Advances in Experimental Medicine and Biology

Volume 682

For other titles published in this series, go to
www.springer.com/series/5584

Dilson E. Rassier
Editor

Muscle Biophysics

From Molecules to Cells

Springer

Editor
Dilson E. Rassier
McGill University
Montreal
Canada
dilson.rassier@mcgill.ca

ISBN 978-1-4419-6365-9 e-ISBN 978-1-4419-6366-6
DOI 10.1007/978-1-4419-6366-6
Springer New York Dordrecht Heidelberg London

Library of Congress Control Number: 2010934106

Printed on acid-free paper

Springer is part of Springer Science+Business Media (www.springer.com)

Preface

Muscle contraction has been the focus of scientific investigation for more than two centuries, and major discoveries have changed the field over the years. Early in the twentieth century, Fenn (1924, 1923) showed that the total energy liberated during a contraction (heat + work) was increased when the muscle was allowed to shorten and perform work. The result implied that chemical reactions during contractions were load-dependent. The observation underlying the "Fenn effect" was taken to a greater extent when Hill (1938) published a pivotal study showing in details the relation between heat production and the amount of muscle shortening, providing investigators with the force-velocity relation for skeletal muscles.

Subsequently, two papers paved the way for the current paradigm in the field of muscle contraction. Huxley and Niedergerke (1954), and Huxley and Hanson (1954) showed that the width of the A-bands did not change during muscle stretch or activation. Contraction, previously believed to be caused by shortening of muscle filaments, was associated with sliding of the thick and thin filaments. These studies were followed by the classic paper by Huxley (1957), in which he conceptualized for the first time the cross-bridge theory; filament sliding was driven by the cyclical interactions of myosin heads (cross-bridges) with actin. The original cross-bridge theory has been revised over the years but the basic features have remained mostly intact. It now influences studies performed with molecular motors responsible for tasks as diverse as muscle contraction, cell division and vesicle transport.

Since the proposal of the cross-bridge theory, the field of muscle physiology has expanded in many laboratories around the world. The structure of the myosin molecule and the interactions between myosin-actin and ATP kinetics are well described (Rayment et al. 1993a, b). New techniques to investigate isolated muscle cells (fibers) and sub-cellular structures have provided scientists with the capability to study contraction in ways never imagined previously.

This book gathers studies performed by scientists who have used diverse muscle techniques, and who have shaped the field of muscle contraction throughout the past years. Some of these studies are, in fact, intrinsically associated with early studies looking at the mechanisms underlying the Fenn effect and load dependence of force production, the relation between force and filament overlap, and the nature of cross-bridge rotation. Although some studies confirm the concepts of the cross-bridge theory, others bring considerable controversies over the mechanisms of contraction. These challenging results and interpretations will be clear to the readers as they read through the chapters.

The organization of the book was flexible; while some authors concentrated mostly on experimental details from the work performed in their laboratories, others reviewed some of the aspects considered of major importance in their work and in the field of muscle contraction. The chapter sequence does not obey a predefined organization, but it generally follows the title of the book: from molecules to cells. After a short introduction on different techniques used in muscle physiology and a historical perspective provided by Dr. K.A.P. Edman, whose scientific contributions are broadly recognized by the scientific community, the chapters deal with muscle mechanics at different levels of analyses. It starts with theoretical approaches to understand myosin molecule function and energetics, and it is followed by experimental work performed with single molecules. Then, results from studies investigating basic mechanisms of contraction, using (mostly in order of appearance) isolated sarcomeres, myofibrils, and fibers are presented. The last chapters summarize studies investigating the effects of acute and chronic adaptations, including weakness and muscle disease.

The book does not intend to cover all aspects of muscle contraction, a field that has expanded significantly over the years. Techniques of molecular biology, computational biology, and genetics, among others, are used in some studies but are not the main focus of investigation. Instead, the book centers mostly on mechanical studies, dealing with force production and regulation; an option clearly indicated by the list of contributing authors. Therefore, the book represents an excellent source of information for readers intending to understand the mechanics of muscle contraction.

Editing a book with several contributors may be a daunting task. However, given the excellence of the research produced by the invited authors, organizing this book was a pleasure, with a minimal effort made by the editor. The result is a fascinating collection of chapters that will hopefully stimulate young investigators to pursue research in this exciting field.

Montreal, QC Dilson E. Rassier, Ph.D.

References

1. Fenn WO (1923) A quantitative comparison between the energy liberated and the work performed by the isolated sartorius muscle of the frog. J Physiol 58:175–203
2. Fenn WO (1924) The relation between the work performed and the energy liberated in muscular contraction. J Physiol 58:373–395
3. Hill AV (1938) The heat of shortening and the dynamic constants of muscle. Proc R Soc Lond B 126:136–195
4. Huxley AF (1957) Muscle structure and theories of contraction. Prog Biophys Biophys Chem 7:255–318
5. Huxley AF, Niedergerke R (1954) Structural changes in muscle during contraction; interference microscopy of living muscle fibres. Nature 173:971–973

6. Huxley H, Hanson J (1954) Changes in the cross-striations of muscle during contraction and stretch and their structural interpretation. Nature 173:973–976
7. Rayment I, Holden HM, Whittaker M, Yohn CB, Lorenz M, Holmes KC, Milligan RA (1993a) Structure of the actin-myosin complex and its implications for muscle contraction. Science 261:58–65
8. Rayment I, Rypniewski WR, Schmidt-Base K, Smith R, Tomchick DR, Benning MM, Winkelmann DA, Wesenberg G, Holden HM (1993b) Three-dimensional structure of myosin subfragment-1: a molecular motor. Science 261:50–58

Contents

Striated Muscles: From Molecules to Cells 1
Dilson E. Rassier

Contractile Performance of Striated Muscle 7
K.A.P. Edman

Energy Economy in the Actomyosin Interaction: Lessons from Simple Models 41
Steven L. Lehman

A Strain-Dependency of Myosin Off-Rate Must Be Sensitive to Frequency to Predict the B-Process of Sinusoidal Analysis 57
Bradley M. Palmer

Electron Microscopic Visualization of the Cross-Bridge Movement Coupled with ATP Hydrolysis in Muscle Thick Filaments in Aqueous Solution, Reminiscences and Future Prospects 77
Haruo Sugi

Role of Titin in Skeletal Muscle Function and Disease 105
Coen A.C. Ottenheijm and Henk Granzier

Contractile Characteristics of Sarcomeres Arranged in Series or Mechanically Isolated from Myofibrils 123
Dilson E. Rassier and Ivan Pavlov

The Force–Length Relationship of Mechanically Isolated Sarcomeres 141
W. Herzog, V. Joumaa, and T.R. Leonard

Extraction and Replacement of the Tropomyosin–Troponin Complex in Isolated Myofibrils 163
Beatrice Scellini, Nicoletta Piroddi, Corrado Poggesi, and Chiara Tesi

Stretch and Shortening of Skeletal Muscles Activated Along the Ascending Limb of the Force–Length Relation 175
Dilson E. Rassier and Clara Pun

Cross-Bridge Properties in Single Intact Frog Fibers Studied by Fast Stretches 191
Barbara Colombini, Marta Nocella, Giulia Benelli, Giovanni Cecchi, and M. Angela Bagni

Crossbridge and Non-crossbridge Contributions to Force in Shortening and Lengthening Muscle 207
K.W. Ranatunga, H. Roots, G.J. Pinniger, and G.W. Offer

Short-Range Mechanical Properties of Skeletal and Cardiac Muscles 223
Kenneth S. Campbell

Crossbridge Mechanism(s) Examined by Temperature Perturbation Studies on Muscle 247
K.W. Ranatunga and M.E. Coupland

Efficiency of Cross-Bridges and Mitochondria in Mouse Cardiac Muscle 267
C.J. Barclay and C. Widén

Mechanisms of Skeletal Muscle Weakness 279
Håkan Westerblad, Nicolas Place, and Takashi Yamada

Stretch-Induced Membrane Damage in Muscle: Comparison of Wild-Type and *mdx* Mice 297
David G. Allen, Bao-ting Zhang, and Nicholas P. Whitehead

Cellular and Whole Muscle Studies of Activity Dependent Potentiation 315
Brian R. MacIntosh

Index 343

Contributors

David G. Allen
Bosch Institute and School of Medical Sciences, University of Sydney, F13, Sydney, NSW 2006, Australia

M. Angela Bagni
Dipartimento di Scienze Fisiologiche and Istituto Interuniversitario di Miologia, Università degli Studi di Firenze, Viale G.B. Morgagni 63, I-50134 Firenze, Italy

C.J. Barclay
School of Physiotherapy and Exercise Science, Griffith University, Gold Coast, QLD, Australia

Giulia Benelli
Dipartimento di Scienze Fisiologiche and Istituto Interuniversitario di Miologia, Università degli Studi di Firenze, Viale G.B Morgagni 63, I-50134 Firenze, Italy

Kenneth S. Campbell
Department of Physiology & Center for Muscle Biology, University of Kentucky, Lexington, KY 40503, USA

Giovanni Cecchi
Dipartimento di Scienze Fisiologiche and Istituto Interuniversitario di Miologia, Università degli Studi di Firenze, Viale G.B Morgagni 63, I-50134 Firenze, Italy

Barbara Colombini
Dipartimento di Scienze Fisiologiche and Istituto Interuniversitario di Miologia, Università degli Studi di Firenze, Viale G.B Morgagni 63, I-50134 Firenze, Italy

K.A.P. Edman
Department of Experimental Medical Science, Biomedical Centre, University of Lund, Lund S-221 84, Sweden

Henk Granzier
Department of Physiology, University of Arizona, Tucson, AZ 85724, USA

W. Herzog
Faculty of Kinesiology, University of Calgary, Calgary, AB, Canada T2N1N4

V. Joumaa
Faculty of Kinesiology ,University of Calgary, Calgary, AB, Canada T2N1N4

Steven. L. Lehman
Department of Integrative Biology, University of California Berkeley, Berkeley, CA 94720, USA

T.R. Leonard
Faculty of Kinesiology, University of Calgary, Calgary, AB, Canada T2N1N4

Brian R. MacIntosh
Faculty of Kinesiology, University of Calgary, 2500 University Dr NW, Calgary, AB, Canada T2N1N4

Marta Nocella
Dipartimento di Scienze Fisiologiche and Istituto Interuniversitario di Miologia, Università degli Studi di Firenze, Viale G.B Morgagni 63, I-50134 Firenze, Italy

G. W. Offer
Muscle Contraction Group, Department of Physiology & Pharmacology, School of Medical Sciences, University of Bristol, Bristol BS8 1TD, UK

Coen A. C. Ottenheijm
Laboratory for Physiology, Institute for Cardiovascular Research, VU University Medical Center, Amsterdam1081 BT, The Netherlands

Bradley M. Palmer
Department of Molecular Physiology and Biophysics, University of Vermont, 122 HSRF – 149 Beaumont Ave, Burlington, VT 05405, USA

Ivan Pavlov
Department of Kinesiology and Physical Education, McGill University, Pine Avenue West 475, Montreal, QC, Canada H2W1S4

G. J. Pinniger
School of Biomedical, Biomolecular & Chemical Sciences, The University of Western Australia, Crawley, WA, 6009, Australia

Nicoletta Piroddi
Dipartimento di Scienze Fisiologiche, Università di Firenze, Viale Morgagni 63, Firenze I-50134, Italy

Nicolas Place
Institute of Movement Sciences and Sport Medicine, Université de Genève, Rue du Conseil Général 10, 1205, Genève, Switzerland

Corrado Poggesi
Dipartimento di Scienze Fisiologiche, Università di Firenze, Viale Morgagni 63, Firenze I-50134, Italy

Clara Pun
Department of Kinesiology and Physical Education, McGill University,
Pine Avenue West 475, Montreal, QC, Canada H2W1S4

K.W. Ranatunga
Muscle Contraction Group, Department of Physiology and Pharmacology,
School of Medical Sciences, University of Bristol, Bristol BS8 1TD, UK

Dilson E. Rassier
Department of Kinesiology and Physical Education, McGill University,
Pine Avenue West 475, Montreal, QC, Canada H2W1S4

H. Roots
Muscle Contraction Group, Department of Physiology & Pharmacology,
School of Medical Sciences, University of Bristol, Bristol BS8 1TD, UK

Beatrice Scellini
Dipartimento di Scienze Fisiologiche, Università di Firenze,
Viale Morgagni 63, Firenze I-50134, Italy

Haruo Sugi
Department of Physiology, School of Medicine, Teikyo University,
Itabashi-ku, Tokyo 173-8605, Japan

Chiara Tesi
Dipartimento di Scienze Fisiologiche, Università di Firenze, Viale Morgagni 63,
Firenze I-50134, Italy

Håkan Westerblad
Department of Physiology and Pharmacology, Karolinska Institutet,
171 77 Stockholm, Sweden

Nicholas P. Whitehead
Bosch Institute and School of Medical Sciences, University of Sydney, F13
Sydney, NSW 2006, Australia

C. Widén
Department of Plant Breeding and Biotechnology, Swedish University of
Agricultural Sciences, Fjälkestadsvägen 459, SE-291 94, Kristianstad, Sweden

Takashi Yamada
Department of Physiology & Pharmacology, Karolinska Institutet,
171 77 Stockholm, Sweden

Bao-ting Zhang
Muscle Physiology Laboratory, Department of Rehabilitation Sciences,
Hong Kong Polytechnic University, Hung Hom, Kowloon, Hong Kong

Striated Muscles: From Molecules to Cells

Dilson E. Rassier

Abstract Striated muscles are complex structures, composed of many cells and different molecules that work together to produce contraction and force. Advanced techniques have allowed scientists to investigate all layer of muscle organization. Forces ranging from piconewtons (pN) to millinewtons (mN) and displacements ranging from nanometers (nm) to millimetres (mm) can be measured in different muscle structures. The integration of different layers of analyses improves significantly our understanding of the mechanisms of contraction.

Keywords Single molecules • Myofilaments • Laser trap • Nanolevers

1 Striated Muscle Organization

Striated muscles are highly ordered structures. They consist of several cylindrical cells (fibers) which contract homogeneously, regulated by the central nervous system. Each muscle fiber is composed of thousands of parallel cylindrical units, the myofibrils, which are formed by longitudinal building blocks – the sarcomeres. The sarcomere is the functional unit of the muscle, aggregating filaments made up of different molecules, most prominently myosin, actin and titin. Within the subtleties of the sarcomere and filaments architecture lie the essentials of the contractile process. The action of many myosin molecules upon actin filaments causes shortening of the sarcomeres, beginning an amplified response that passes through myofibrils, cells and ultimately muscles. Simultaneously, titin molecules stabilize myosin-actin filaments in the center of the sarcomeres, and provide most of the passive forces present in striated muscles.

D.E. Rassier (✉)
Department of Kinesiology and Physical Education, McGill University,
Pine Avenue West 475, Montreal, QC, Canada H2W1S4
e-mail: dilson.rassier@mcgill.ca

D.E. Rassier (ed.), *Muscle Biophysics: From Molecules to Cells*,
Advances in Experimental Medicine and Biology 682,
DOI 10.1007/978-1-4419-6366-6_1,

Understanding the mechanisms of contraction and force production requires understanding the action of all muscle structures. While classic studies on muscle physiology have used mostly whole muscles or isolated fibers – a preparation still used with irreplaceable applications - advances in technology have changed the field considerably. Experimental apparatuses that allow the evaluation of tasks within nanonewton (nN) ranges and piconewtons (pN) forces have enabled researchers to undercover the molecular world inside muscle fibers. Studies have been performed on several layers of structure to understand how they interact and produce force. Each layer of organization brings original contribution to our knowledge; each associated technique allows the testing of specific hypotheses and mechanisms involved in contraction.

1.1 Single Muscle Fibers

Mechanical measurements in intact muscle fibers (Fig. 1a) have been performed for many years, allowing the testing of the cross-bridge theory, establishing the details

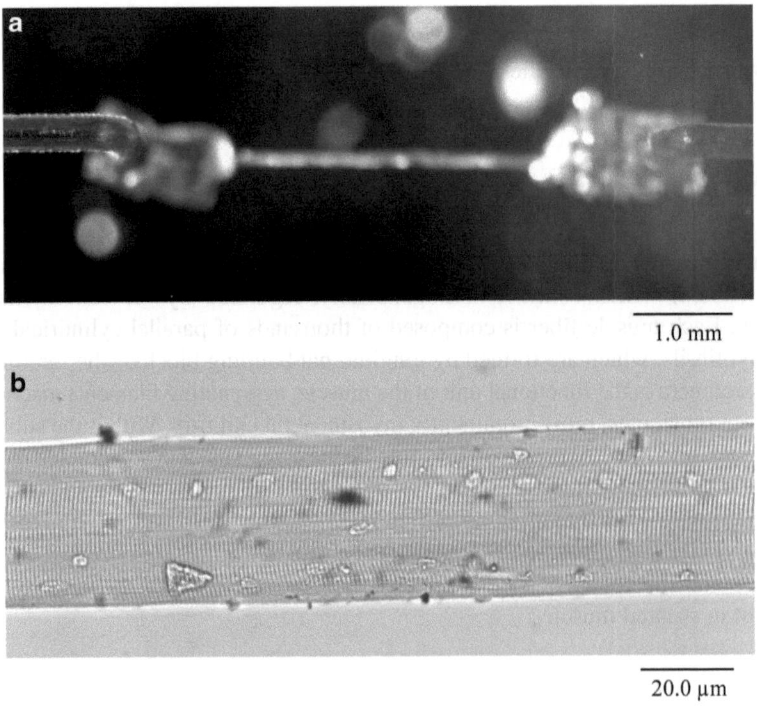

Fig. 1 Single fibers isolated from skeletal muscles. (**a**) Intact fiber from the frog lumbrical muscle, attached between two hooks, which are connected to a motor arm and a force transducer, respectively. (**b**) Permeabilized fiber from the rabbit psoas muscle. Note that the striation pattern is clear, allowing measurements of average sarcomere length

of the force-length and force-velocity relations, among several characteristic features of contraction. The first studies – and most mechanical studies – have been performed with fibers isolated from amphibians (Huxley and Simmons 1971; Gordon et al. 1966). Experiments with fibers isolated from mammalian muscles were subsequently developed with important physiological implications (Lannergren and Westerblad 1987). In muscle fibers, the striation pattern produced by the myosin and actin filaments (mostly visible in amphibian fibers) is used for measurements of the average sarcomere length, and the degree of filament overlap can be estimated during experiments.

Muscle fibers can also be isolated after being permeabilized (Fig. 1b), and the media surrounding the fibers can be manipulated (pH, Ca^{2+}, etc.). Permeabilized fibers are not as resistant as intact fibers, but still provide reliable force measurements. This preparation has enabled scientists to evaluate the detail of the force-pCa^{2+} relation, an important feature of muscle regulation associated with the sensitivity of the contractile apparatus in generating force.

Single fibers represent robust preparations in which repeated force measurements are made in an environment where all structures involved in contraction are present.

1.2 Single Myofibrils and Sarcomeres

Depending on the goal of the study, fibers may present limitations: (1) the diffusion time inside the fibers may be slow, and (2) the behaviour of individual sarcomeres cannot be evaluated. Sarcomere length measurements are based on the average profile of millions of sarcomeres arranged in series and in parallel, which may vary considerably during contractions. Single myofibrils, on the other hand, allow the visualization of individual sarcomeres and half-sarcomere in-series (Rassier et al. 2003; Telley et al. 2006) (Fig. 2a). As a result, the analysis of force and sarcomere dynamics can be performed directly. Giving their small diameter and diffusion time, myofibrils allow quick activation and relaxation, and investigators can determine precisely the kinetics of force development and relaxation. These measurements will ultimately correspond to the kinetics of cross-bridge attachment and detachment rates.

Myofibrils provide consistent force measurements for the investigation of sub-cellular muscle mechanics. Nevertheless, sarcomeres in a given myofibril may present complex behaviours. Significant sarcomere length non-uniformities have been observed, as sarcomeres may change their length continuously during activation and relaxation (Telley et al. 2006). There is an inherent inter-sarcomere coordination, suggesting that sarcomeres in a myofibril are all dependent on each other (Shimamoto et al. 2009). Recently, individual sarcomeres have been mechanically isolated from myofibrils (Pavlov et al. 2009) (Fig. 2b). Forces of the same magnitude as those produced by myofibrils are observed, and the length dependence of force resembles that observed in single fibers. The preparation may open new opportunities for the investigation of sarcomere mechanics in diverse situations.

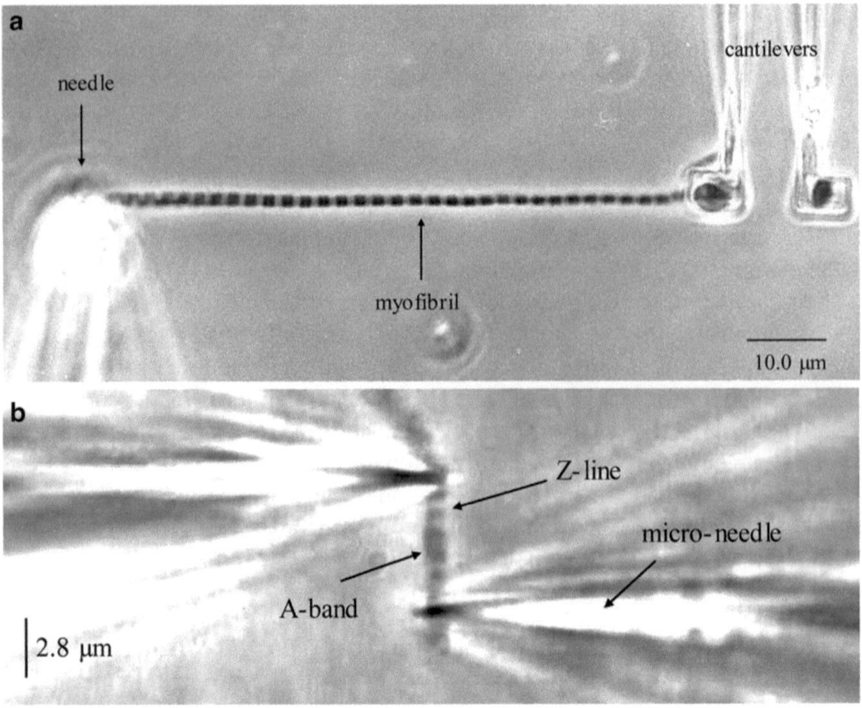

Fig. 2 (**a**) Myofibril isolated from the rabbit psoas muscle attached between a pair of cantilevers at one end and a glass needle at the other end. The cantilevers are used for force measurements, and the glass needle is connected to a motor arm. (**b**) Sarcomere isolated from a rabbit psoas myofibril. The sarcomere is caught between two pre-calibrated micro-needles that allow force measurements during contractions

1.3 Single Filaments and Molecules

Sarcomeres maintain the three dimensional structure of the striated muscles intact, and thus represent the basic unit of contraction. However, each sarcomere has many muscle filaments, and molecular interpretations may be limited. Single filaments, on the other hand, provide a direct evaluation of the interactions between myosin and actin. Experimenting with isolated filaments is challenging, as they have to be either co-polymerized after single molecule preparation (Hikikoshi et al. 2005), or isolated directly from muscles (usually from mussel, Mytilus) (Liu and Pollack 2004).

Force measurements during myosin-actin filament interactions are performed with the "laser-trap" essay (Hikikoshi et al. 2005) (Fig. 3a), or with the use of micro-fabricated cantilevers (Liu and Pollack 2004) (Fig. 3b). The laser trap essay has also allowed scientists to investigate the mechanics of single myosin molecules (Finer et al. 1994) (Fig. 3c), although other methods can be used (e.g. force probes (Kitamura et al. 1999) and in-vitro motility essays (VanBuren et al. 1994)).

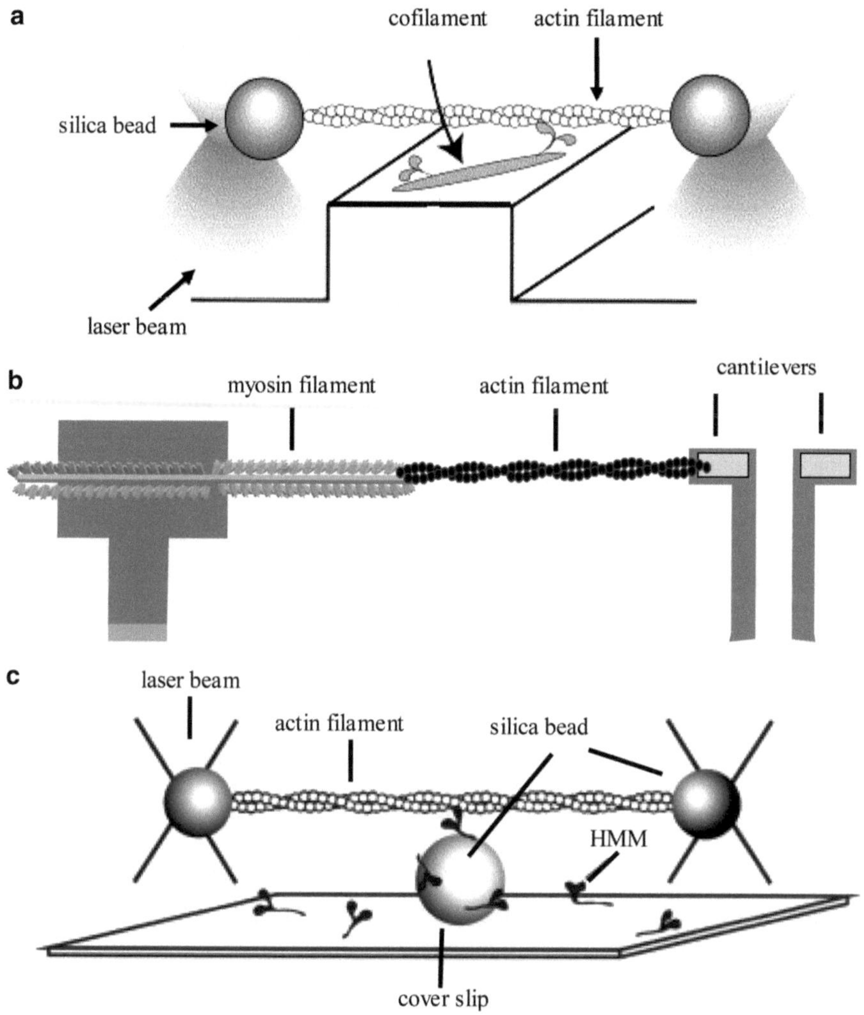

Fig. 3 Molecular experimentation during myosin-actin interactions. (**a**) Actin filament caught between two beads interacts with a myosin filament (adapted from Hikikoshi et al. (2005). (**b**) Actin and myosin filaments attached between cantilevers. (**c**) Actin filament caught between two beads interacts with a myosin molecule on the top of a third bead. Adapted from Finer et al. (1994)

With the laser trap, one myosin molecule supported in a pedestal attaches to actin; when actin slides as a result of myosin power stroke, the force can be measured. Such approach has provided investigators with a precise estimation of basic events underlying contraction, i.e. the magnitude of the myosin power stroke, the force and stiffness of individual myosin molecules, and the relation between ATP consumption and force.

Lately, incredible advancements have been made towards understanding the structure responsible for passive forces, most notably titin molecules. Scientists have isolated titin molecules and also specific regions of titin, to then attach them to atomic force probes (e.g. Kellermayer et al. 2001). With such approach the mechanical characteristics of titin, including its extensibility, persistence length and length-dependent stiffness have been studied in details – new physiological roles for the molecules emerge all the time.

Each technique used in the field of muscle contraction has advantages and disadvantages. Molecular and sub-cellular preparations allow scientists to understand the basic mechanisms of contraction and force production, while larger preparations approximate the experimental conditions to the physiological environment. Although all experimental techniques have provided valuable information, the mechanism of contraction and its applications to varying situations is still under investigation. The field will certainly see more changes in the near future, which promises to bring new approaches and interpretations.

References

Finer JT, Simmons RM, Spudich JA (1994) Single myosin molecule mechanics: piconewton forces and nanometre steps. Nature 368:113–119

Gordon AM, Huxley AF, Julian FJ (1966) The variation in isometric tension with sarcomere length in vertebrate muscle fibres. J Physiol 184:170–192

Hikikoshi IA, Tanaka H, Morimoto S, Ishijima A, Yanagida T (2005) The neck domain of myosin II primarily regulates the actomyosin kinetics, not the stepsize. J Mol Biol 353:213–221

Huxley AF, Simmons RM (1971) Proposed mechanism of force generation in striated muscle. Nature 233:533–538

Kellermayer MS, Smith SB, Bustamante C, Granzier HL (2001) Mechanical fatigue in repetitively stretched single molecules of titin. Biophys J 80:852–863

Kitamura K, Tokunaga M, Iwane AH, Yanagida T (1999) A single myosin head moves along an actin filament with regular steps of 5.3 nanometres. Nature 397:129–134

Lannergren J, Westerblad H (1987) The temperature dependence of isometric contractions of single, intact fibres dissected from a mouse foot muscle. J Physiol 390:285–293

Liu X, Pollack GH (2004) Stepwise sliding of single actin and Myosin filaments. Biophys J 86:353–358

Pavlov I, Novinger R, Rassier DE (2009) The mechanical behavior of individual sarcomeres of myofibrils isolated from rabbit psoas muscle. Am J Physiol Cell Physiol 297:C1211–C1219

Rassier DE, Herzog W, Pollack GH (2003) Dynamics of individual sarcomeres during and after stretch in activated single myofibrils. Proc Biol Sci 270:1735–1740

Shimamoto Y, Suzuki M, Mikhailenko SV, Yasuda K, Ishiwata S (2009) Inter-sarcomere coordination in muscle revealed through individual sarcomere response to quick stretch. Proc Natl Acad Sci U S A 106:11954–11959

Telley IA, Denoth J, Stussi E, Pfitzer G, Stehle R (2006) Half-sarcomere dynamics in myofibrils during activation and relaxation studied by tracking fluorescent markers. Biophys J 90:514–530

VanBuren P, Work SS, Warshaw DM (1994) Enhanced force generation by smooth muscle myosin in vitro. Proc Natl Acad Sci U S A 91:202–205

Contractile Performance of Striated Muscle

K.A.P. Edman

Abstract The single muscle fiber preparation provides an excellent tool for studying the mechanical behaviour of the contractile system at sarcomere level. The present article gives an overview of studies based on intact single fibers from frog and mouse skeletal muscle. The following aspects of muscle function are treated: (1) The length–tension relationship. (2) The biphasic force–velocity relationship. (3) The maximum speed of shortening, its independence of sarcomere length and degree of activation. (4) Force enhancement *during* stretch, its relation to sarcomere length and myofilament lattice width. (5) Residual force enhancement *after* stretch. (6) Force reduction after loaded shortening. (7) Deactivation by active shortening. (8) Differences in kinetic properties along individual muscle fibers.

Keyword Contractile performance of striated muscle

1 Introduction

A study published in Nature 1939 by Engelhardt and Ljubimova can be said to form the beginning of a new era in muscle physiology. It was demonstrated in this article that "myosin", known at the time to be one of the principal components of striated muscle (Bailey 1937), had the ability to split the high-energy phosphate compound adenosine triphosphate (ATP). The Engelhardt-Ljubimova discovery inspired the Hungarian biochemist Albert Szent-Györgyi to further elucidate the "myosin"-ATP reaction and its relation to force development and movement of muscle. In the years to follow (1940–1946), while Europe was ravaged by war, Szent-Györgyi and his school made the most fundamental discoveries ever in muscle physiology.

K.A.P. Edman (✉)
Department of Experimental Medical Science, Biomedical Centre, University of Lund, Lund S-221 84, Sweden
e-mail: paul.edman@farm.lu.se

D.E. Rassier (ed.), *Muscle Biophysics: From Molecules to Cells*,
Advances in Experimental Medicine and Biology 682,
DOI 10.1007/978-1-4419-6366-6_2,

What was known as "myosin" at the time was found to be a complex of two proteins: the true myosin (initially named myosin A) and a newly identified protein, named actin. The discovery of actin was made by F.B. Straub (1942), a member of Szent-Györgyi's team. The actin-myosin complex (actomyosin) could be extracted from muscle tissue in a high-salt solution and finally shaped into threads at physiological ionic strength. The threads so produced were shown to contract heavily on addition of ATP, a phenomenon that Albert Szent-Györgyi in later years described as the most impressive phenomenon he had experienced in his scientific career. Significant contributions in the early studies of the contractile machinery also came from the laboratory of H.H. Weber. It was demonstrated by Weber and co-workers (Weber and Portzehl 1952) that ATP has two important actions on reconstituted actomyosin fibers: it induces contraction and at the same time makes the fibers plastic, the latter effect being demonstrable by blocking the myosin ATPase activity by means of an SH-group reagent. It should be noted at this point that actomyosin lacking the regulatory system, the troponin–tropomyosin complex, does not require calcium to be contracted by ATP.

Near the end of the war, before the borders of Hungary were effectively closed for immigration, Albert Szent-Györgyi left Budapest to spend half a year at the Karolinska Institute in Stockholm, Sweden. During this time he was invited to republish a review article of his seminal work on muscle in Acta Physiologica Scandinavica (Szent-Györgyi 1944). Before proceeding to the US, Albert Szent-Györgyi arranged for two of his previous pupils in Budapest to come to work at Uppsala University. One of them, T. Erdös, had previously demonstrated, at the suggestion of Szent-Györgyi, that the state of rigor in muscle is due to depletion of ATP, i.e. loss of the plasticizing action of ATP. My own interest in muscle function came in at this point as I happened to be a research student at the Department of Pharmacology at Uppsala University where one of Szent-Györgyi's previous pupils (A. Czapó) came to work. My first papers (Edman 1950, 1951, 1953) describe the effects of cardiac glycosides on the ATPase activity of cardiac myosin. In parallel with this work and together with my mentor E.H. Bárány, an electronic technique was developed for measuring the time course of dissociation of actomyosin in solution after addition of ATP. Our main interest in this project was to establish how calcium and magnesium and variations in the electrolyte composition of the medium might affect the rate of dissociation of actomyosin in response to ATP (Bárány et al. 1951a, b, 1952). This was followed by detailed studies of the relaxing effects of zinc in glycerinated (skinned) muscle fibers (e.g. Edman 1959a, b). The essential role of the sarcoplasmic reticulum in the relaxation process was unknown at the time, and the possible role of zinc as a relaxing agent became a most intriguing possibility after observing that there is as much zinc in muscle as there is calcium. My subsequent work on muscle has mainly been performed on intact preparations of smooth muscle (Edman and Schild 1962, 1963), myocardium (Edman and Nilsson 1968, 1972) and intact skeletal muscle fibers. The latter preparations have enabled studies of the contractile machinery at sarcomere level using various mechanical and optical techniques. In the following account some basic properties of the contractile system will be described as elucidated in the single fiber preparation.

2 The Length–Tension Relationship in Striated Muscle

It was demonstrated in early studies on isolated frog muscle by M. Blix (professor of Physiology at the University of Lund; Blix 1894) that the force producing capability of skeletal muscle is critically dependent on the degree of extension of the muscle. Blix demonstrated that, after account had been made for passive tension, there was a bell-shaped relationship between active force and muscle length and that maximum force was produced near the length assumed by the muscle at rest in the body. It soon became clear that the length–tension relationship contains information of great relevance to our understanding of the contractile process, and the observations presented by Blix (1894) were confirmed and extended in several subsequent studies on both whole muscle (Evans and Hill 1914) and single muscle fibers of the frog (Ramsey and Street 1940) and on partially isolated voluntary muscles of human subjects (Ralston et al. 1947). With the presentation of the sliding-filament hypothesis of muscle contraction (Huxley and Hanson 1954; Huxley and Niedergerke 1954) it became vital to relate force production and shortening velocity to sarcomere length, i.e. to relate the mechanical performance of the muscle to the state of overlap between the thick (myosin) and thin (actin) filaments that make up the sarcomeres. The first two studies with this particular aim, both carried out on intact muscle fibers (M. semitendinosus) of the frog, were published in the same journal by Gordon et al. (1966) and Edman (1966). Both studies demonstrated that maximum tetanic force increased with increasing filament overlap reaching a maximum at a sarcomere length where the thin filaments would reach the central, inert, zone of the thick filaments. The paper by Edman (1966) was specifically focused on the question whether the contractile state at a given sarcomere length might depended on how this length is being reached, by passive movement at rest or by active shortening from a prestretched position. It was found that, provided the fiber is allowed to shorten against a light load, the starting length does not affect the force producing capability at the end of the movement. That is, the isometric force was found to be determined by the actual sarcomere length only, in full accordance with the sliding filament hypothesis.

The studies reported by Gordon et al. (1966) and Edman (1966) were both carried out on fairly long fiber preparations in which intrinsic, non-uniform sarcomere movements might affect force recording. A renewed study of the length–tension relationship was therefore carried out (Edman and Reggiani 1987), again on isolated frog muscle fibers, using a technique by which isometric force and shortening velocity could be performed on short (0.5–0.7 mm) marked segments along the length of the fiber. By this technique, described in detail by Edman and Reggiani (1984), a limited population of sarcomeres (circa 150–300 sarcomeres in series) could be held at constant overall length during tetanic stimulation or, alternatively, be allowed to shorten actively against a constant, preset load. The length–tension relationship depicted by the latter technique (Fig. 1) differs in certain respects from the polygonal length–tension curve proposed by Gordon et al. (1966). The new data showed that the length–tension relationship has a smoother shape. It exhibits no distinct plateau

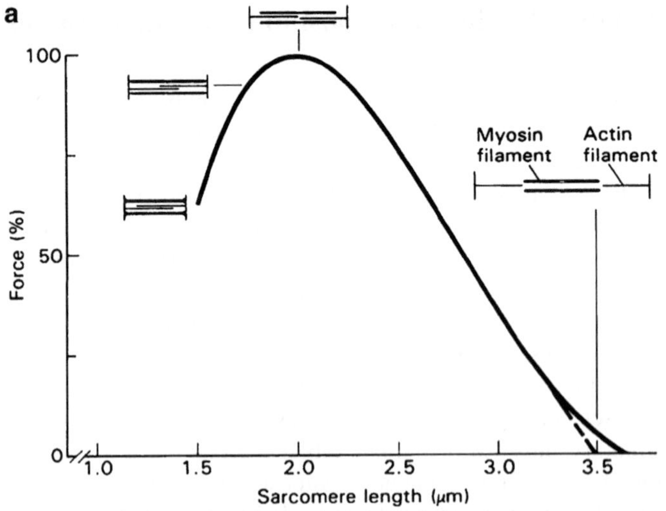

Fig. 1 (**a**) Variation of maximum tetanic force with sarcomere length. Curve from Edman and Reggiani (1987) based on measurements from short, length-controlled segments of single muscle fibers. The *dashed line* shows extrapolation to abscissa of the steep portion of the length–tension relation. The intersection of the *dashed line* with the abscissa marks the sarcomere length at which the majority of the A- and I-filaments are in end-to-end position. *Insets* show degree of filament overlap at four different sarcomere lengths. (**b**) The length–tension relationship shown in A (*full line*) compared with the polygonal length–tension curve (*dashed line*) described by Gordon et al. (1966)

between 2.0 and 2.2 μm sarcomere length, and the descending limb can be seen to have a slightly sigmoid shape with the main portion extrapolating to a distinctly shorter sarcomere length (3.49 μm) than that (3.65 μm) postulated by Gordon et al. (1966). As demonstrated by Edman and Reggiani (1987) the length–tension relationship, even in its revised form, is fully explainable in terms of the sliding-filament hypothesis.

The results are consistent with the view that: (1) there is a substantial variability (SD 0.2 μm) in the degree of overlap between the thick (A) and thin (I) filaments and (2) the average functional length of the A and I filaments is 1.55 and 1.94 μm, respectively, in frog striated muscle. These data agree closely with the revised values of the A and I filament lengths (1.55 and 1.92–1.96 μm, respectively) that were presented by Page 1968 and Huxley 1973. Several factors may contribute to the variability in filament overlap predicted by the length–tension curve: (1) non-uniformity in length of the thin filaments, (2) longitudinal misalignment (staggering) of the thick and thin filaments within the A and I bands of the sarcomere and (3) differences in sarcomere length (mean M-Z line spacing) among individual myofibrils within a given cross-section of the fiber. It is essential to point out that a randomized variation of filament overlap is entirely consistent with the finding that the sarcomere pattern remains constant during length-clamp recording of a short fiber segment. This has to do with the fact that elastic structures connecting adjacent myofibrils across the fiber will be strained and will support those sarcomeres, or sarcomere halves, that have less capacity to develop force than has a neighboring region (see further Edman and Tsuchiya 1996).

The first study of the length–tension relationship performed on mammalian intact single muscle fibers was recently presented (Edman 2005). The experimental approach was essentially the same as that employed in the earlier frog fiber experiments, i.e. tetanic force was recorded from a short (ca. 0.3 mm) marked segment that was held at constant length during stimulation. The fibers dissected from the *flexor digitorum brevis* muscle of the mouse were merely 25–40 μm wide and 0.5–0.7 mm in length between the tendon insertions. As illustrated in Fig. 2, the

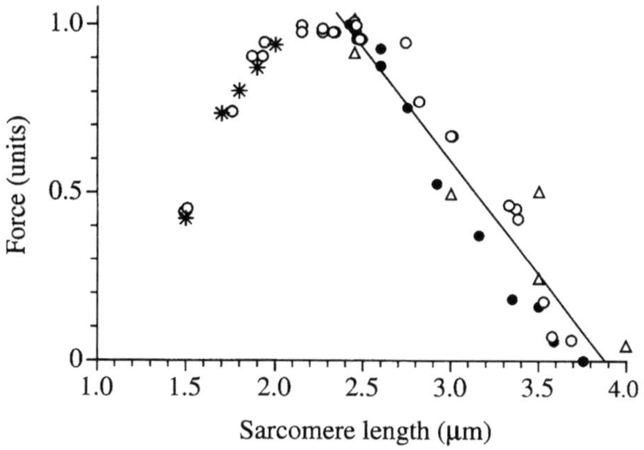

Fig. 2 Tetanic force versus sarcomere length in mouse single muscle fibers. Force is expressed in units of the tetanic force recorded near 2.4 μm sarcomere length. The *straight line* is linear regression of force upon sarcomere length for values between 2.35 and 4.00 μm sarcomere length. The *asterisks* along the ascending limb are calculated values based on the assumption that force is reduced below optimum length in proportion to the occurrence of double filament overlap and, at more extreme degrees of shortening, to compression of the thick filaments against the Z disks. For further details, see Edman (2005)

length–tension relationship has a shape that, similar to the situation in amphibian muscle fibers, can be fitted well with the sliding-filament hypothesis. The slope of the descending limb predicts that the average length of the thin filaments is 1.10 μm and the length of the thick filaments is 3.88 μm, both values being in excellent agreement with recent electron microscopical measurements of the two filaments in rabbit psoas muscle (Sosa et al. 1994).

3 The Force–Velocity Relationship. Maximum Velocity of Shortening

In contrast to a simple elastic body active muscle has the remarkable property to adjust its force to precisely match the load that is experienced during shortening. It does so by continuously adjusting the speed at which it moves. Thus, when the load is small, the force produced by the contractile system is made correspondingly small by increasing the speed of shortening appropriately. Conversely, when the muscle is heavily loaded, the active force is raised to an equivalent level by reducing the shortening velocity sufficiently. Fenn and Marsh (1935) were first to formulate the inverse relationship between load and velocity of shortening. Hill (1938) further characterized the force–velocity relationship and described it in terms of a rectangular hyperbola. Hill also pointed out the important role that the force–velocity relationship is likely to have for understanding the basic mechanisms of muscle contraction.

A detailed study on single muscle fibers (Edman et al. 1976) revealed that the force–velocity relationship deviates from Hill's hyperbolic equation in the high-force range. Improved recording techniques, including measurements from short length-controlled segments of intact fibers, made it possible to characterize this phenomenon in considerable detail (Edman 1988). As illustrated in Fig. 3 the force–velocity relationship exhibits two distinct curvatures, both with an upward concavity on either side of a breakpoint near 80% of the isometric force (P_0). The breakpoint is most clearly illustrated by plotting the force–velocity data in a semilogarithmic diagram as shown in the insets of Fig. 3. The biphasic shape of the force–velocity relationship, with a breakpoint near 0.8 P_0, also appears in a similar way in *mammalian* muscle fibers (Edman 2005; Devrome and MacIntosh 2007) and in isolated *muscle spindles* of the frog (Edman et al. 2002). It is worth pointing out that the high-force deviation of the force–velocity curve has been shown to exist also in skinned muscle fibers and to be unrelated to the state of activation of the contractile system (Lou and Sun 1993). The biphasic shape of the force–velocity relation thus seems to reflect a basic feature of the sliding-filament process. The following empirical equation provides an excellent fit to the experimental data (Fig. 4):

$$V = \frac{\left(P_0^* - P\right)b}{P + a}\left(1 - \frac{1}{1 + e^{-k_1\left(P - k_2 P_0\right)}}\right) \tag{1}$$

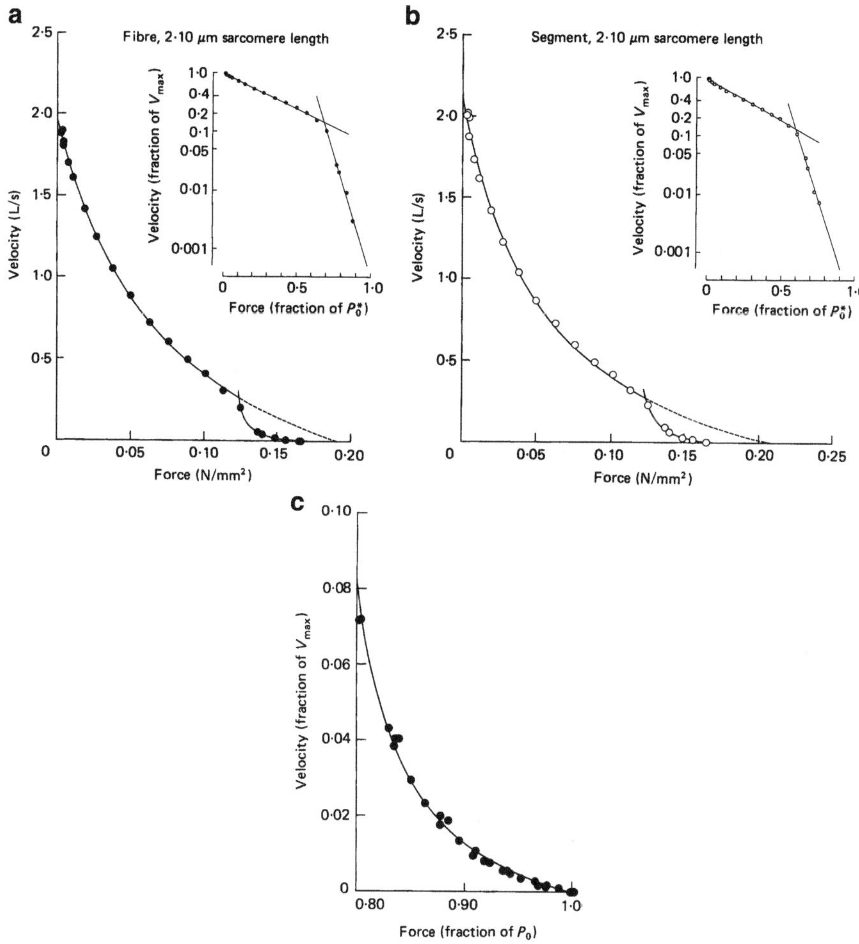

Fig. 3 Force–velocity relationship based on measurement in a single muscle fiber. Velocity of shortening recorded in the fiber as a whole (**a**) and in a short (ca. 0.6 mm) segment of the same fiber (**b**). Note difference in curvature of force–velocity relation on either side of a break point near 0.8 P_0. *Continuous lines*: Hill's hyperbolic equation fitted to data points above and below the break point. *Insets*: semilogarithmic plots of the force–velocity data given in the main diagrams. *Lines* in the insets, linear regressions of velocity upon force on either side of the break point. Sarcomere length: 2.10 μm. Temperature: 2.2°C. (**c**), Force–velocity data (from whole fiber) derived in the high-force range plotted in diagram in which the ordinate has been placed at 0.80 P_0. Hyperbola fitted by least-squares method. From Edman (1988)

V denotes the velocity of shortening, *P* the load on the muscle fiber and P_0 is the measured isometric force. The first term of the equation expresses the force-velocity relation below 0.8 P_0 and represents a rectangular hyperbola in the form described by Hill (1938). P_0^* is the isometric force predicted from this hyperbola and *a* and *b* are constants with dimensions of force and velocity, respectively.

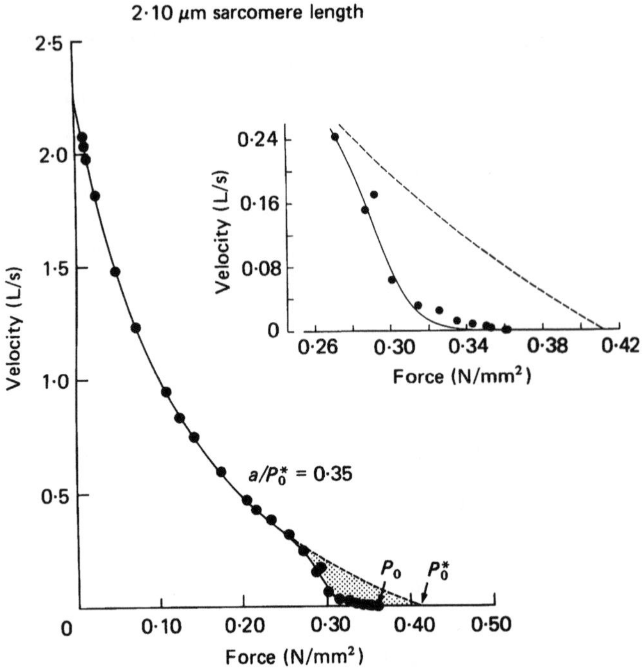

Fig. 4 Equation (1) fitted to force–velocity data derived from a single muscle fiber. Principle of curve fitting: The first term in eqn. (1) (Hill's hyperbolic equation) was initially fitted, by least-squares method, to data truncated at 0.78 P_0 in this way deriving the numerical values of constants a and b. Constants k_1 and k_2 were thereafter determined by fitting of eqn. (1) to all data points, i.e. including loads greater than 0.78 P_0. *Inset*: detail of high-force portion of the force–velocity relation shown in main diagram. *Continuous line*, fit of eqn. (1). *Dashed line*, extrapolation of hyperbola derived from data below 0.78 P_0. From Edman (1988)

The second term within parenthesis (referred to as the 'correction term'; Edman 1988) reduces V at high loads to fit the distinct upward-concave curvature at loads greater than approximately 0.8 P_0. The constant k_1 in the correction term has the dimension of 1/force, whereas k_2 is dimensionless. k_1 determines the steepness of the high-force curvature and k_2 the relative force at which the correction term reaches its half value. The two constants are useful parameters when comparing data from different studies.

Several attempts have been made to simulate the characteristic change of the force–velocity relation in the high-force range using different cross-bridge models (Mitsui and Chiba 1996; Edman et al. 1997; Duke 1999; Nielsen 2003). The model presented by Edman et al. (1997) simulates the experimental force–velocity and stiffness-velocity relationships exceedingly well. According to this model there is a Gaussean position-dependence of the attachment rate constant along the thin filament. This will provide a region early during the power stroke where the cross-bridge attachment is slow. This feature of the model leads to a marked decrease in the number of pulling cross-bridges during shortening in the high-force range,

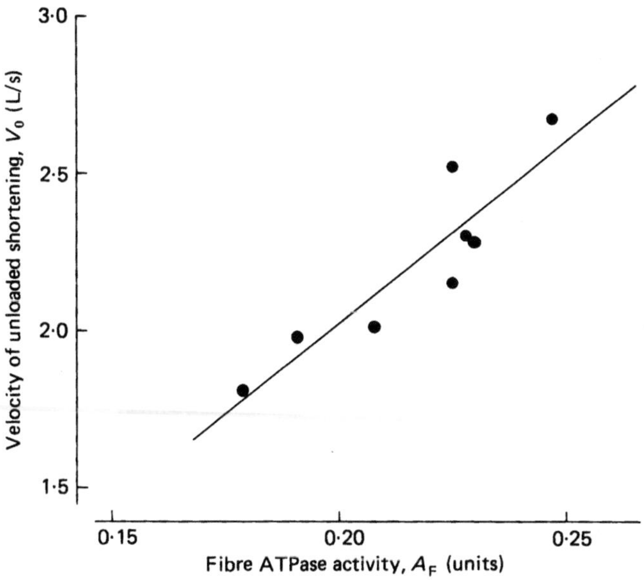

Fig. 5 Relationship between velocity of unloaded shortening (V_0, lengths/s) and myofibrillar ATPase activity (A_F, units) in eight anterior tibialis muscle fibers from frogs. *Line*: least-squares regression of V_0 upon A_F. From Edman et al. (1988)

and to a lower force output per bridge. Taken together, these changes account for the upwards-concave shape of the force–velocity relation at high loads ($0.8\ P_0 - P_0$) according to the model.

There is reason to believe that the maximum speed of shortening (V_{max}), which is attained when the muscle is free to shorten at zero load, can be presumed to express the maximum cycling rate of the myosin cross-bridges. Supporting this view V_{max} has been found to correlate well with the maximum rate of splitting of ATP within the contractile system. This was first indicated in studies of whole muscles (Bárány 1967), and later in a more quantitative way (Edman et al. 1988), by correlating V_{max} and myofibrillar ATPase activity in isolated single muscle fibers (Fig. 5).

A critical test of the sliding-filament model is establishing whether V_{max} varies with the number of cross-bridges that interact with the thin filament. If V_{max} represents the maximum speed at which the cross-bridges are able to cycle, V_{max} may be presumed to be independent of the actual number of interacting bridges. That is, maximum speed of shortening would be expected to stay constant at different degrees of overlap between the thick and thin filaments and, also, at different degrees of activation of the contractile system. The results illustrated in Fig. 6 clearly demonstrate that these predictions hold true. Figure 6a shows that, over a wide range of sarcomere lengths (1.6–2.7 μm), the maximum speed of shortening remains very nearly constant in the individual muscle fiber while, over the same range of sarcomere lengths, there is a great variation in the fiber's capacity to produce force. Thus, in contrast to the fiber's ability to produce force, the maximum speed of

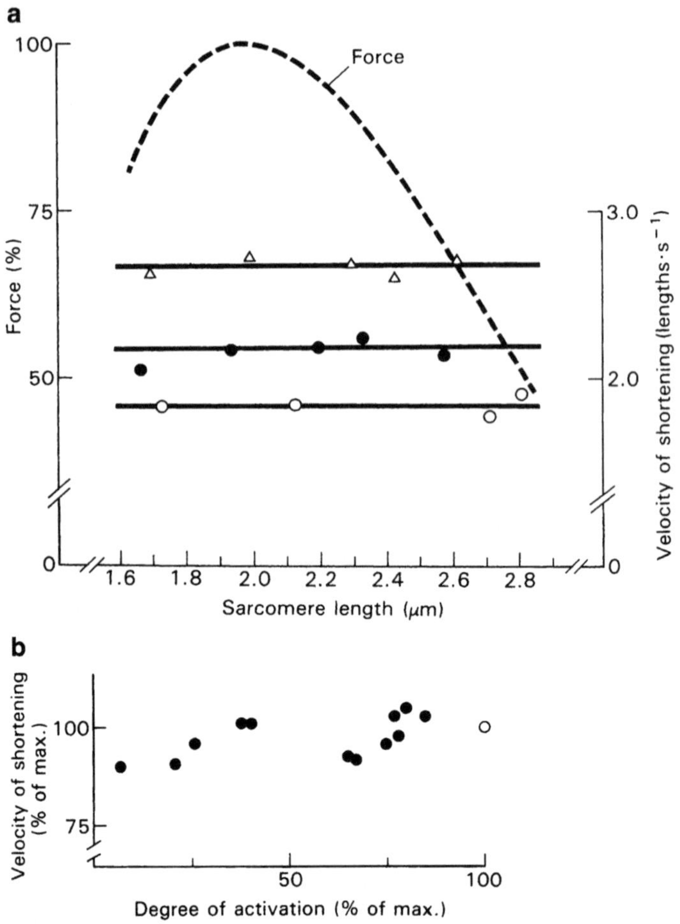

Fig. 6 (**a**) Maximum velocity of shortening (V_{max}, right ordinate) measured at various sarcomere lengths in three different fibers (indicated by different symbols). V_{max} can be seen to remain very nearly constant as sarcomere length is changed. The *dashed line* shows, for comparison, the variation in isometric force within the range of sarcomere lengths considered (1.6–2.8 μm). Replotted data from Fig. 5 of Edman (1979). (**b**) Maximum speed of shortening (V_{max}) recorded at different degrees of activation of single muscle fibers. *Open circle*, V_{max} during tetanus, i.e. at full activation in the respective fiber. *Filled circles*, Measurements of V_{max} during twitch contractions representing various degrees of activation as indicated on abscissa. V_{max} can be seen to remain quite constant as activation is changed. (Plotting of data from Table 1 of Edman 1979)

shortening does not depend on the number of myosin bridges that are able to interact with the thin filaments. The results illustrated in Fig. 6b provide evidence that V_{max} is also independent of the degree of activation of the contractile system. V_{max} during a twitch is thus virtually the same as during tetanus, even after depressing the twitch force to less than 10% of the control isometric tetanus by dantrolene, an agent known to depress the release of activator calcium from the sarcoplasmic reticulum.

4 The Slack Test Method. Braking Force of Cross-Bridges at Negative Strain

A standard approach in the past to gain information on the maximum speed of shortening of muscle has been to extrapolate from velocity measurements performed at finite loads. However, the steepness of the force–velocity curve in the low-force range makes this approach uncertain and typically leads to underestimation of the highest shortening speed that the muscle is capable of. A convenient method of measuring the speed of shortening at zero load was originally described for use on intact single muscle fibers (Edman 1979) but the approach has been found useful in other situations as well, both in intact and skinned preparations of skeletal, cardiac and smooth muscles (see further Reggiani 2007).

The method, here referred to as the slack test method, has certain features that are of interest to point out. The preparation is quickly released during activity (Fig. 7) so as to rapidly slacken the preparation to cause an abrupt fall in tension to zero, thereafter allowing the preparation to shorten actively to take up the slack. Three or more amplitudes of release (ΔL) are used, care being taken to ensure that the smallest release is more than sufficient to slacken the series elasticity. The time (Δt) measured from the onset of release to the onset of force redevelopment is plotted against the amplitude of the release movement (ΔL) as shown in Fig. 8. The slope of the straight line so derived (regression analysis) provides a measure of the speed of shortening at zero load. By this approach it is possible to exclude any contribution from the recoil of series elastic components that occurs when the

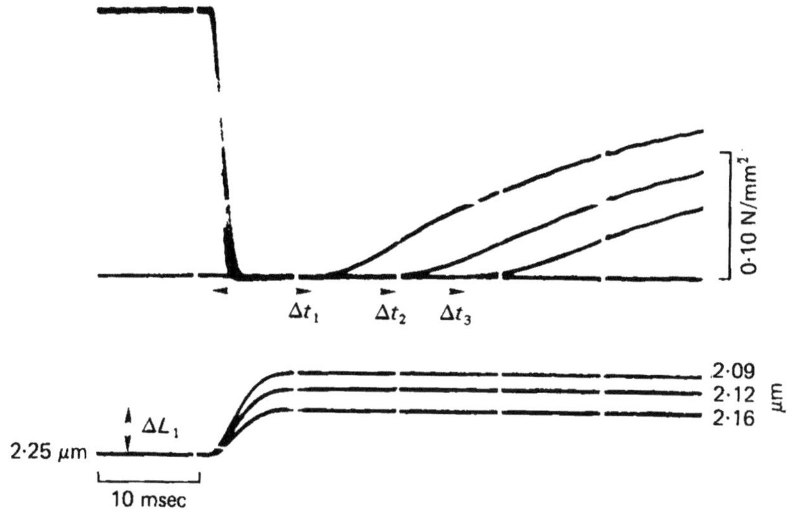

Fig. 7 Superimposed records from a single muscle fiber showing three releases of different amplitudes during plateau of tetanus. *Upper traces*: tension. *Lower traces*: release steps calibrated in µm sarcomere length. From Edman (1979)

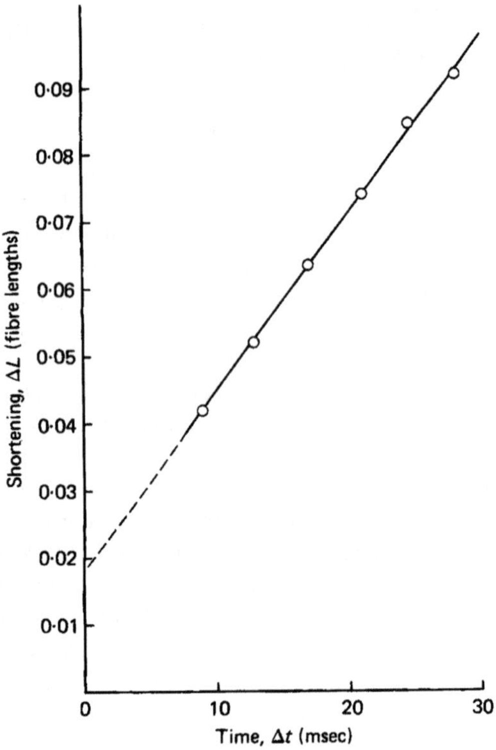

Fig. 8 Relationship between amplitude of shortening (ΔL) and time from onset of release to beginning of force redevelopment (Δt). Releases performed from 2.25 μm sarcomere length during plateau of tetanus as illustrated in Fig. 7. Each data point is the mean of five release recordings. *Straight line*: least-squares regression of Δt upon ΔL. Intersection of line with ordinate: total series compliance of preparation during isometric tetanus. From Edman (1979)

muscle is released and the tension drops to zero. The magnitude of the total series compliance is provided in the slack test measurement by extrapolating the line relating ΔL and Δt to zero time (Fig. 8).

The slack test method performed at an extended fiber length provides compelling evidence that the myosin cross-bridges are able to exert a braking force when bridges are brought into negative strain. This is demonstrated in Fig. 9, which shows, superimposed, three release steps of different amplitudes (lower panel) ending up at a given sarcomere length (2.95 μm) in each case. The resulting changes in force are shown in Fig. 9a when the release steps are carried out *at rest*. The results show that each release step results in a rapid fall in resting tension and that the tension is maintained quite constant at the new level. By contrast, as shown in Fig. 9b, when the same releases are carried out *during activity* the tension drops to zero, i.e. the response is similar to that recorded at a short sarcomere length where no resting tension exists (cf. Fig. 7).

The finding that tension is reduced to zero in the active fiber in spite of the fact that the parallel elastic elements are still under strain after the release steps (Fig. 9a), provides evidence that the passive force exerted by the parallel elastic elements is absorbed by the myofibrils: A fraction of active cross-bridges may thus be presumed to go into negative strain after the release in this way neutralizing the compressive force produced by the parallel elastic elements.

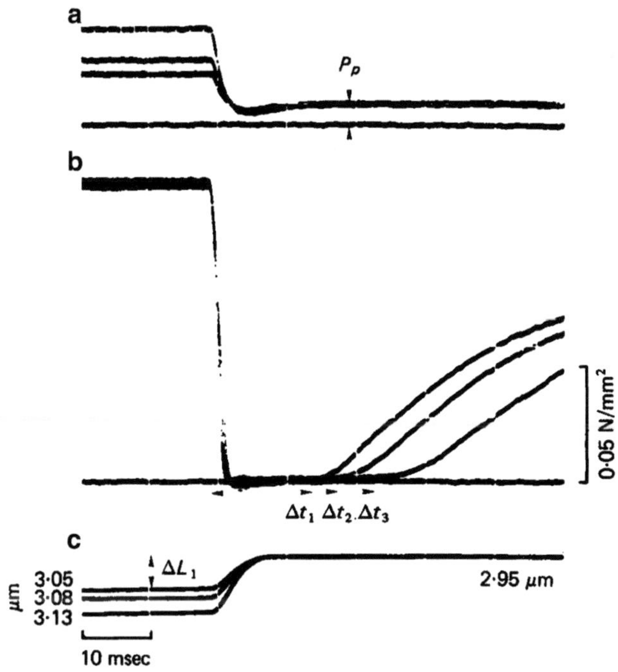

Fig. 9 Superimposed quick release records illustrating changes in force when the release steps are carried out at rest (**a**) and during the plateau of tetanus (**b**) in a prestretched fiber. The release steps (**c**) indicate the sarcomere length before and after release. Note that the fiber holds a finite resting tension (**a**) after the release steps whereas tension is reduced to zero when the fiber is quickly released during tetanus (**b**). See text for further explanation. From Edman (1979)

5 Force Enhancement by Stretch

When striated muscle is stretched during activity there is an initial rapid rise of force that reaches a breakpoint after which the force remains essentially constant or, if the muscle is stretched beyond its optimal length, increases slowly during the remainder of the stretch period. This phenomenon is referred to as 'force enhancement *during* stretch' and has been investigated in several previous studies, both in whole muscle (Fenn 1924; Abbott and Aubert 1952; Hill and Howarth 1959; Cavagna and Citterio 1974) and in single fiber preparations (Sugi 1972; Edman et al. 1978; Julian and Morgan 1979; Edman 1999). The amplitude of force attained during stretch increases with the velocity of stretch, and a plot of the enhanced force against the stretch velocity exhibits the well-known 'negative' branch of the force–velocity relationship in skeletal muscle (Katz 1939; Aubert 1956; Edman 1988). The force enhancement during stretch gradually disappears after the end of stretch and reaches a stable level within 4–5 s in a single muscle fiber at

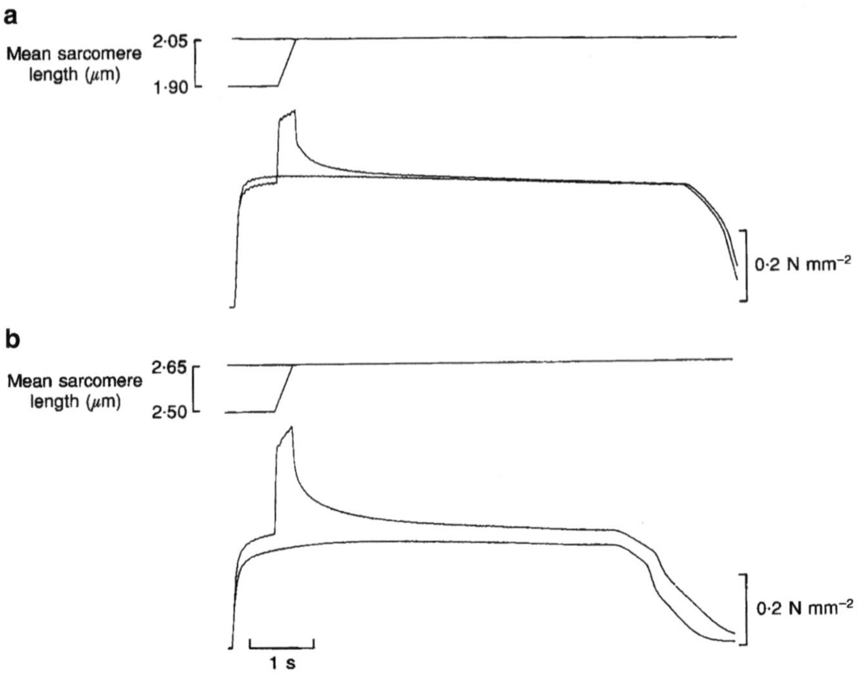

Fig. 10 Force enhancement by stretch during tetanus of a single muscle fiber at two different sarcomere lengths. (**a**), Stretch during activity from 1.90 to 2.05 μm sarcomere length compared with ordinary isometric tetanus at 2.05 μm. (**b**), Stretch from 2.50 to 2.65 μm sarcomere length compared with isometric tetanus at 2.65 μm. Note that the same force level is finally reached in both records in (a), whereas force after stretch remains above the isometric control level throughout the activity period in (b). From Edman and Tsuchiya (1996)

low temperature (Fig. 10a, b). It is noteworthy, however, that if the stretch has been performed above optimal length, the final tension remains clearly higher than the control isometric tetanus at the stretched length. This remaining component of force increase is independent of the stretch velocity and is referred to as 'residual force enhancement *after* stretch' (Edman et al. 1978).

5.1 Force Enhancement During Stretch, Its Relation to Sarcomere Length and Myofilament Lattice Width

The force enhancement during stretch is likely to mark the point at which the cross-bridges can no longer withstand the external pull and therefore start to slide along the thin filament. In a study of length controlled segments of intact muscle fibers it was demonstrated (Edman 1999) that the amplitude of force enhancement during stretch varies with sarcomere length but in a way that is markedly different from that of the plain isometric force.

Figure 11 illustrates oscilloscope records from a length-controlled segment of an intact muscle fiber that is subject to stretch during the plateau of an isometric tetanus. The stretching can be seen to cause a rapid rise of tension to a breakpoint after which the tension remains constant during the remainder of the stretch period. The force enhancement considered here is the rise in force above the isometric level ending at the breaking of the force myogram (marked *b* in Fig. 11).

As demonstrated in Fig. 12a (filled symbols), force enhancement during stretch increases with sarcomere length above 1.8 μm reaching a maximum at approximately 2.4 μm, at which point the enhanced force can be seen to be equally

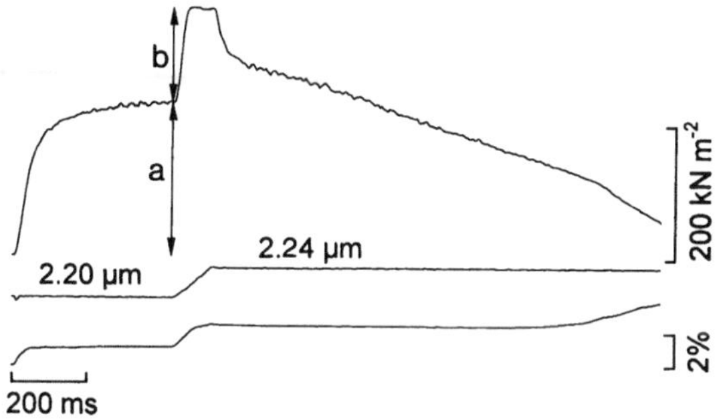

Fig. 11 Force and displacement records from a length-controlled segment of an intact muscle fiber subjected to stretch on the plateau of a fused tetanus. *Upper trace*, force; *middle trace*, segment length calibrated in μm per sarcomere; *lower trace*, change of puller position as required to adjust the segment length. Note the rapid rise of force after the onset of stretch ending at a breakpoint after which the force remains quite constant during the rest of the stretch ramp. Stretch amplitude at the breakpoint of the force myogram is approximately 8 nm per half sarcomere. (**a**), Isometric force; (**b**), Measured force enhancement. From Edman (1999)

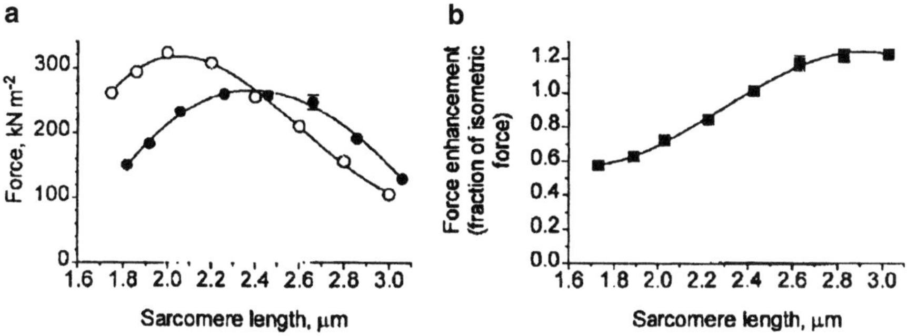

Fig. 12 Isometric tetanic force and force enhancement during stretch in relation to sarcomere length. (**a**) Measurements of tetanic force (*Open circles*) and force enhancement during stretch (*filled circles*) recorded from a given length-controlled segment of a single muscle fiber. Mean values of five or six repeated recordings at each sarcomere length. Error bars (if exceeding symbols) indicate SEM From Edman (1999). (**b**) Force enhancement during stretch normalized to the tetanic force recorded at respective sarcomere length. From Edman (1999)

large as the isometric force. With further increase in sarcomere length the force enhancement during stretch becomes gradually smaller. Figure 12b shows the data from Fig. 12a replotted as a fraction of the control isometric force recorded at the various sarcomere lengths in this way normalizing the force enhancement to the amount of filament overlap at each sarcomere length. Pooled data from seven experiments are shown in Fig 13. The force enhancement during stretch can be seen to increase steadily, relative to the isometric force, as the sarcomere length is changed from approximately 1.8 to 2.8 μm sarcomere length. This finding strongly suggests that the *width* of the myofilament lattice is an important determinant of the force enhancement during stretch given the fact that an intact muscle fiber behaves as a closed volume (see further below). The relative fiber width at the different sarcomere lengths is indicated above the abscissa in Fig. 13 and has been calculated on the assumption that the myofilament lattice maintains a constant volume within the range of sarcomere lengths considered (Elliott et al. 1963; Huxley 1969; Millman 1998).

The data in Fig. 13 show that a decrease in relative fiber width below the value (1.05) existing at 1.8 μm sarcomere length is associated with a steady increase of the normalized force enhancement during stretch. Maximum force enhancement is attained near 2.8 μm sarcomere length at which point the relative fiber width is estimated to be merely 0.85 of the value at 2.0 μm sarcomere length. The tighter packing of the myofilament lattice at the greater sarcomere length apparently makes the myosin cross-bridges more capable to withstand stretch; the resistive force can be seen to increase by approximately 65% by increasing the sarcomere length from approximately 1.8 to 2.8 μm (Fig. 13). This width dependence of the force enhancement during stretch was confirmed in the same study by varying the fiber width over a similar range by altering the tonicity of the extracellular medium (for details, see Edman 1999).

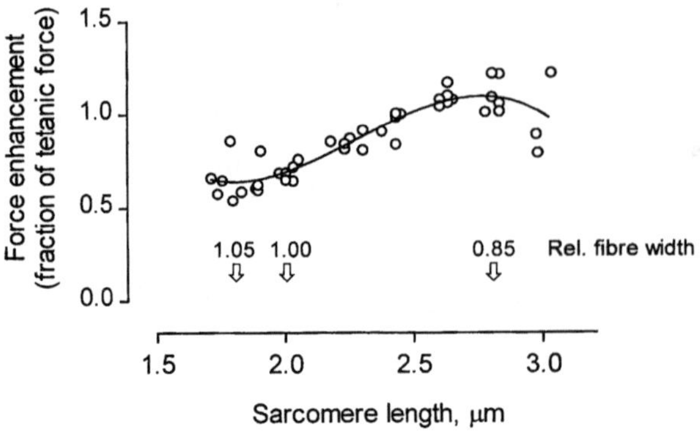

Fig. 13 Force enhancement during stretch (normalized to tetanic force) related to sarcomere length and fiber width. Pooled data from seven experiments similar to that illustrated in Fig. 12b. The relative fiber width at different sarcomere length is indicated above the abscissa. From Edman (1999)

The above results suggest that the strength by which the myosin bridges bind to the thin filaments during activity is greatly dependent on the width of the myofilament lattice. It is of interest to note in this connection that the myosin head (approximately 17 nm long and 6 nm wide; Elliott and Offer 1978; Rayment et al. 1993) is large enough to fill the entire gap between the thick and thin filaments near slack fiber length, the side-to-side distance between the two filaments being 17.2 nm at 1.8 μm sarcomere length. (The estimation of the interfilament spacing is based on published data on the 1.0 lattice spacing of the X-ray diffraction pattern and on electron microscopic measurements of filament diameters; for details see Edman 1999). The free space available for the myosin head will become drastically reduced when the sarcomere length is increased above slack length. At 2.8 μm sarcomere length the side-to-side distance between the thick and thin filaments can be estimated to be merely 11.9 nm. The tighter packing of the myofilaments at greater sarcomere lengths can be presumed to increase the contact area between the actin and myosin molecules. This may increase the opportunities for bond formation between the two structures and lead to firmer attachment of the bridge. Another corollary of the limited space between the filaments is that the shaft of the myosin bridge (the S2 region) will stay virtually parallel with the actin filament over the entire working range of sarcomere lengths. This is fully in line with the earlier conclusion that the force producing capability of striated muscle is a function of the length of filament overlap per se with no relation to the myofilament lattice width (Edman and Andersson 1968).

5.2 *Residual Force Enhancement After Stretch*

The long lasting component of force enhancement by stretch (Fig. 10b) that remains at the end of a long (5–10 s) tetanus was originally observed on isolated whole muscles of the frog (Fenn 1924; Abbott and Aubert 1952; Hill and Howarth 1959) and has later been subjected to detailed studies on isolated single muscle fibers (Edman et al. 1978, 1982; Sugi 1972; Edman and Tsuchiya 1996; Sugi and Tsuchiya 1988; Julian and Morgan 1979; Rassier and Herzog 2004a, b; Herzog and Rassier 2006) and, recently, on isolated cardiac and skeletal muscle myofibrils (Telley et al. 2006; Joumaa et al. 2008; Rassier et al. 2003). The continued interest in this phenomenon is based on the fact that the long lasting force enhancement by stretch does not seem to be readily explainable on the basis of the sliding-filament hypothesis. The muscle fiber is, however, a complex mechanical structure, to say the least, and there is so far no convincing evidence that the residual force enhancement after stretch represents a phenomenon that is not ultimately based on the sliding-filament model.

In early studies on intact single muscle fibers (Edman et al. 1978, 1982) a number of salient features of the phenomenon ‘force enhancement after stretch’ were described. The magnitude of force enhancement was found to be independent of the stretch velocity, to increase with the amplitude of stretch and to occur on the descending limb of the length–tension relationship only. It was found that the absolute magnitude of the residual force enhancement increased with sarcomere length to a maximum at ~2.9 μm and then declined. Particular attention was paid to the question of whether the residual force enhancement after stretch might exceed the control isometric force at optimal sarcomere length. Our results showed that stretches causing a marked force enhancement above optimal length, did not clearly increase the force when performed on the plateau of the isometric length–tension relationship (Edman et al. 1982).

The notion that force enhancement after stretch is based on non-uniform sarcomere behaviour in response to stretch is supported by electron microscopic studies in which muscle fibers have been rapidly fixed after a stretch ramp. Brown and Hill (1991) were first to demonstrate marked irregularities of myofilament overlap within myofibrillar sarcomeres along the muscle fiber after a stretch ramp. That is, in spite of an overall elongation of the sarcomere pattern in response to the stretch the degree of filament overlap may differ between individual sarcomere halves. Thus, the amount of filament overlap in one half of the sarcomere may actually become larger during and after stretch at the expense of the filament overlap in the opposite half of the sarcomere which becomes correspondingly smaller. The uneven distribution of filament overlap after stretch has been confirmed in other similar studies (Edman and Lou, unpublished data based on a rapid-freezing technique; Edman and Lou 1992; Talbot and Morgan 1996).

An essential component of the mechanism of residual force enhancement after stretch is the development of strained elastic elements during the stretch process. This particular problem was addressed by Edman and Tsuchiya (1996) in a study of frog muscle fibers with simultaneous measurements of length and force from the fiber as a whole and from marked segments of the same fiber. The existence of strained elastic elements along the fiber during tetanic contraction manifested itself in a damped initial transient when the fiber was released to shorten against a very small load at different times after a stretch ramp. The initial transient increased linearly with the amplitude of the preceding stretch and with the force enhancement existing at the moment of release (for details, see Fig. 4 in Edman and Tsuchiya 1996). The evidence presented by Edman and Tsuchiya suggests that strain of passive elastic elements along the fiber does not in itself explain the extra force recorded during and after stretch. The extended elastic elements are likely to be dispersed within the fiber volume and merely serve to support the weakened parts of the myofibrils (parts with reduced filament overlap) to enable them to match the stronger parts of the fiber during and after stretch. According to this view the measured force enhancement after stretch is thus actively produced by those sarcomeres, or sarcomere halves, that have acquired a more favorable degree of filament overlap during stretch than other regions in series.

Consistent with this view no residual force enhancement after stretch would be expected when the fiber is stretched over the "plateau" of the length–tension relation. The strongest regions after stretch would already be at optimal filament overlap and so would not achieve a greater force after stretch than that produced in the isometric control (however, see below). Results reported by Edman et al. (1982), based on a large number of experiments, support this prediction by showing that the residual force after stretch was not clearly above the plateau of the length–tension curve. This problem has recently been revisited by Lee and Herzog (2008) who conclude that a stretch ramp carried out within the plateau of the length–tension relation may result in a small (4.3%) but distinct residual force enhancement.

The effect described by Lee and Herzog (2008) by no means disqualifies the idea that force enhancement after stretch is based on non-uniform sarcomere behaviour. Firstly the length–tension relationship in frog muscle fibers does not have a perfectly flat maximum between 2.0 and 2.2 μm sarcomere length as early studies seemed to suggest (Edman 1966; Gordon et al. 1966). In reality there is a steady increase of the tetanic force of approximately 7%, as the overall sarcomere length is changed from 2.2 to 1.95 μm as demonstrated in an extensive study on short, length-controlled segments of isolated frog muscle fibers (Edman and Reggiani 1987). The absence of a real plateau of the length–tension relation can be explained by the fact that the thick and thin filaments are not constant in length and, furthermore, that there is normally some staggering (longitudinal misalignment) of the filaments within the A and I bands in a given fiber cross-section (see further Edman and Reggiani 1987). To this may be added that the force producing capacity may differ by several (up to 10) % along the length of the muscle fiber (Edman et al. 1985). Taken together these findings imply that one or more regions of the fiber are actually capable of producing a somewhat greater force than that recorded from the fiber as a whole within the "plateau" of the length–tension relation (see Edman and Reggiani 1987). During stretch, on the other hand, it is the *strongest* regions of the muscle fiber that alone determine the force recorded from the fiber. The weaker parts in series will match the weaker segments by recruiting supportive elastic components as described in some detail by Edman and Tsuchiya (1996).

6 Force Reduction After Loaded Shortening

It has been known for a long time (Abbott and Aubert 1952) that striated muscle that is released to shorten against a high load during tetanus produces less force after the end of shortening than during a fixed-end tetanus at the corresponding length (Maréchal and Plaghki 1979). Figure 14 illustrates the force depression after loaded shortening in a single muscle fiber (a) and, for comparison, the lack of force depression when the same amount of shortening occurs without load (b). Non-uniform sarcomere behaviour was early considered as one possible cause of the force deficit, an explanation that gained direct experimental support by Julian and Morgan (1979) in studies of single muscle fibers. However, there has been a continuous search for alternative explanations and several imaginative mechanisms have been proposed (e.g. Maréchal and Plaghki 1979; Sugi and Tsuchiya 1988; Granzier and Pollack 1989; Schachar et al. 2004). The phenomenon of force depression by loaded shortening was revisited by Edman, Caputo and Lou (1993) in frog single muscle fibers using a combination of supplementary techniques to elucidate the possible role of non-uniform sarcomere behaviour.

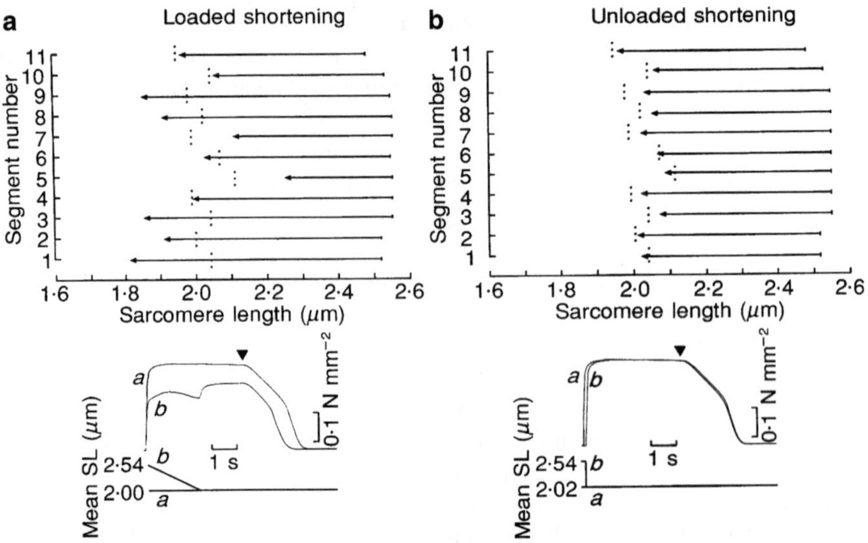

Fig. 14 Differential length changes along a single muscle fiber during tetanus with loaded (**a**) and unloaded (**b**) shortening, respectively. The change in sarcomere length from rest to the end of the tetanus plateau (*triangle* above myograms) is indicated by *horizontal arrows* for 11 consecutive segments. *Dotted vertical lines* indicate the sarcomere length in the same segments during fixed-end tetanus at the short fiber length. The superimposed records below the diagrams show force and puller position during control tetanus at short length (myograms a) and during tetanus with shortening (myograms b). Note that loaded shortening leads to a marked force depression and a wide distribution of sarcomere length. By contrast, unloaded shortening does not result in force depression and the sarcomere pattern after unloaded shortening is not markedly different from that of a fixed-end tetanus. From Edman et al. (1993)

The following observations were made:

1. The force deficit after loaded shortening is quantitatively related to the variation in sarcomere shortening along the length of the fiber. The interrelationship between force depression and sarcomere non-uniformity after loaded shortening is exemplified in Fig. 14 and is presented statistically in Fig. 15.
2. The force depression after loaded shortening disappears after imposing a brief (10 s) period of relaxation (Fig. 16). Furthermore, as illustrated in Fig. 16, the recovery of force after relaxation was associated with the reappearance of a quite uniform sarcomere pattern along the fiber.
3. No force deficit occurs after loaded shortening when recording is made from the short length-controlled segment of a single muscle fiber. This finding holds true either the loaded shortening occurs above or below slack length (Fig. 17).
4. A computer model of a typical fiber with non-uniform sarcomere behaviour simulates the force deficit after loaded shortening exceedingly well. For details, see Edman et al. (1993).

Taken together the above results would seem to leave little doubt that non-uniform sarcomere behaviour is the main cause of the force deficit after loaded shortening

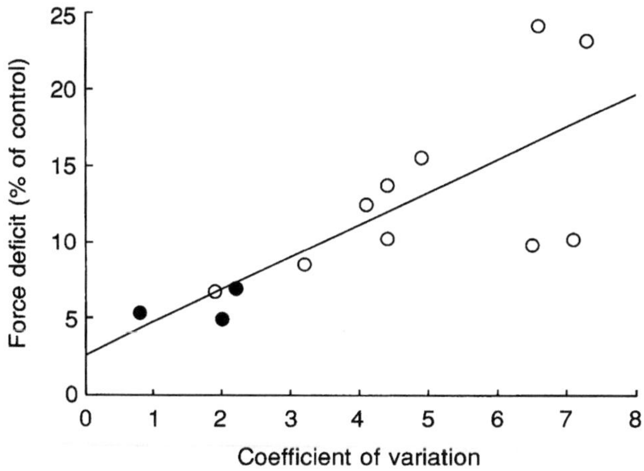

Fig. 15 Relation between force deficit after loaded shortening (ΔF) and coefficient of variation (K_v) of sarcomere length along the muscle fiber. K_v is the standard deviation expressed as percentage of the mean. *Open circles*, shortening above slack length; *filled circles*, shortening below slack length. Regression line ($P<0.005$): $\Delta F = 2.10\ K_v + 2.55$. From Edman et al. (1993)

in striated muscle. It is essential to point out that the force depression so produced is distinct from the transitory deactivation by active shortening to be described next.

7 Deactivation by Active Shortening

When striated muscle is allowed to shorten during activity it loses temporarily some of its capacity to produce force. This depressant effect of shortening has been demonstrated in both isolated muscle preparations (Jewell and Wilkie 1960; Edman and Kiessling 1971; Briden and Alpert 1972) and muscles in situ in the body (Joyce et al. 1969). This phenomenon, which is unrelated to fatigue, has previously been studied in considerable detail on both intact and skinned muscle fibers (Edman 1975, 1980, 1981; Ekelund and Edman 1982).

The depressant effect of active shortening is demonstrated in Fig. 18 which illustrates two superimposed records of a partially fused tetanus in a frog single muscle fiber. Record A (control) shows the development of force at 2.05 μm sarcomere length when the fiber ends are fixed throughout the entire stimulation period. In record B the same stimulation sequence is repeated, but here the fiber is initially set at a longer sarcomere length, 2.55 μm, and the fiber allowed to shorten to 2.05 μm during the first twitch period. The fiber is arrested at this length to develop force during the remainder of the stimulation period, i.e. at the same length as in the control run. The records in Fig. 18 clearly demonstrate that force development is depressed after the shortening phase. Twitch No. 2 in myogram B can thus be seen to be considerably lower than the first twitch in the control run.

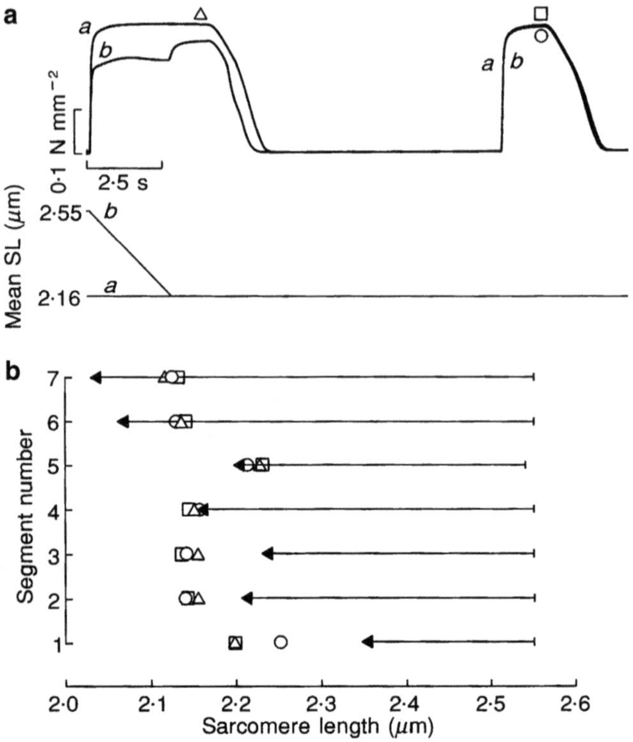

Fig. 16 Disappearance of force depression after loaded shortening by interposing a brief period of relaxation. The superimposed myograms in (**a**) show the following stimulation sequence which was repeated every 4 min: 4 s tetanic stimulation followed by a 10 s pause and a new, 1.5 s period of tetanic stimulation. In myogram b the fiber was allowed to shorten against a high load from a mean sarcomere length of 2.55–2.16 μm during the 4 s tetanus. In myogram a (control) the fiber was held at the short length throughout. Note that the force deficit produced by loaded shortening during the first tetanus is no longer seen when tetanization is resumed after relaxation. The *horizontal arrows* in (**b**) show the change in sarcomere length of successive segments along the fiber during the 4 s tetanus with loaded shortening. *Open triangles*, *open squares* and *Open circles* indicate the sarcomere length in the same segments at the times shown at respective force myogram in (**a**). Note that the force deficit after loaded shortening is associated with a grossly irregular sarcomere pattern along the fiber and that the length pattern returns to normal, together with restoration of the tetanic force, as the fiber is restimulated after a brief period of relaxation. From Edman et al. (1993)

This is significant since tension starts from zero level and occurs at the same sarcomere length in both cases. The lower force recorded in twitch period No. 2 in myogram B thus represents a true depression of the fiber's capacity to produce force, due to the preceding shortening. It is clear, however, that the depressant effect of shortening disappears with time during continued stimulation. Full strength can be seen to be regained 1–2 s after the end of the shortening phase.

There is evidence to suggest that shortening temporarily reduces the degree of activation of the contractile system. In line with this view the depressant effect of

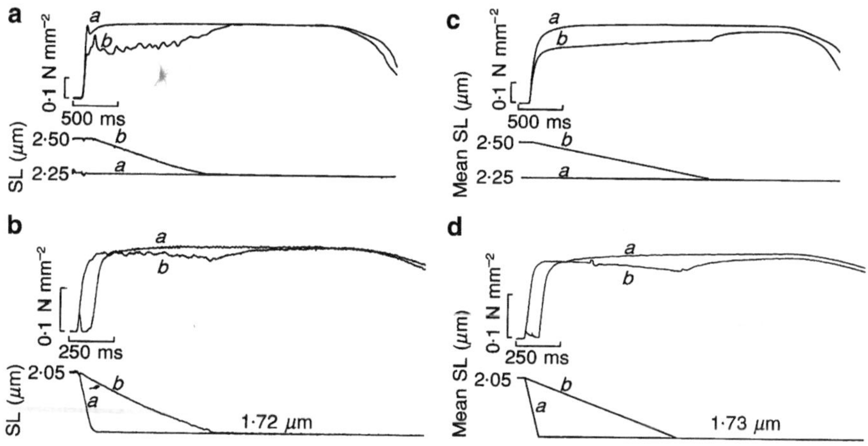

Fig. 17 Lack of force deficit after loaded shortening of length-controlled segment of single muscle fibers above (**a**) and below (**b**) slack length. (**c**) and (**d**) show, for comparison, the effects of loaded shortening when recording is made from the fiber as a whole. The same fiber in (**a**) and (**c**) and in (**b**) and (**d**), respectively. Myograms a, control tetani; myograms b, tetani with loaded shortening. Control tetani below slack length contain an initial phase of unloaded shortening. From Edman et al. (1993)

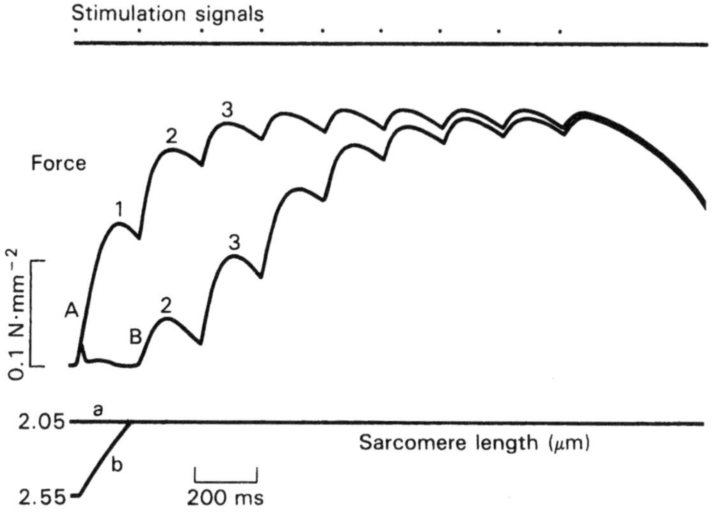

Fig. 18 Transitory depressant effect of shortening during partially fused tetanus of a frog muscle fiber. The superimposed myograms A and B show force development at 2.05 μm sarcomere length. In myogram B contraction is initiated at 2.55 μm sarcomere length and the fiber is allowed to shorten to 2.05 μm during the first twitch cycle as indicated by the length record underneath. The first few twitch cycles in the myograms are numbered for identification in the text

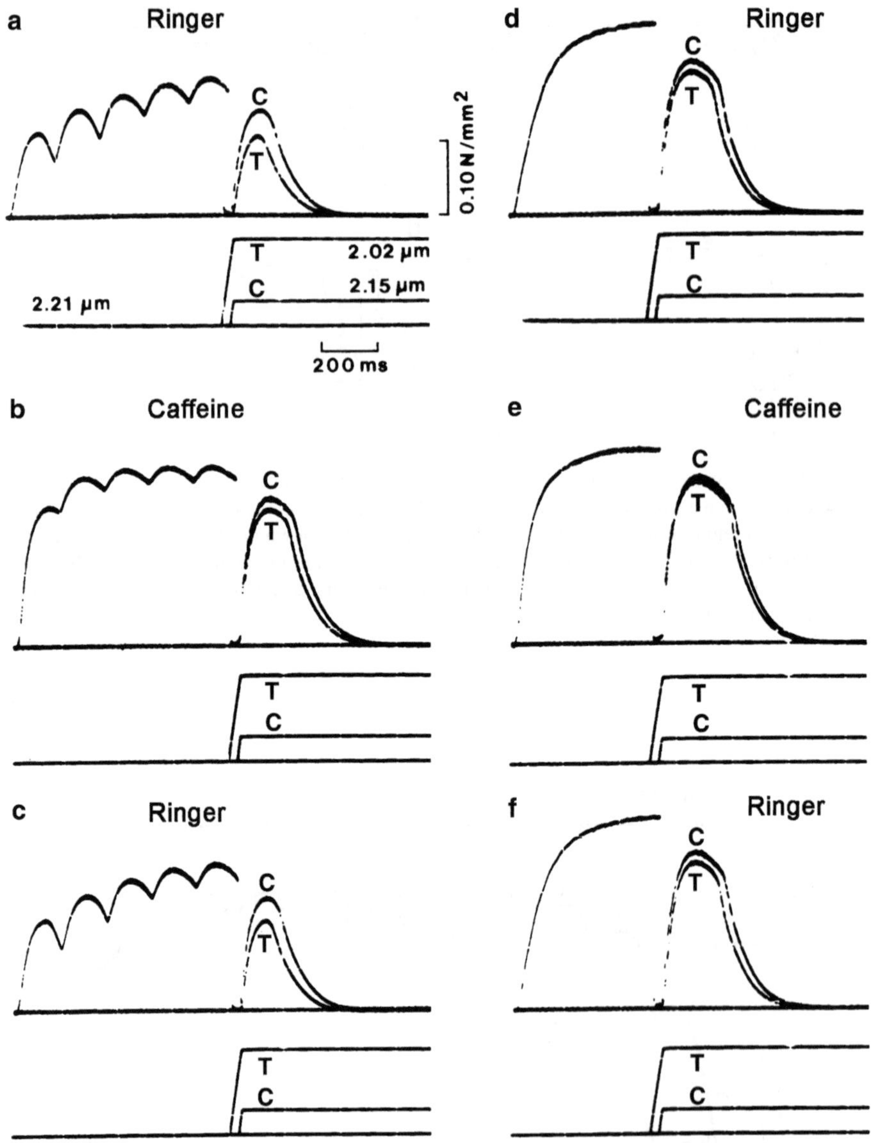

Fig. 19 Depressant effect of active shortening tested at different states of activity of a single muscle fiber. Control (C) and test (T) movements performed during the last cycle of partially fused tetanus (**a**–**c**) and prior to the last stimulus of fused tetanus (**d**–**f**). a, d: fiber immersed in ordinary Ringer solution. b, e: 45–65 min after addition of 0.5 mM caffeine to the bathing fluid. c, f: 50–65 min after change to ordinary Ringer solution. *Lower traces* in each *panel* indicate sarcomere length before and after the shortening. Temperature: 2.5°C. From Edman (1980)

shortening is diminished if the state of activity in the fiber is raised. The top left panel (a) of Fig. 19 illustrates the depressant effect in a partially fused tetanus. Myogram *a* (control) of Fig. 19 shows the redevelopment of force after a small shortening which is just sufficient to produce a drop in tension and an immediate redevelopment of force. In myogram *b* the amplitude of shortening is larger, and the release is timed so as to make the redevelopment of force start at the same time as in the control. The additional amount of shortening produced in myogram *b* can be seen to depress the amplitude of the redeveloped force by approximately 25%. Panel d shows, for comparison, the corresponding effect of shortening during a *fused* tetanus in the same fiber, i.e. under conditions when the contractile system can be presumed to be fully activated. The shortening is here produced during the plateau of the tetanus, before the last stimulus. The resulting depression of force (approximately 7%) can now be seen to be considerably smaller than during the partially fused tetanus.

The middle panels (b and e) of Fig. 19 shows that the depressant effect of shortening is much reduced in the presence of caffeine in a concentration that is known to increase the release of calcium from the sarcoplasmic reticulum in amphibian muscle (Blinks et al. 1978). The depressant effect of shortening can be seen to be greatly reduced in the presence of caffeine, both in absolute and relative terms, during the partially fused tetanus, and the movement effect is almost abolished during the fused tetanus. Finally, the depressant effect of shortening is fully restored after removal of caffeine from the bathing fluid (panels c and f).

The above results support the view that active shortening temporarily reduces the state of activation of the contractile system, most likely by reducing the affinity of calcium at its binding-sites on troponin (see further below). In line with this view, the movement effect is large when the contractile system is submaximally activated as is the case during a twitch and a partially fused tetanus. This interpretation of the movement effect is also strongly supported by the finding that there is virtually no force depression by shortening when the contractile system is maximally, or supramaximally, activated which is the case during a fused tetanus in the presence of caffeine. Under these circumstances a decrease in calcium affinity at the binding sites would be compensated by a higher free calcium concentration in the myofibrillar space.

Experiments on skinned muscle fibers from frog and mouse provide further support for the view that the movement effect is based on a change in the myofilament system that leads to a temporary decrease in mechanical activity. The approach used is illustrated in Fig. 20 which shows the redevelopment of force after a small (control) and large (test) release step (C and T, respectively) during a calcium induced contraction of a skinned muscle fiber of the mouse. The records readily show that the rate of force redevelopment is lower after the larger shortening step. However, the same final tension was finally attained after both amplitudes of release. For quantification of the depressant effect of shortening the *rate* of force redevelopment, at 25% of maximum force, was used as an index as shown in Fig. 21 (inset). The fact that the measurements refer to the same tension level in test and control implies that the load on the contractile system was the same in both cases. As illustrated in Fig. 21 (main diagram), an increase in free calcium concentration from 1.5 to 6–12 μM greatly reduced the depressant effect of shortening.

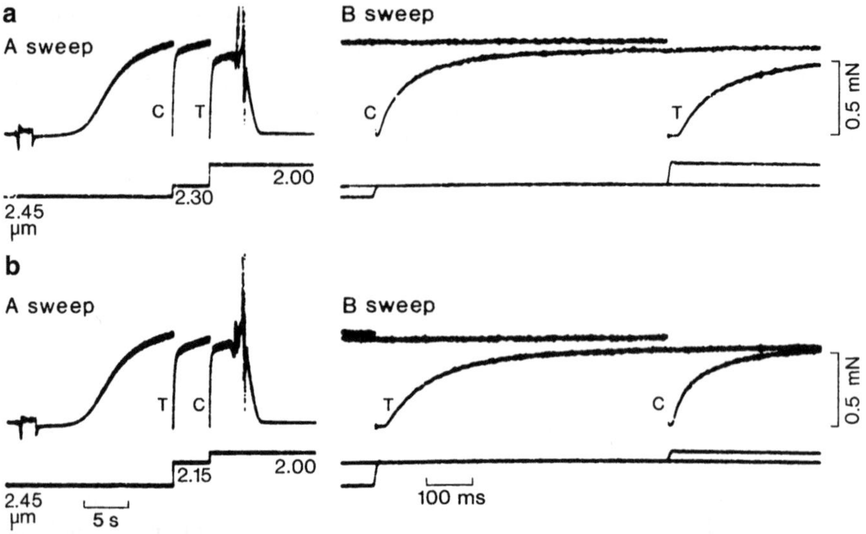

Fig. 20 Depressant effect of active shortening during calcium induced contraction of chemically skinned fiber of the mouse. Small (control) and large (test) release steps, designated C and T, respectively, carried out in sequence during the same contraction. The order of C and T reversed in (**a**) and (**b**). The *right-hand traces* show the same release recordings on a faster time base. Note that the rate of force redevelopment is smaller after the large (test) release. From Ekelund and Edman (1982)

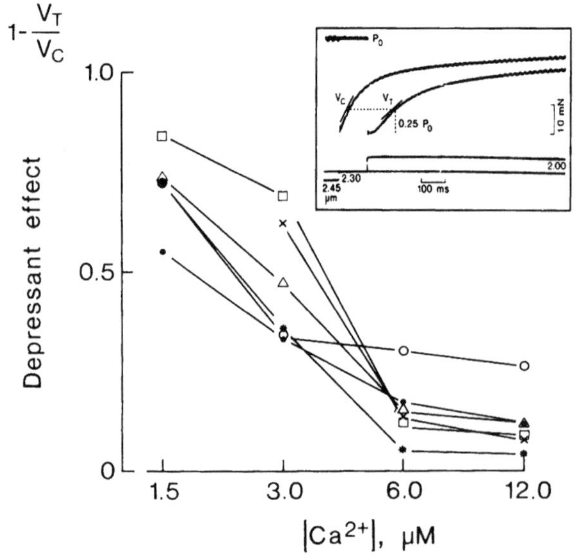

Fig. 21 Depressant effect of active shortening of skinned muscle fibers of the mouse related to the free calcium concentration. The depressant effect was measured as the decrease in the rate of force redevelopment at 25% of maximum isometric force as illustrated in the *inset*. Each data set denoted by the *same symbol* is from a given fiber preparation. From Ekelund and Edman (1982)

The evidence suggests that active sliding of the myofilaments leads to a decrease in activation of the contractile system. Activity is needed for the depressant effect to occur. That is, the two filaments have to interact during sliding for the movement effect to occur (Edman 1980). These findings support the view (Edman 1975; Ekelund and Edman 1982) that the myosin bridges interfere with the regulatory proteins as the bridges interact with the thin filaments during muscle shortening. The movement effect may actually be a manifestation of a cooperative activity between the proteins of the calcium-regulated thin filament leading to a transitory decrease in calcium affinity and, therefore, to a transient deactivation of the contractile machinery. In line with this idea, there is a transitory increase in the myoplasmic calcium ion concentration during the shortening process, suggesting that calcium previously bound to troponin is released due to the lowered affinity for calcium at the binding sites (Fig. 22).

There is reason to believe that shortening induced deactivation plays a part in normal life. Mammalian motor units are activated to produce partially fused tetani and under such conditions the movement effect may be expected to be pronounced as demonstrated in Figs. 18 and 22. The movement effect may set a limit to what

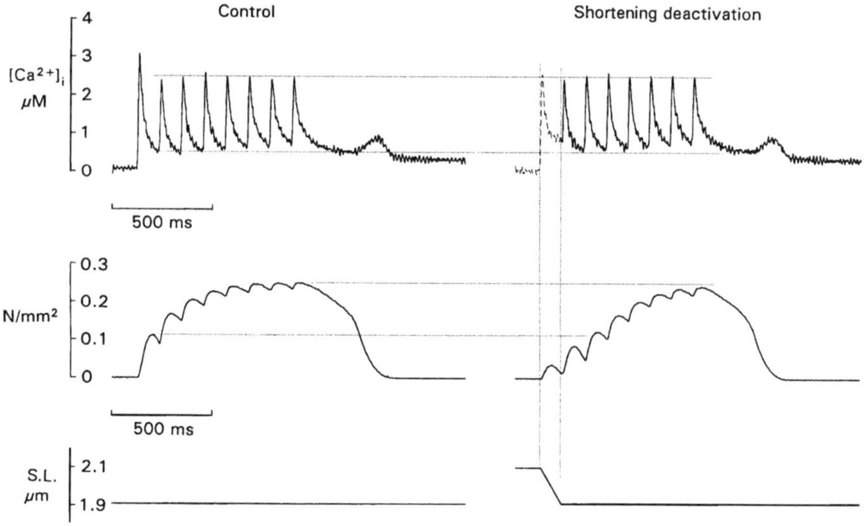

Fig. 22 Simultaneous recordings of intracellular free calcium ($[Ca^{2+}]$i) and force (*middle traces*) during a partially fused tetanus of a single muscle fiber of Rana temporaria. The *bottom trace* shows puller position calibrated in μm sarcomere length. *Left panel*: isometric contraction at 1.91 μm sarcomere length; *right panel*: the contraction is initiated at 2.1 μm sarcomere length and the fiber is allowed to shorten to 1.91 μm during the first twitch period. The transitory force depression by shortening is quite similar to that described in Fig. 18. Note that $[Ca^{2+}]$i is substantially higher immediately after the end of shortening as compared with the corresponding time in the control. However, from the second stimulus onwards, the $[Ca^{2+}]$i transient is almost identical to that recorded in the control run, while the force remains depressed and only slowly recovers after the shortening. The approach used for the intracellular calcium measurement has been described by Caputo, Edman, Lou and Sun (1994)

the body muscles are able to achieve, for example during running, weightlifting etc. This might in fact be the main purpose of the effect, to serve as a safety mechanism to prevent overuse of the muscles.

8 Differences in Kinetic Properties Along Individual Muscle Fibers

It is well established that individual muscles in the body vary with respect to their maximum speed of shortening (Close 1972; Buchthal and Schmalbruch 1980). Marked differences in kinetic properties are also known to exist among individual fibers in a muscle, and there is abundant evidence to suggest that these differences in mechanical performance among muscles and individual muscle fibers are based on differences in isomyosin composition of the myofilament system (Schiaffino and Reggiani 1996; Bottinelli and Reggiani 2000). Marked differences in shortening characteristics may exist among fibers in a given muscle. This applies in particular to mammalian and avian skeletal muscles, but substantial differences in shortening kinetics, although less pronounced than those observed among mammalian muscle fibers (Julian et al. 1981), are also observed among twitch fibers of amphibian skeletal muscles (Edman and Hwang 1977; Edman 1979).

There is now evidence that the kinetics of shortening may even differ substantially from one part to another *within* individual fibers in vertebrate striated muscles. Experimental data illustrating this point are shown in Fig. 23. In these experiments (Edman et al. 1985) opaque markers were placed on the fiber surface, 0.5–0.8 mm apart, along the length of the preparation. The change in distance between two adjacent markers (one segment) was monitored by means of a photoelectric recording system, while the fiber was released to shorten isotonically against a load close to zero during tetanic stimulation. The speed of shortening (V_0) was recorded during repeated tetani from each of the segments, the accuracy of these measurements being better than 4% in all parts of the fiber. As exemplified in Fig. 23a the maximum velocity of shortening differed substantially along the fibers, each fiber exhibiting a unique pattern of velocities. In general there was a clear trend towards an increase or decrease in V_0 over a sequence of segments, but in some cases an abrupt, large change in velocity was seen between two adjacent segments. The observed differences in velocity along a given fiber remained the same (within the accuracy of the measurement) over several hours of experimentation (Fig. 23b). It is noteworthy that the segmental differences in V_0 did not bear any relation to the passive elastic properties of the muscle fibers (see further Edman et al. 1985) supporting the view that the observed differences in shortening velocity do reflect true differences in the kinetic properties of the myofilament system along the fiber.

The segmental differences in V_0 were found to be associated with specific differences in the shape of the force–velocity relationship. Segments having a lower V_0 were consistently found to be less curved, i.e. to have a higher value of a/P_0 in terms of Hill's (1938) hyperbolic equation. This finding has an interesting

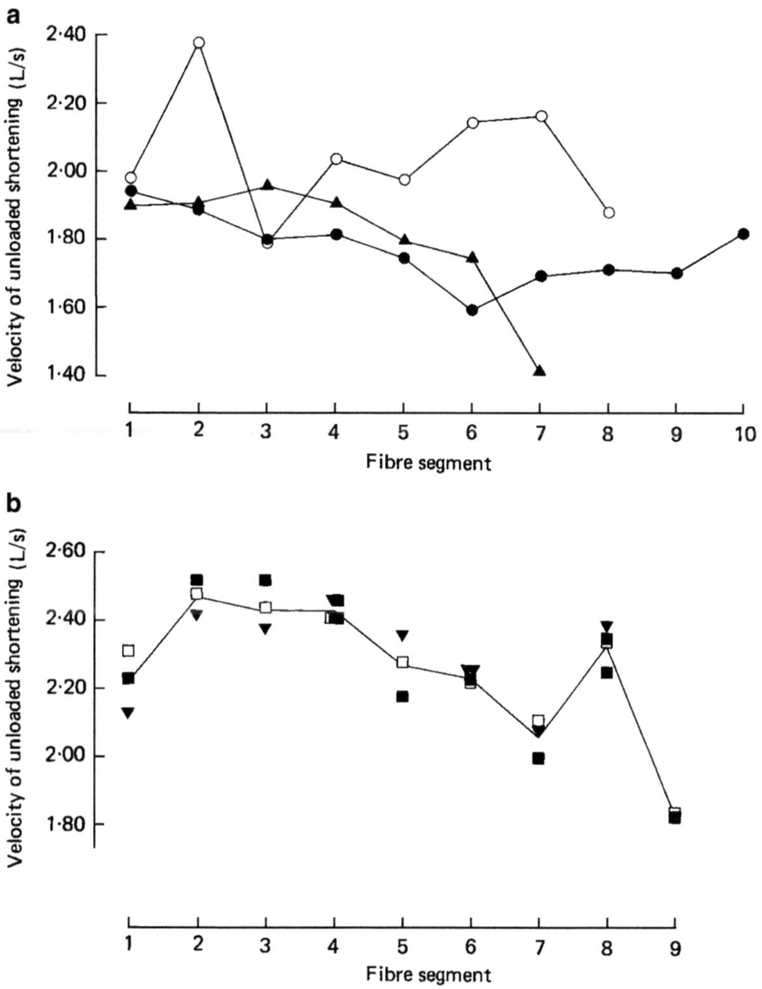

Fig. 23 (**a**) Velocity of unloaded shortening (V_0) recorded in consecutive segments of three different muscle fibers. Length of segments, 0.6–0.8 mm. Total fiber length at 2.25 μm sarcomere length: *Open circles*, 7.65 mm; *filled circles*, 8.40 mm; *filled triangles*, 6.25 mm. Note that each fiber has a unique V_0 pattern and that, in certain regions of the fiber, there is a steady increase or decrease in V_0 over several segments in succession. Data points are values of two or more recordings in each segment. (**b**) Repeated V_0 measurements along a single muscle fiber during different time intervals after the onset of the experiment illustrating consistency of the measurements over several hours. Time interval after onset of experiment: *open squares*, 30–65 min; *filled squares*, 70–105 min; *inverted filled triangles*, 200–230 min. From Edman et al. (1985)

similarity with the original (two-step) cross-bridge model of muscle contraction (Huxley 1957) according to which an increase in V_0 is predicted to be associated with an increased curvature of the force–velocity relationship. It should be noted, however, that this relation between V_0 and the curvature of the force–velocity relation

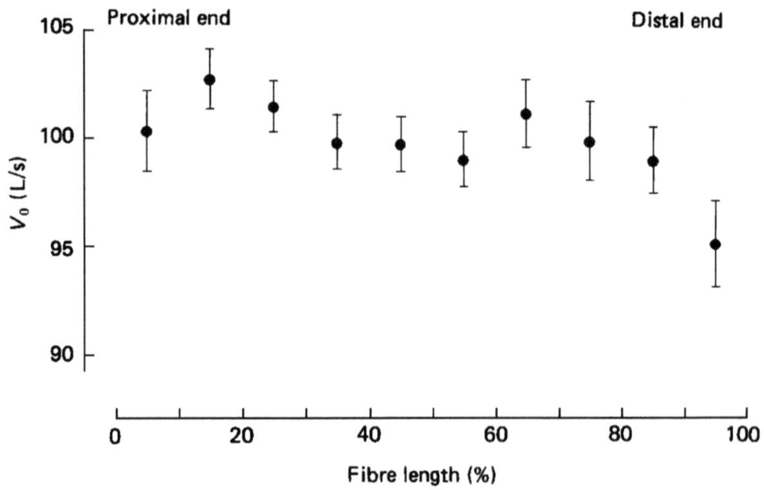

Fig. 24 V_0 of individual fiber segments related to the fiber's orientation in the body. The length of each fiber has been normalized to 100%. V_0 estimated for ten successive segments of equal length from proximal (*left*) to distal (*right*) end of the fiber. Data points are mean values (±SEM) based on measurements in 14 fibers. From Edman et al. (1985)

is opposite to what is generally found when 'fast' and 'slow' muscles are compared. This apparent disagreement is not understood at the present time and requires further studies to be explained.

It was realized that the segmental differences in contractile properties along individual muscle fibers might somehow be related to the fiber's orientation in the body. A series of experiments was therefore carried out in which the fiber ends were identified as 'proximal' and 'distal' with respect to the fiber's attachment between the knee and ankle of the frog's leg. Figure 24 summarizes the results obtained from 14 fibers in which V_0 was determined in ten successive segments in each preparation. The orientation of the fiber in the body does indeed seem to have some influence upon the V_0 pattern. Thus, there was a clear trend towards higher velocities in the proximal part of the fiber and towards lower velocities in the most distal part (for details concerning the statistical treatment, see Edman et al. (1985)).

A variety of isomyosins are now identified in striated muscle and their existence has been found to correlate with specific kinetic properties of individual fibers in both mammalian and amphibian muscles (e.g. Lännergren and Hoh 1984; Spurway and Rowlerson 1989; Schiaffino and Reggiani 1996; Bottinelli and Reggiani 2000). Co-existence of isomyosins of different kinetic types has been demonstrated in individual muscle fibers (Edman et al. 1988) and appears to be a common phenomenon. A non-uniform isomyosin composition along the muscle fiber may, at least partly, account for the segmental differences in V_0.

Inhomogeneities of the isomyosin pattern along the fiber do not seem unreasonable in view of the multicellular origin of the muscle fiber. A skeletal muscle fiber, which in every respect functions as a single cell, is in reality a syncytium of several myoblasts that have fused (Bloom and Fawcett 1975). The genetic information contained in a given fiber thus stems from multiple sources, and the genetic instruction may therefore differ along the fiber. As a result certain features of the contractile system may become more pronounced in one region than in another. However, as demonstrated in Fig. 24, the orientation of the muscle fiber in the body does seem to play a part as well. The differences in contractile performance along the fiber may therefore not be entirely accidental but somehow related to the close environment in which the fiber and the muscle operates in the body.

References

Abbott BC, Aubert XM (1952) The force exerted by active striated muscle during and after change of length. J Physiol 117:77–86

Aubert X (1956) Le couplage energetique de la contraction musculaire. These d'agregation l'enseignement superieur. Editions Arsaia, Bruxelles

Bailey K (1937) Composition of the myosins and myogen of skeletal muscle. Biochem J 31:1406–1413

Bárány M (1967) ATPase activity of myosin correlated with speed of muscle shortening. J Gen Physiol 50:197–218

Bárány EH, Edman KAP, Palis A (1951a) The influence of electrolytes on the rate of viscosity drop in ATP-actomyosin mixtures. Acta Physiol Scand 24:361–367

Bárány EH, Edman KAP, Högberg F, Rosner H (1951b) A recording viscosimeter for study of rapid viscosity changes. Acta Physiol Scand 23:128–136

Bárány EH, Edman KAP, Palis A (1952) The influence of potassium chloride concentration on the rate of drop of viscosity in ATP-actomyosin mixtures. Acta Soc Med Ups 56:269–272

Blinks JR, Rudel R, Taylor SR (1978) Calcium transients in isolated amphibian skeletal muscle fibres: detection with aequorin. J Physiol 277:291–323

Blix M (1894) Die Länge und die Spannung des Muskels. Skand. Arch. Physiol 5:173–206

Bloom W, Fawcett DW (1975) A Textbook of Histology, p. 328. Philadelphia: W.B. Saunders Company

Bottinelli R, Reggiani C (2000) Human skeletal muscle fibres: molecular and functional diversity. Prog Biophys and Mol Biol 73:195–262

Briden KL, Alpert NR (1972) The effect of shortening on the time-course of active state decay. J Gen Physiol 60:202–220

Brown LM, Hill L (1991) Some observations on variations in filament overlap in tetanized muscle fibres and fibres stretched during a tetanus, detected in the electron microscope after rapid fixation. J Muscle Res Cell Motil 12:171–182

Buchthal F, Schmalbruch H (1980) Motor unit of mammalian muscle. Physiol Rev 60:90–142

Caputo C, Edman KAP, Lou F, Sun YB (1994) Variation in myoplasmic Ca^{2+} concentration during contraction and relaxation studied by the indicator flyo-3 in frog muscle fibres. J Physiol 478:137–148

Cavagna GA, Citterio G (1974) Effect of stretching on the elastic characteristics and the contractile component of frog striated muscle. J Physiol 239:1–14

Close R (1972) Dynamic properties of mammalian skeletal muscles. Physiol Rev 52:129–197

Devrome AN, MacIntosh BR (2007) The biphasic force–velocity relationship in whole rat skeletal muscle in situ. J Appl Physiol 102:2294–2300

Duke TA (1999) Molecular model of muscle contraction. Proc Nat Acad Sci U S A 96:2770–2775

Edman KAP (1950) The action of ouabain on heart actomyosin. Acta Physiol Scand 21:230–237

Edman KAP (1951) The action of ouabain on actomyosin from striated musculature. Acta Physiol Scand 23:137–142
Edman KAP (1953) Action of cardiac glycosides on the ATP-induced contraction of glycerinated muscle fibers. Acta Physiol Scand 30:69–79
Edman KAP (1959a) Relaxation of glycerol-extracted muscle fibre bundles induced by zinc in the presence of ATP and other polyphosphates. Acta Physiol Scand 46:62–87
Edman KAP (1959b) The binding of zinc to glycerol-extracted muscle, and its relaxing effect. Acta Physiol Scand 46:209–227
Edman KAP (1966) The relation between sarcomere length and active tension in isolated semitendinosus fibres of the frog. J Physiol 183:407–417
Edman KAP (1975) Mechanical deactivation induced by active shortening in isolated muscle fibres of the frog. J Physiol 246:255–275
Edman KAP (1979) The velocity of unloaded shortening and its relation to sarcomere length and isometric force in vertebrate muscle fibres. J Physiol 291:143–159
Edman KAP (1980) Depression of mechanical performance by active shortening during twitch and tetanus of vertebrate muscle fibres. Acta Physiol Scand 109:15–26
Edman KAP (1981) Deactivation of the contractile system induced by shortening of striated muscle. In The regulation of muscle contraction: excitation-contraction coupling. Eds.: Grinnell AD, Brazier, MAB. Academic Press, New York, pp. 281–296
Edman KAP (1988) Double-hyperbolic force–velocity relation in frog muscle fibres. J Physiol 404:301–321
Edman KAP (1999) The force bearing capacity of frog muscle fibres during stretch: its relation to sarcomere length and fibre width. J Physiol 519:515–526
Edman KAP (2005) Contractile properties of mouse single muscle fibers, a comparison with amphibian muscle fibers. J Exp Biol 208:1905–1913
Edman KAP, Andersson KE (1968) The variation in active tension with sarcomere length in vertebrate skeletal muscle and its relation to fibre width. Experientia 24:134–136
Edman KAP, Hwang JC (1977) The force–velocity relationship in vertebrate muscle fibres at varied tonicity of the extracellular medium. J Physiol 269:255–272
Edman KAP, Kiessling A (1971) The time course of the active state in relation to sarcomere length and movement studied in single skeletal muscle fibres of the frog. Acta Physiol Scand 81:182–196
Edman KAP, Lou F (1992) Myofibrillar fatigue versus failure of activation during repetitive stimulation of frog muscle fibres J Physiol 457:655–673
Edman KAP, Mulieri LA, Scubon-Mulieri B (1976) Non-hyperbolic force-velocity relationship in single muscle fibres. Acta Physiol Scand 98:143–156
Edman KAP, Nilsson E (1968) The mechanical parameters of myocardial contraction studied at a constant length of the contractile element. Acta Physiol Scand 72:205–219
Edman KAP, Nilsson E (1972) Relationship between between force and velocity of shortening in rabbit papillary muscle. Acta Physiol Scand 85:488–500
Edman KAP, Reggiani C (1984) Redistribution of sarcomere length during isometric contraction of frog muscle fibres and its relation to tension creep. J Physiol 351:169–198
Edman KAP, Reggiani C, teKronnie G (1985) Differences in maximum velocity of shortening along single muscle fibres of the frog. J Physiol 365:147–163
Edman KAP, Reggiani C (1987) The sarcomere length–tension relation determined in short segments of intact muscle fibres of the frog. J Physiol 385:709–732
Edman KAP, Schild HO (1962) The need for calcium in the contractile responses induced by acetylcholine and potassium in rat uterus. J Physiol 161:424–441
Edman KAP, Schild HO (1963) Calcium and the stimuland and inhibitory effects of adrenaline in depolarized smooth muscle. J Physiol 169:404–411
Edman KAP, Tsuchiya T (1996) Strain of passive elements during force enhancement by stretch in frog muscle fibres. J Physiol 490:191–205
Edman KAP, Elzinga G, Noble MIM (1978) Enhancement of mechanical performance by stretch during tetanic contractions of vertebrate skeletal muscle fibres. J Physiol 281:139–155

Edman KAP, Elzinga G, Noble MIM (1982) Residual force enhancement after stretch of contracting frog single muscle fibers. J Gen Physiol 80:769–784

Edman KAP, Reggiani C, Schiaffino S, te Kronnie G (1988) Maximum velocity of shortening related to myosin isoform composition in frog skeletal muscle fibres. J Physiol 395:679–694

Edman KAP, Caputo C, Lou F (1993) Depression of tetanic force induced by loaded shortening of frog muscle fibres. J Physiol 446:535–552

Edman KAP, Månsson A, Caputo C (1997) The biphasic force–velocity relationship in frog muscle fibres and its evaluation in terms of cross-bridge function. J Physiol 503:141–156

Edman KAP, Radzyukevich T, Kronborg B (2002) Contractile properties of isolated muscle spindles of the frog. J Physiol 541:905–916

Ekelund MC, Edman KAP (1982) Shortening induced deactivation of skinned fibres of frog and mouse striated muscle. Acta Physiol Scand 116:189–199

Elliott A, Offer G (1978) Shape and flexibility of the myosin molecule. J Mol Biol 123:505–519

Elliott GF, Lowy J, Worthington CR (1963) An X-ray and light-diffraction study of the filament lattice of striated muscle in the living state and in rigor. J Mol Biol 6:295–305

Engelhardt WA, Ljubimowa MN (1939) Myosine and Adenosinetriphosphatase. Nature 144:668–669

Evans CL, Hill AV (1914) The relation of length to tension development and heat production on contraction in muscle. J Physiol 49:10–16

Fenn WO (1924) The relationship between work performed and the energy liberated in muscular contraction. J Physiol 58:373–395

Fenn WO, Marsh BS (1935) Muscular force at different speeds of shortening. J Physiol 85:277–297

Gordon AM, Huxley AF, Julian FJ (1966) The variation in isometric tension with sarcomere length in vertebrate muscle fibres. J Physiol 184:170–192

Granzier HL, Pollack GH (1989) Effect of active pre-shortening on isometric and isotonic performance of single frog muscle fibres. J Physiol 415:299–327

Herzog W, Lee EJ, Rassier DE (2006) Residual force enhancement in skeletal muscle. J Physiol 574:635–642

Hill AV (1938) The heat of shortening and the dynamic constants of muscle. Proc R Soc Lond B 126:136–195

Hill AV, Howarth JV (1959) The reversal of chemical reactions in contracting muscle during an applied stretch. Proc R Soc Lond B 151:169–193

Huxley HE (1969) The mechanism of muscular contraction. Science 164:1356–1366

Huxley HE (1973) Molecular basis of contraction in cross-striated muscles. In The Structure and Function of Muscle, vol. 1, 2nd edn., ed Bourne, G., pp. 301–387. New York: Academic Press

Huxley HE, Hanson J (1954) Changes in the cross-striations of muscle during contraction and stretch and their natural interpretation. Nature 173:973–977

Huxley AF (1957) Muscle structure and theories of contraction. Progr Biophys biophys Chem 7:255–311

Huxley AF, Niedergerke R (1954) Structural changes in muscle during contraction: Interference microscopy of living muscle fibres. Nature 173:971–973

Jewell BR, Wilkie DR (1960) The mechanical properties of relaxing muscle. J Physiol 152:30–47

Joumaa V, Leonard TR, Herzog W (2008) Residual force enhancement in myofibrils and sarcomeres. Proc Biol Sci 275:1411–1419

Joyce GC, Rack PMH, Westbury DR (1969) The mechanical properties of cat soleus muscle during controlled lengthening and shortening movements. J Physiol 204:461–474

Julian FJ, Morgan DL (1979) The effect on tension of non-uniform distribution of length changes applied to frog muscle fibres. J Physiol 293:379–392

Julian FJ, Moss RL, Waller GS (1981) Mechanical properties and myosin light chain composition of skinned muscle fibres from adult and new-born rabbits. J Physiol 311:201–218

Katz B (1939) The relation between force and speed in muscular contraction. J Physiol 96:54–64

Lännergren J, Hoh JFY (1984) Myosin isoenzymes in single muscle fibres of Xenopus laevis: analysis of five different functional types. Proc R Soc Lond B 222:401–408

Lee EJ, Herzog W (2008) Residual force enhancement exceeds the isometric force at optimal sarcomere length for optimized stretch conditions. J Appl Physiol 105:457–462

Lou F, Sun Y-B (1993) The high-force region of the force–velocity relation in frog skinned muscle fibres. Acta Physiol Scand 148:243–252
Maréchal G, Plaghki L (1979) The deficit of the isometric tetanic tension redeveloped after a release of frog muscle at constant velocity. J Gen Physiol 73:453–467
Millman BM (1998) The filament lattice of striated muscle. Physiol Rev 78:359–391
Mitsui T, Chiba H (1996) Proposed modification of the Huxley-Simmons model for myosin head motion along an actin filament. J Theor Biol 182:147–159
Nielsen BG (2003) Unfolding transitions in myosin give rise to the double hyperbolic force–velocity relation in muscle. J Phys Condens Matter 15:1759–1765
Page SG (1968) Fine structure of tortoise skeletal muscle. J Physiol 197:709–715
Ralston HJ, Inman VT, Strait LA, Shaffrath MD (1947) Mechanics of human isolated voluntary muscle. Am J Physiol 151:612–620
Ramsey RW, Street SF (1940) The isometric length tension diagram of isolated skeletal muscle fibres of the frog. J Cell Comp Physiol 15:11–34
Rassier DE, Herzog W (2004a) Effects of shortening on stretch-induced force enhancement in single skeletal muscle fibers. J Biomech 37:1305–1312
Rassier DE, Herzog W (2004b) Considerations on the history dependence of muscle contraction. J Appl Physiol 96:419–427
Rassier DE, Herzog W, Pollack GH (2003) Dynamics of individual sarcomeres during and after stretch in activated single myofibrils. Proc Biol Sci 270:1735–1740
Rayment I, Rypniewsky WR, Schmidt-Bäse K, Smith R, Tomchick DR, Benning MM, Winkelmann DA, Wesenberg G, Holden HM (1993) Three-dimensional structure of myosin subfragment-1: a molecular motor. Science 261:50–58
Reggiani C (2007) When fibres go slack and cross bridges are free to run: a brilliant method to study kinetic properties of acto-myosin interaction. J Physiol 583:5–7
Schachar R, Herzog W, Leonard TR (2004) The effects of muscle stretching and shortening on isometric forces on the descending limb of the force-length relationship. J Biomech 37:917–926
Schiaffino S, Reggiani C (1996) Molecular diversity of myofibrillar proteins: gene regulation and functional significance. Physiol Rev 76:371–423
Sosa H, Popp D, Ouyang G, Huxley HE (1994) Ultrastructure of skeletal muscle fibers studied by a plunge quick freezing method: myofilament lengths. Biophys J 67:283–292
Spurway NC, Rowlerson AM (1989) Quantitative analysis of histochemical and immunohistochemical reactions in skeletal muscle fibres of Rana and Xenopus. Histochem J 21:461–471
Straub FB (1942) Actin. Studies from the Institute of Medical Chemistry University Szeged, 2:3–15
Sugi H (1972) Tension changes during and after stretch in frog muscle fibers. J Physiol 225:237–253
Sugi H, Tsuchiya T (1988) Stiffness changes during enhancement and deficit of isometric force by slow length changes in frog skeletal muscle fibres. J Physiol 407:215–229
Szent-Györgyi A (1944) Studies on muscle. Acta Physiol Scand 9, suppl. 25:1–128.
Talbot JA, Morgan DL (1996) Quantitative analysis of sarcomere non-uniformities in active muscle following a stretch. J Muscle Res Cell Motil 17:261–268
Telley IA, Stehle R, Ranatunga KW, Pfitzer G, Stüssi E, Denoth J (2006) Dynamic behaviour of half-sarcomeres during and after stretch in activated rabbit psoas myofibrils: sarcomere asymmetry but no 'sarcomere popping'. J Physiol 573:173–185
Weber HH, Portzehl H (1952) Kontraktion, ATP-Cyclus und fibrilläre Proteine des Muskels. Ergeb Physiol 47:369–468

Energy Economy in the Actomyosin Interaction: Lessons from Simple Models

Steven L. Lehman

Abstract The energy economy of the actomyosin interaction in skeletal muscle is both scientifically fascinating and practically important. This chapter demonstrates how simple cross-bridge models have guided research regarding the energy economy of skeletal muscle. Parameter variation on a very simple two-state strain-dependent model shows that early events in the actomyosin interaction strongly influence energy efficiency, and late events determine maximum shortening velocity. Addition of a weakly-bound state preceding force production allows weak coupling of cross-bridge mechanics and ATP turnover, so that a simple three-state model can simulate the velocity-dependence of ATP turnover. Consideration of the limitations of this model leads to a review of recent evidence regarding the relationship between ligand binding states, conformational states, and macromolecular structures of myosin cross-bridges. Investigation of the fine structure of the actomyosin interaction during the working stroke continues to inform fundamental research regarding the energy economy of striated muscle.

Keywords Energy cost • Tension economy • Efficiency • ATP turnover

1 Introduction

The energy cost of maintaining and using skeletal muscle is a large part of the human body's energy budget (about 20% at rest; over 90% during hard exercise). Skeletal muscle accounts for about 40% of the lean body mass of a healthy adult (28 kg of a formerly average 70 kg person). Its average energy use is only about 0.5 W/kg at rest, but can exceed 50 W/kg during hard exercise. As the major portion of this large and variable energy flux is due to the myosin ATPase,

S.L. Lehman (✉)
Department of Integrative Biology, University of California Berkeley, Berkeley, CA 94720, USA
e-mail: slehman@berkeley.edu

D.E. Rassier (ed.), *Muscle Biophysics: From Molecules to Cells*,
Advances in Experimental Medicine and Biology 682,
DOI 10.1007/978-1-4419-6366-6_3,

the energy economy of the actomyosin interaction is both scientifically fascinating and practically important.

The energetic aspects of skeletal muscle contraction have been a vital component of muscle science throughout the twentieth and early twenty-first centuries, and early findings on the thermodynamics of whole muscle have been greatly extended by studies of work and heat production by single muscle fibers (see Woledge et al. 1985 for a comprehensive review). In the past two decades, new techniques have enabled a multitude of new energetic results. New assays have enabled measurements of ATP consumption by single permeabilized fibers, and energy use by the actomyosin ATPase has been distinguished from that by other muscle ATPases. Photorelease experiments in which ATP and phosphate are introduced into skinned fibers with fully activated thin filaments have clarified some of the connections between the myosin ATPase cycle and cross-bridge mechanics. The atomic structures of myosin have been solved for several putative states in the cross-bridge cycle. Increasingly sophisticated optical tweezer techniques have begun to characterize the effects of load on the actomyosin interaction. Investigation of signaling pathways and protein modification mechanisms in skeletal muscle analogous to those that enable modulations of contractility and energy use in cardiac muscle have begun.

New techniques have led to answers to many long-standing questions, but many fundamental questions remain: Can recent findings regarding energy economy of the myosin ATPase and other ATPases be integrated to achieve an understanding of energetic properties of living fibers, whole muscles, and ultimately, whole animals? In the language of the American Physiological Society, is it possible to go 'from molecule to man'? How far have we progressed in the integration of mechanical and biochemical cycles? How does the very fine scale structural information we now have regarding myosin relate to function, including free energy transduction? How are the energetic properties of skeletal muscle modulated by load and the biochemical composition of the myofibrillar space? Are the mechanical and energetic properties of skeletal muscle, like cardiac muscle, modulated?

From the earliest days of muscle mechanics and energetics, mathematical models have been useful. A.V. Hill's phenomenological models of the force–velocity relationship and rates of enthalpy liberation (Hill 1938) led to bold hypotheses, sometimes disproved (Hill 1964). A.F. Huxley's brilliant two-state cross-bridge model (Huxley 1957) showed that the relationship between velocity of shortening and steady-state force can emerge from a model with just two cross-bridge states and reasonable assumptions regarding cross-bridge attachment and detachment rates. Simple lumped-parameter models of biochemical kinetics have often been useful in figuring out the relationship between ligand binding states and force production in the cross-bridge cycle. In the era of molecular motors, quite detailed models of the geometry of actin and myosin filaments are the obvious tool for synthesizing structural data and relating it to mechanics.

Some simple models (e.g. Hill 1938; Huxley 1957) continue to be useful, both didactically and as the first steps in understanding new data. This chapter will demonstrate how simple cross-bridge models have guided and continue to inform research regarding the energy economy of skeletal muscle. Parameter variation on a very simple two-state strain-dependent model will distinguish between events early and late in a cross-bridge's working stroke, and show that early events influence energy efficiency, whereas late events determine maximum shortening velocity. Addition of a weakly-bound state preceding force production allows uncoupling of cross-bridge detachment from completion of the ATPase cycle. Discussion of the evolution of our understanding of the coupling between the biochemical myosin ATPase and mechanical cross-bridge cycles is centered around a simple three-state model that can simulate the non-monotonic velocity-dependence of ATP turnover. Consideration of this model leads to a review of recent evidence regarding the relationship between ligand binding states, conformational states, and macromolecular structures of myosin heads that shows how investigation of the fine structure of the working stroke continues to inform fundamental research regarding the energy economy of striated muscle.

2 Lessons from a Two State Model

Impressively, the first cross-bridge model (Huxley 1957) simulated the known force–velocity relationship, and consequently the power-velocity relationship for shortening muscle. The model was based on a few simple assumptions: that only one actin site is available to a cross-bridge at any given time, that one ATP is hydrolyzed in each cross-bridge cycle, that attached cross-bridges produce force proportional to extension of an elastic element, that the rate constant for cross-bridge attachment (f) increases with positive strain out to a maximum distension, and that the rate constant for detachment (g) is large for negative strains (drag). Using the empirical relationships between shortening speed and force and rate of heat production to fit the constants of his model, Huxley obtained excellent agreement with A.V. Hill's experimental force–velocity results and Hill's 1938 results indicating a linear dependence of the rate of total enthalpy (heat + work) liberation on shortening velocity.

In his wonderful book on mechanics of motor proteins, Howard (2001) developed an even simpler didactic two-state strain-dependent model, in which he assumed that a cross-bridge can bind only in a very limited range of positions relative to the thick filament (the "strike zone"); and that there can be no detachment until strain drops to zero. For negative strains, the detachment rate increases abruptly to a large value. The narrow strike zone and high off rate at negative strains ensure a small fraction of attached cross-bridges. Howard used this ultimately simple model to develop the important concept of 'duty ratio', the fraction of time a cross-bridge spends in its attached state. He showed that for models with just one attached state the duty ratio is also equal to the ratio of the distance traveled by a cross-bridge while attached (the 'working stroke') to the distance moved by the sliding filaments during one ATP hydrolysis (the 'path distance'). Even so simple a model predicts monotonically declining force as shortening velocity increases – Howard points out that the shape of the force–velocity relationship depends on the details of the dependence of attachment and detachment rate constants on strain.

This simple model also predicts that the rate of ATP turnover to be a monotonically increasing function of the rate of shortening (i.e. the shortening velocity), as ATP turnover is coupled one to one with attachment and passage through the working stroke.

Two-state strain-dependent models continue to be useful as a first line of attack in research on energetic questions. For example, in their review of skeletal muscle efficiency, Smith et al. (2005) added myofilament compliances to two-state model similar to Huxley's 1957 model, and used it to predict the effects of filament compliance on efficiency.

My didactic two-state model (Fig. 1) has one attached state and one detached state. The free energy of the detached state is assumed to be zero. The attached state exerts zero force at a cross-bridge distension defined to be x = 0, and exerts a force proportional to x as it is stretched in the positive direction (force = dG/dx, positive force being defined as force in the direction of shortening). I assume that at any instant the cross-bridge has available to it only one actin site, and that it acts independently of any other cross-bridges.

For shortening, imagine a single actin site appearing on the right side of Fig. 1a and moving to the left, at velocity v, past the single cross-bridge. The x axis should extend to +18 nanometers on the right and to −18 nanometers on the right, as 36 nm is the distance between equivalent actin sites, but no attachment can occur over most of this range, as the reach of a myosin cross-bridge is limited. As the actin site passes x = 7, the free energy of the attached state becomes lower than the free energy of the detached state, and attachment becomes energetically favorable. The attachment rate constant f(x) is assumed constant for x in the interval [0,7] nm (Fig. 1b). The detachment rate constant g(x) is related to the attachment rate constant f by the Boltzmann factor (Hill 1989; Howard 2001):

$$g(x) = f(x) * e^{-(Gd-Ga)/kT}$$

Due to this relationship, the detachment rate constant g is much smaller than f over most of the interval [0,7] and becomes equal to f at x = 7 nm. An attached cross-bridge follows the free energy down the parabola Ga(x), doing work on the actin filament as it shortens toward x = 0, where the rate constant g− for detachment is

Fig. 1 A simple strain-dependent cross bridge model. (**a**) Free energies of the detached (Gd, Gd′) and a strongly-bound attached state (Ga) as functions of cross-bridge distension (x). Free energy available from hydrolysis of a molecule of ATP (Gd–Gd′ in this diagram) is 100 pNnm; maximum work from a cross-bridge is 80 pNnm; power stroke length is 7 nm. (**b**) Rate constants for this model, chosen to fit force-velocity data for rabbit psoas fibers (Fig. 2a). Attachment rate constant f (*solid line*) is 175 s^{-1} for x in the interval [0,7] nanometers, zero elsewhere. Detachment rate constant g (*dashed line*) for x > 0 is determined by the Boltzmann relationship between f, g, Ga and Gd. Detachment rate constant g− = 1,500 s^{-1} for cross-bridges distended into the negative force-producing region [−7,0]. A very high detachment rate is fixed for x < −7 nm. (**c**) Probability that a cross-bridge is attached, comparing isometric contraction (*solid line*) to maximum shortening velocity (*dashed line*)

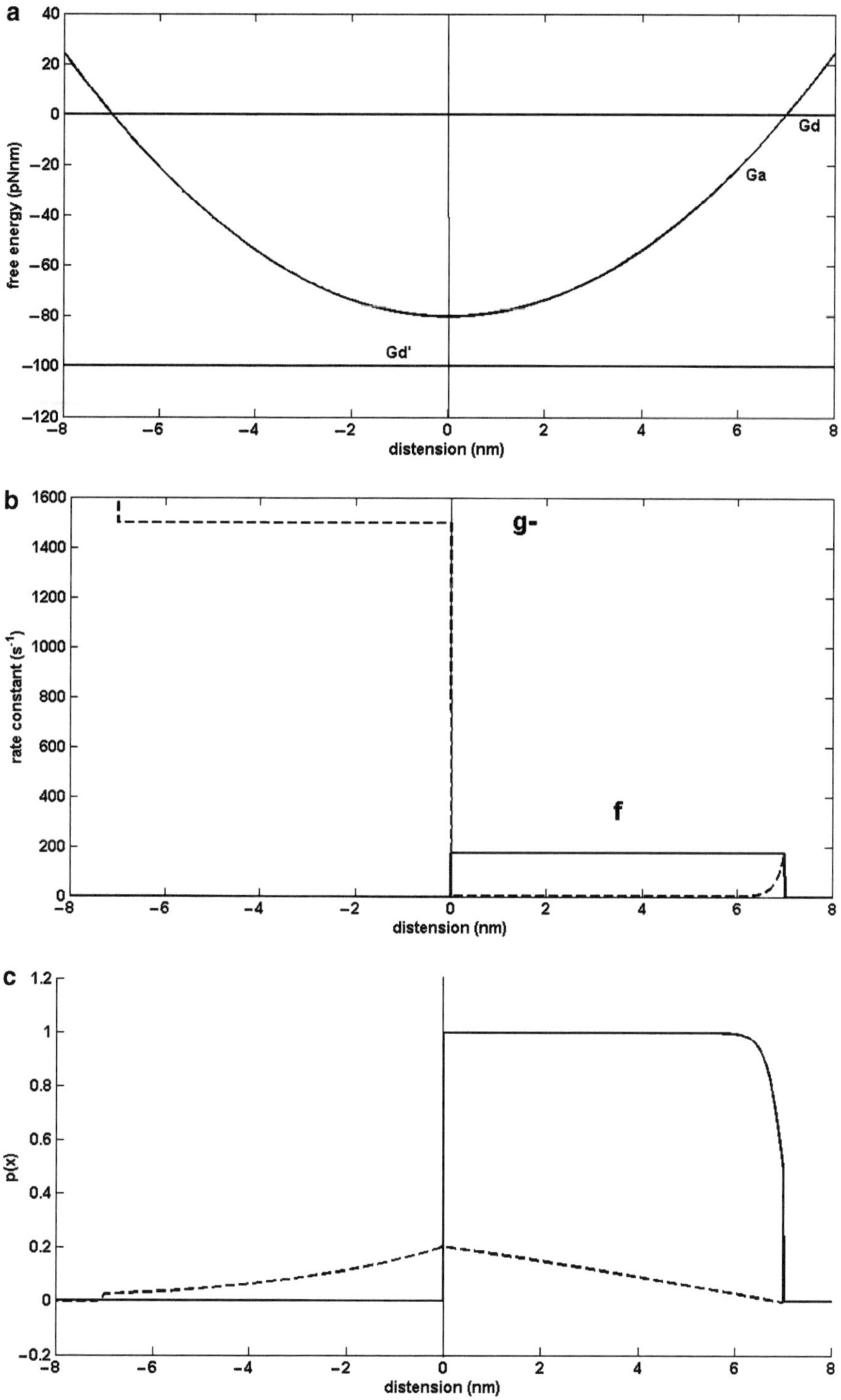
a
free energy (pNnm)
distension (nm)
Gd
Ga
Gd'
b
rate constant (s-1)
distension (nm)
g-
f
c
p(x)
distension (nm)

assumed to become large. The horizontal line at Gd′ = −100 pNnm (~25 kT) represents the free energy available from hydrolysis of a single ATP.

The probability of a cross-bridge being attached at each distension x, which is determined by the free energies and the rate constants, is shown for the isometric case (v = 0) and for the maximum velocity of shortening (vmax) in Fig. 1c. As an actin sites moves across the figure from right to left, the probability of attachment increases throughout the 'power stroke' region until the cross-bridge reaches x = 0. Attached cross-bridges are dragged into the 'drag stroke' region [−7,0] by the actin filament. Detachment becomes increasingly likely as drag force (dG/dx) increases. Cross-bridges are assumed to be broken off at distensions to the left of x = −7 nm, consistent with the size of myosin's S1 fragment. At higher velocities, the probability of attachment becomes much smaller throughout the 'power stroke' region, and higher in the 'drag stroke' region. At v_{max} (dashed line) the force due to the distribution of dragging cross-bridges (x < 0) becomes equal to the force due to the distribution of cross-bridges doing positive work on the actin filament (x > 0).

This simple model produces a remarkably nice fit to force–velocity data (Fig. 2a). Variation of the two parameters (Fig. 2b, c) shows that the attachment rate constant f influences the T(V) curve at low velocities, and the detachment rate constant g– influences the curve at high velocities. In this simple model, this means that the attachment rate constant influences the curvature of T(V) – increasing f maintains a larger population of cross-bridges in the positive-force-producing region (x > 0), decreasing curvature. The detachment rate constant influences the maximum shortening velocity – decreasing g– allows more cross-bridges to drag into the negative-force-producing region, decreasing vmax. As the power stroke begins on the right end of [0,7], where the f becomes non-zero and the drag stroke begins at the right end of [−7,0], where g becomes non-zero, Fig. 2b and c suggest that events early in the power stroke affect curvature of T(V), and events late in the power stroke affect v_{max}.

These sensitivities of T(V) to the two rate constants relate to the efficiency of skeletal muscle. Woledge (1998) derived expressions for the work rate (power), heat rate, and mechanical efficiency (work/(work+heat)) of muscle from the empirical equations of Hill (1938) and used them to fit work and heat rate data. He showed (his Fig. 2) that the parameter G = P0/a, which strongly influences the curvature of the force–velocity curve, also affects both power and heat rate, and therefore the efficiency – muscles with flatter force–velocity curves (smaller G) are predicted to produce more power but also larger heat rates, and thus have lower efficiencies. Fast muscles generally have larger parameters a/P0 (smaller G) than slow muscles of the same animal (Woledge et al. 1985, Table 2.II). The maximum power output of fast-twitch fibers is often two to three times the maximum power output of slow fibers from the same species, and peak cross-bridge efficiencies of fast fibers are generally lower than of slow fibers from the same species (reviewed in Smith et al. 2005).

The assumption that each attachment-work-detachment cycle of a cross-bridge is coupled to hydrolysis of an ATP implies that the rate of enthalpy release is a monotonically increasing function of shortening velocity. While Hill's early

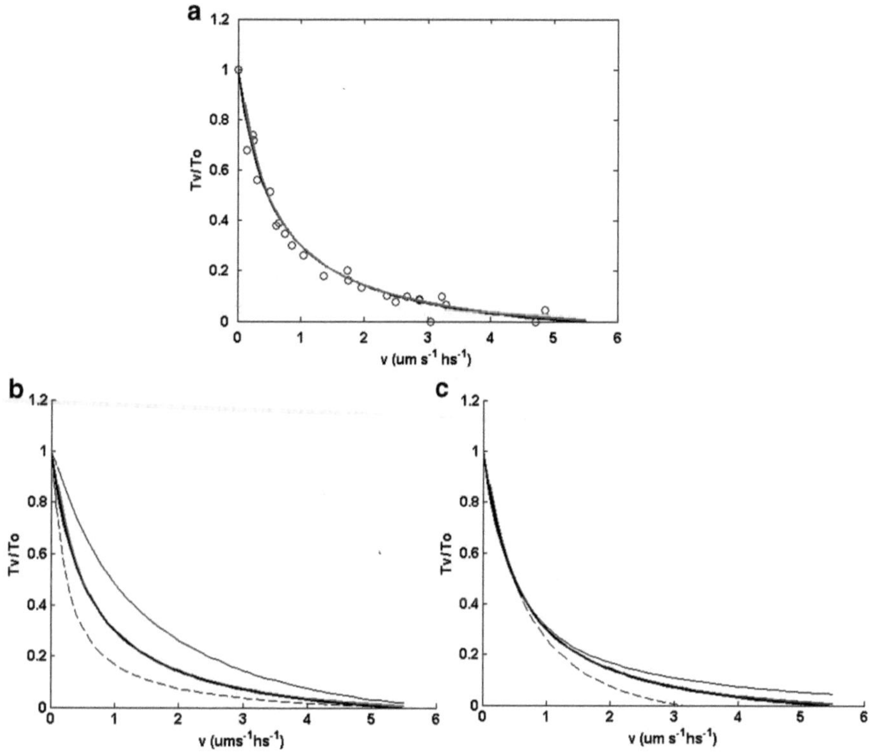

Fig. 2 Model fit to data and effects of parameter variations. (**a**) Steady state force produced by the model in Fig. 1 (*grey line*) chosen to fit force-velocity data for rabbit psoas muscle fibers at 12° (data points redrawn from Caremani et al. 2009). *Black line* shows Hill curve best fit to data points (v_{max}=5.3 μm hs^{-1} s^{-1}, a/To=0.11). Each force is normalized to isometric tension T0. (**b**) Sensitivity of the fit to variation of attachment rate constant (f). Doubling f (*thin solid line*) produces less curvature by maintaining attachment of force-producing cross-bridges (x>0), and halving f (*dashed line*) produces greater curvature. *Heavy grey* and *black curves* show simulation of the same model with best-fit attachment rate constant (f=175 s^{-1}) and Hill curve. (**c**) Sensitivity of the fit to variation of detachment rate constant (g–). Halving g– (*dashed line*) decreases vmax by extending the range of dragging cross-bridges, and doubling g– (*thin solid line*) increases vmax

data suggested a monotonic increase, Hill's data of 1964 and many subsequent experiments indicated a decline in enthalpy liberation rate with shortening velocity at moderate to high velocities. To address this discrepancy while maintaining the single attached state of the 1957 model, A.F. Huxley (1973) amended the 1957 model to include two stages of cross-bridge attachment. Attachment in the first stage was hypothesized to occur over a narrow range of displacements, but detachment from this stage could occur at any displacement, without ATP hydrolysis. Attachment and detachment from the second stage, with ATP hydrolysis, followed the same rules as in the 1957 model. Thus the 1973 model solved the energy release rate problem by adding detail to the strain dependence of attachment.

An alternative solution to the same energetic problem that also deals with the existence of multiple actomyosin states in the ATPase cycle is to insert a second attached state in the cross-bridge cycle.

3 Lessons from a Three State Model

Study of the biochemistry of the actomyosin interaction (reviewed in Szent-Györgyi 2004) has a long history, often independent of the mechanical studies. The biochemical and cross-bridge worlds interacted strongly at a 1972 meeting at Cold Spring Harbor (Cooke 2004). Just before that meeting, Lymn and Taylor published a mechanism for ATP hydrolysis by actomyosin that began to connect ATP hydrolysis to cross-bridge cycling (Lymn and Taylor 1971):

$$M + T \rightleftarrows A + MT \rightleftarrows MDP \rightleftarrows AMDP \rightleftarrows AM'$$

where A stands for actin, M for myosin, M′ for myosin in a force-producing conformation, T for ATP and D for ADP. According to this scheme, ATP binds to actomyosin, causing the actomyosin bond to break. ATP is then hydrolyzed on the detached myosin head, and the myosin, with the hydrolysis products, binds to actin. There is then a structural change in the myosin head, coupled to release of the hydrolysis products, which produces force and relative movement of the actin and myosin filaments.

The release of two ATP hydrolysis products, ADP and Pi, is coupled to the structural change that brings about force production and shortening. In the 1970s, muscle biochemists resolved the order of release and found energetic data that suggested that it was Pi release that was coupled to the conformational change in myosin. Trentham et al. (1976) found that Pi was released before ADP (AMDP → AMD → AM′), and White and Taylor (1976) showed that much of the free energy from ATP release occurs in the part of the ATPase cycle that corresponds to Pi release. Thus Pi release might be a trigger a transition from a non-force-producing attached state (AMDP) to a force-producing state (AM′D and AM′). Later, Siemankowski and White (1984) measured the kinetics of ADP dissociation from myosin S1 and concluded that the rate constant of ADP dissociation was slow enough to limit the rate of myosin S1 dissociation from actin. Siemankowski et al. (1985) found evidence that ADP release could limit the rate of cross-bridge detachment in striated muscle, and hypothesized that ADP release might be a general mechanism for limiting maximum shortening velocity (regarding the generality of this mechanism, see Nyitrai et al. 2006). In a decade, the domain of the actomyosin interaction in the cross-bridge cycle expanded from a single state to multiple actomyosin states, from Pi release, possibly triggering a structural change that brings about force production, to ADP release, possibly rate limiting cross-bridge release.

Evidence from studies of muscle fibers supported the existence of a weakly-bound, non-force-producing attached state (AMDP) that led to a force producing state or states by release of Pi. Pate and Cooke (1989) probed the reaction thought to be at the crux of energy release in the power stroke by manipulating [Pi] in permeabilized fibers. Measuring the decrease in T0 caused by increasing [Pi], they found a linear relationship between log([Pi]) and T0. Using a quite general model of force generation by a cross-bridge, they showed that a logarithmic relationship is consistent with models in which added Pi populates non-force-producing (AMDP) states that precede force-producing cross-bridge states.

In measurements of ATP turnover by single muscle fibers, Potma and co-investigators also found evidence consistent with a non-force-producing AMDP state – ATP turnover during shortening saturated at moderate velocities (Potma et al. 1994a), and high [Pi] depressed isometric force and ATP turnover to different degrees (Potma et al. 1995).

Potma et al. (1994a) used a very simple three-state model of cross-bridge kinetics to simulate force production and myofibrillar ATP turnover. Consistent with the concept of force production after Pi release, the model had a single detached state (MDP), and a non-force-producing Pi -bearing state (AMDP) that preceded a force-producing (AMD) state (denoted 'd', 'w' and 's' in Fig. 3a). Rate constants for attachment to and detachment from the 'weakly-bound' state (fw and gw) agreed well with parameter identifications using sinusoidal analysis (Kawai and Zhao 1993) and with measurements of relaxation after photorelease of caged Pi (Dantzig et al. 1992). Simulation of ATP turnover during rapid shortening required large increases in the rate constants for detachment from the attached states (gw and gs), by factors of 4 and 10, respectively. The much faster off rate from the AMD state depopulated the AMD state (35% occupancy to 8%), reducing force. The faster off rate from the AMDP state allowed detachment without product release (weak coupling), and thus a two- to threefold increase in ATP turnover during a tenfold increase in detachment rate.

Variation of parameters in this simple model allowed Potma and co-investigators to simulate the effects of pH on ATPase activity in fast and slow fibers (Potma et al. 1994b) and the effects of added Pi on ATP consumption during isometric contraction (Potma et al. 1995) and during shortening (Potma and Stienen 1996). Simulations of slow fiber ATP turnover required an order of magnitude slower rate constant for detachment from the AMD state, as well as slower attachment and isomerization rates. Isometric ATPase changed little when pH declined from 7.3 to 6.4, but a decline in force, simulated by increasing the detachment rate from the force-producing state and decreasing the isomerization rate, caused an increase in tension cost. The model accounted for the effects of added Pi on force and ATP turnover by mass action on the transition from the AMD to the AMDP state.

A strain-dependent version of this simple, effective kinetics model obviates the necessity of an ad hoc increase in the rate constant g_s in order to simulate shortening. Addition of a non-force-producing weakly-bound attached state (Gw in Fig. 3b) converts the two-state strain-dependent model (Fig. 1) to a strain-dependent three-state model. Addition of the non-force-producing attached state that isomerizes to a force-producing state adds fine structure to the working stroke region (Fig. 3c) – an area to the right of $x = 7$ nm populated by weakly-bound cross-bridges and populations of non-force-producing cross-bridges throughout the working stroke. The probability of strong attachment is lower throughout the power

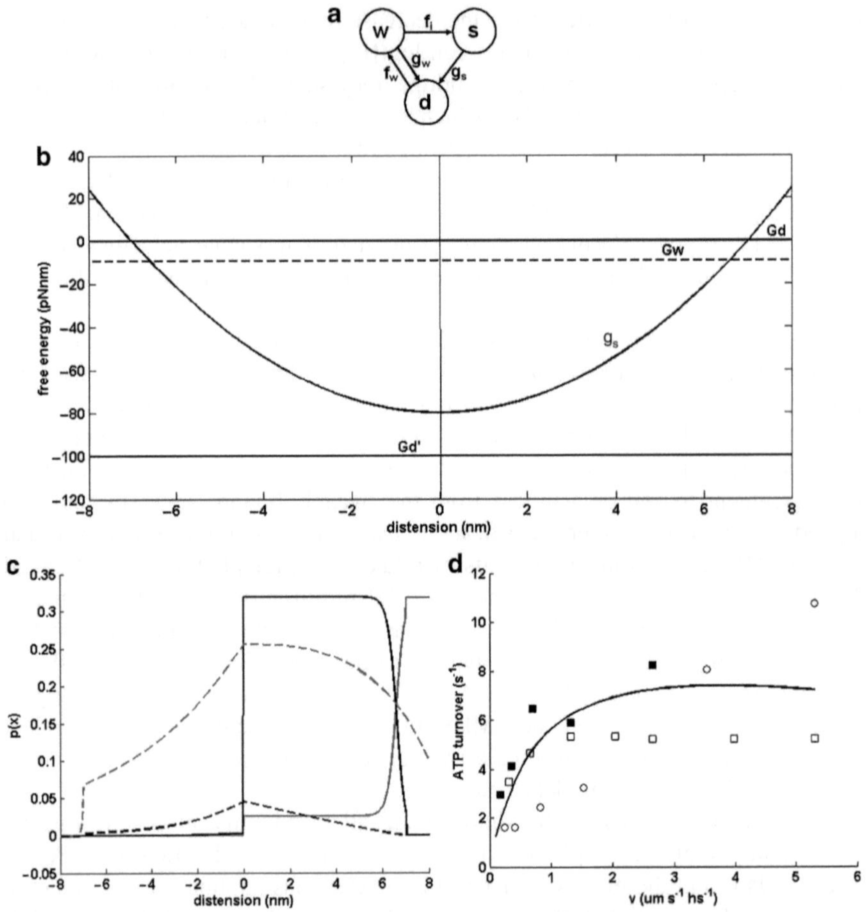

Fig. 3 (**a**) Strain-independent model with one detached state (d), a weakly-bound attached state (w) and a strongly-bound attached state (s), after Potma et al. 1995. Arrows indicate transitions in the indicated directions. (**b**) Free energies of detached (Gd, Gd′), weakly-bound (Gw) and strongly-bound (g_s) states of a strain-dependent three-state model inspired by the model in part (**a**), as functions of cross-bridge distension (x). Rapid equilibrium is assumed between the weakly-bound state and the detached state. Isomerization from weakly- to strongly-bound attachment is at a rate similar to the attachment rate in the two-state model. Detachment from the strongly-bound state is slow for $x>0$, and similar to g− in Fig. 1b for $x<0$. (**c**) Probability that a cross-bridge is attached in the weakly-bound, non-force-producing state (*grey lines*) or the strongly-bound force-producing state (*black lines*), in an isometric contraction (*solid lines*) or during steady-state shortening at maximum velocity (*dashed lines*). Attachment of non-force-producing cross-bridges is allowed for x in [0,8] so that this attachment precedes isomerization to the force-producing state. Note change of scale from Fig. 1c. (**d**) ATP turnover as a function of velocity for the three-state strain-dependent model with parameters chosen to fit the force–velocity curve for rabbit psoas shown in Fig. 2a (*solid curve*). Data are ATP turnover rates derived from Sun et al. 2001, Figure 6 (*open circles*), Potma et al. 1994a, Figure 6b (*open squares*) and Potma and Stienen 1996, Figure 5 (*filled squares*), with vmax for each data **set** scaled to match vmax in Fig. 2a

stroke region (Fig. 3c) because detachment from the force-producing state and the slower ATP hydrolysis step are assumed to be separate events, so that the majority of myosin heads are in the MT to MDP transition. Like the two-state model, the three-state version produces a nice fit to force–velocity data relationship. The three-state model also produces a reasonable relationship between ATP turnover and shortening velocity (Fig. 3d). Data on ATP turnover as a function of shortening velocity are sparse and varied (all data for ATP turnover by rabbit psoas during constant velocity shortening are shown in Fig. 3d, scaled to the same vmax), but this simplest of three-state strain-dependent models allows weak coupling and strain-dependent detachment rates (a much faster detachment rate for $x < 0$), and thus simulates the plateau or even decline in ATP turnover indicated by some of the data.

While investigating the kinetics of Pi binding and release, Dantzig et al. (1992) found evidence that adds further detail to the early part of the power stroke. When they photolyzed caged Pi in fully activated rabbit psoas fibers, they observed a decline in force that was consistent with Pi binding to an AMD state. However, the force transient included a surprising initial lag, followed by the expected nearly-exponential fall of force. Also, the rate constant of the exponential phase increased with Pi concentration, and saturated at high Pi. These kinetics were not consistent with a single transition from a force-producing AM′D state to a non-force-producing AMDP state. However, the kinetics were consistent with a two-step mechanism – Pi binding to a force-producing AM′D state to form a force-producing AM′DP state, followed by isomerization to a non-force-producing AMDP state:

$$MDP \rightleftarrows AMDP \rightleftarrows AM'DP \rightleftarrows AM'D$$

Simulations of this scheme reproduced the Pi dependence of the rate constant of force decline and saturation at high [Pi]. Placed in the normal shortening cycle, this mechanism predicts force production before Pi release, by force-producing AM′DP cross-bridges that isomerize from non-force-producing AMDP cross-bridges!

Myosin S1 structural evidence throughout the 1990s seemed consistent with release of Pi as the event that causes force production, and atomic structures even suggested an attractive hypothesis for the coupling of Pi release to the swinging of a lever arm, by the action of a loop called Switch II. However, recently-solved myosin S1 structures have led to a new hypothetical switching mechanism, with two switches (I and II) that allow more options for the connection between Pi release and force production, including force-producing and non-force-producing states with phosphate in place. The new switching mechanism has the advantage of explaining and old puzzle – myosin's reciprocal affinity for nucleotides and actin (see Takagi et al. 2004 for a complete description). Useful as atomic-level structure has been for suggesting mechanisms of free energy transduction within mysosin S1, structural data for understanding structure at the actomyosin interface in any attached state are still lacking.

On the functional side, experiments on the mechanics of single myosin motors are beginning to produce new data on the effects of load on actomyosin kinetics. Takagi et al. (2006) used a novel 'isometric' optical clamp version of the three-bead assay to study attachment events for a single myosin. They found that attachment duration was load-dependent. Under high load, attachment events were of much shorter duration than under low load. This observation and data on dependences of attachment duration on [Pi] and [ATP] suggest that an applied load can reverse the work done during a power stroke, so that myosin detaches closer to the beginning of the power stroke. Thus loading would increase the fraction of time spent at the beginning of the power stroke, increasing efficiency during isometric contractions, and unloading would allow cross-bridges a longer stroke, and 'tune myosin efficiency to the immediate task'. Similarly, in living fibers, Piazzesi et al. (2002) found that length of the working stroke depends on load – a larger load makes the working stroke smaller. They also found that mechanical energy of the working stroke (work) increased with load, as force was larger under higher load.

Might tuning also be achieved by effects of the myofibrillar environment on the actomyosin interface? Prolonged tetanus in some fast muscles makes their tension cost decline (Crow and Kushmerick 1982a), an effect originally correlated with phosphorylation of regulatory light chains (Crow and Kushmerick 1982b), though causality has not been demonstrated. Tuning of energy economy to compensate for fatigue (reviewed in Cooke 2007) may require a more complete simulation of fatigue than has generally been tested – Karatzaferi et al. (2008) showed that modulation of unloaded shortening velocity by regulatory light chain phophorylation does occur, but only at higher temperatures, and with a combination of low pH and high [Pi], as occurs in fatigued muscle.

The balance of phosphorylation states of several muscle proteins (e.g. phosphorylation of troponin I by PKC in response to adrenergic stimulation) modulates contractility and energy use of cardiac myocytes (see Solaro 2008 for a brief review). Is skeletal muscle contractility and energy consumption modulated similarly? Most of the controls in cardiac muscle are exerted on thin filament proteins, and are thought to have their effect through activation state of the thin filament. Analogous modulation of activation state of the striated muscle thin filament may not occur during tetanus, but could be significant during everyday (unfused) muscle contractions.

4 Conclusions and Future Directions

In their review of cross-bridge theories in 1985, Woledge et al. (1985) were left with the impression that the biochemical and mechanical sides of cross-bridge theory remained separate – 'biochemical predictions could probably be altered without much effect on the physiological ones, and vice versa.' Since then, structures of myosin cross-bridges corresponding to a few biochemical states have been solved, and mechanical properties of single myosin motors have been measured. Physiological predictions have been tested by biochemical interventions, and new biochemical states have been inserted in the mechanical cycle as a result. Simple models of biochemical kinetics have been useful in identifying new fine structure in the working stroke, and increasingly complicated strain-dependent models have extended the kinetic models to make use of structural data, and to make predictions about transient behaviors and effects of distributions of cross-bridges. Still, the fundamental energetic questions listed in the introduction remain unanswered.

The 'molecule to man' integration to link the energy economies of myosin molecules, muscle fibers, whole muscle and whole humans are at an early stage. The initial step in vertical integration, from single molecule mechanics of the myosin ATPase to actomyosin ATP turnover and mechanics in a skinned fiber, has seen some early experimental breakthroughs, but more data and three-dimensional models will be necessary to complete the correspondence. Recent measurements that parse ATP turnover among the myosin, sarcoplasmic reticulum and ion pump ATPases (e.g. Szentesi et al. 2001) should extend the economics of skinned fibers to living fibers. At the level of whole muscle or whole animals, however, mechanical properties of muscle are still represented by Hill models, and integration that includes the architecture, motor unit diversity, neural control options, and multiple compliances of whole muscle, will be much harder than the problem of representing mechanical properties.

Discoveries of mechanical tuning of the working stroke in single motors and fibers have led to the exciting prospect that muscle is mechanically tuned to its task. The fundamental questions regarding the dependence of efficiency on velocity and load seem to be yielding rapidly to new measurements and simple models. Progress in this area seems to depend on further refinement of the nature of the actomyosin interface, the main conclusion of this chapter. Simple models still seem appropriate in this area, but data under physiological conditions (temperature, ion and protein composition of myofibrillar space) are sparse. The prospect of modulation of mechanical properties and ATP turnover by biochemical changes within a fiber (e.g. fatigue) or by signals from outside the muscle (as in cardiac muscle) have hardly begun in skeletal muscle (Chalovich 2002), but given the progress in cardiac myocytes and the array of techniques available for investigating various signaling pathways, this seems an area with great potential.

References

Caremani M, Lehman S, Lombardi V, Linari M (2009) Orthovanadate slows kinetics of the actomyosin interaction in skinned muscle fibers by competition between actin-myosin-ADP-Pi and actin-myosin-ADP-Vi cross-bridges for actin sites. 59th annual meeting of the Biophysical Society, Boston, February 28–March 4

Chalovich JM (2002) Regulation of striated muscle contraction: a discussion. J Muscle Res Cell Motil 23:353–361

Cooke R (2004) The sliding filament model: 1972–2004. J Gen Physiol 123:643–656

Cooke R (2007) Modulation of the actomyosin interaction during fatigue of skeletal muscle. Muscle Nerve 36:756–777

Crow MT, Kushmerick MJ (1982a) Chemical energetics of slow- and fast-twitch muscles of the mouse. J Gen Physiol 79:147–166

Crow MT, Kushmerick MJ (1982b) Myosin light chain phosphorylation is associated with a decrease in the energy cost for contraction in fast twitch mouse muscle. J Biol Chem 257:2121–2124

Dantzig JA, Goldman YE, Millar NC, Lactis J, Homsher E (1992) Reversal of the cross-bridge force-generating transition by photogeneration of phosphate in rabbit psoas muscle fibres. J Physiol 451:247–278

Hill AV (1938) The heat of shortening and the dynamic constants of muscle. Proc R Soc Lond B 126:136–195

Hill AV (1964) The variation of total heat production in a twitch with velocity of shortening. Proc R Soc Lond B 159:596–605

Hill TL (1989) Free energy transduction and biochemical cycle kinetics. Springer-Verlag, New York

Howard J (2001) Mechanics of motor proteins and the cytoskeleton. Sinauer Associates, Inc., Sunderland, MA

Huxley AF (1957) Muscle structure and theories of contraction. Prog Biophys Biophys Chem 7:255–318

Huxley AF (1973) A note suggesting that the cross-bridge attachment during muscle contraction may take place in two stages. Proc Roy Soc Lond B 183:83–86

Karatzaferi C, Franks-Skiba K, Cooke R (2008) Inhibition of shortening velocity of skinned skeletal muscle fibers in conditions that mimic fatigue. Am J Physiol Regul Integr Comp Physiol 294:R948-R955

Kawai M, Zhao Y (1993) Cross-bridge scheme and force per cross-bridge state in skinned rabbit psoas muscle fibers. Biophys J 65:59–62

Lymn RW, Taylor EW (1971) Mechanism of adenosine triphosphate hydrolysis by actomyosin. Biochemistry 10:4617–4624

Nyitrai M, Rossi R, Adamek N, Pellegrino MA, Bottinelli R, Geeves MA (2006) What limits the velocity of fast-seletal muscle contraction in mammals? J Mol Biol 355:432–442

Pate E, Cooke R (1989) Addition of phosphate to active muscle fibers probes actomyosin states within the powerstroke. Pflugers Arch 414:73–81

Piazzesi G, Lucii L, Lombardi V (2002) The size and the speed of the working stroke of muscle myosin and its dependence on the force. J Physiol 545:145–151

Potma EJ, Stienen GJM (1996) Increase in ATP consumption during shortening in skinned fibres from rabbit psoas muscle: effects of inorganic phosphate. J Physiol 496:1–12

Potma EJ, Stienen GJM, Barends JPF, Elzinga G (1994a) Myofibrillar ATPase activity and mechanical performance of skinned fibres from rabbit psoas muscle. J Physiol 474.2:303–317

Potma EJ, van Graas IA, Stienen GJM (1994b) Effects of pH on myofibrillar ATPase activity in fast and slow skeletal muscle fibers of the rabbit. Biophys J 67:2404–2410

Potma EJ, van Graas IA, Stienen GJM (1995) Influence of inorganic phosphate and pH on ATP utilization in fast and slow skeletal muscle fibers. Biophys J 69:2580–2589

Siemankowski RF, White HD (1984) Kinetics of the interaction between actin, ADP, and cardiac myosin-S1. J Biol Chem 259:5045–5053

Siemankowski RF, Wiseman MO, White HD (1985) ADP dissociation from actomyosin subfragment 1 is sufficiently slow to limit the unloaded shortening velocity in vertebrate muscle. Proc Natl Acad Sci U S A 82:658–662

Smith NC, Barclay CJ, Loiselle DS (2005) The efficiency of muscle contraction. Prog Biophys Mol Biol 88:1–58

Solaro JR (2008) Multiplex kinase signaling modifies cardiac function at the level of sarcomeric proteins. J Biol Chem 283:26829–26833

Sun Y-B, Hilber K, Irving M (2001) Effect of active shortening on the rate of ATP utilization by rabbit psoas muscle fibres. J Physiol 531:781–791

Szentesi P, Zaremba R, van Mechelen W, Stienen GJM (2001) ATP utilization for calcium uptake and force production in different types of human skeletal muscle fibres. J Physiol 531:393–403

Szent-Györgyi AG (2004) The early history of the biochemistry of muscle contraction. J Gen Physiol 123:631–641

Takagi Y, Shuman H, Goldman YE (2004) Coupling between phosphate release and force generation in muscle actomyosin. Phil Trans R Soc B 359:1913–1920

Takagi Y, Homsher EE, Goldman YE, Shuman H (2006) Force generation in single conventional actomyosin complexes under high dynamic load. Biophys J 90:1295–1307

Trentham DR, Eccleston JF, Bagshaw CR (1976) Kinetic analysis of ATPase mechanisms. Q Rev Biophys 9:217–281

White HD, Taylor EW (1976) Energetics and mechanism of actomyosin adenosine triphosphatease. Biochemistry 15:5818–5826

Woledge RC (1998) Possible effects of fatigue on muscle efficiency. Acta Physiol Scand 162:267–273

Woledge RC, Curtin NA, Homsher E (1985) Energetic aspects of muscle contraction. Academic Press, San Diego

A Strain-Dependency of Myosin Off-Rate Must Be Sensitive to Frequency to Predict the B-Process of Sinusoidal Analysis

Bradley M. Palmer

Abstract Muscle force arises as the result of many myosin molecules, each producing a force discrete in magnitude and in time duration. In previous work we have developed a computer model and a mathematical model of many myosin molecules acting as an ensemble and demonstrated that the time duration over which myosin produces force at the molecular level (referred to here as "time-on") gives rise to specific visco-elastic properties at the whole muscle level. That model of the mechanical consequences of myosin-actin interaction predicted well the C-process of small length perturbation analysis and demonstrated that the characteristic frequency $2\pi c$ provided a measure of the myosin off-rate, which is equal to the reciprocal of the mean time-on. In this study, we develop a mathematical hypothesis that a strain-dependence of the myosin off-rate at the single molecule level can result in a negative viscous modulus like that observed at low frequencies, i.e., the B-process. We demonstrate here that a simple monotonic strain-dependency of the myosin off-rate cannot account for the observed B-process. However, a frequency-dependent strain-dependency, as may occur when visco-elastic properties of the myosin head are introduced, can explain the observed negative viscous modulus. These findings suggest that visco-elastic properties of myosin constitute the specific molecular mechanisms that underlie the frequency-dependent performance of many oscillatory muscles such as insect flight muscle and mammalian cardiac muscle.

Keywords Oscillatory muscle • Oscillatory work • Insect fight muscle • Cardiac muscle • Sarcomere

B.M. Palmer (✉)
Department of Molecular Physiology and Biophysics, University of Vermont,
122 HSRF – 149 Beaumont Ave, Burlington, VT 05405, USA
e-mail: Bradley.Palmer@uvm.edu

D.E. Rassier (ed.), *Muscle Biophysics: From Molecules to Cells*,
Advances in Experimental Medicine and Biology 682,
DOI 10.1007/978-1-4419-6366-6_4,

1 Introduction

Pick up a pencil, and the muscles in your arm and fingers casually produce the forces necessary to achieve the required movement and stability. We generally experience muscle force as being easily controlled over a continuous range of forces. At the molecular level, however, muscle force is discrete in both magnitude and time duration (Finer et al. 1994; Spudich 1994; Tyska and Warshaw 2002). Myosin is the specific molecule that produces this discrete force as it goes through the following biochemical-mechanical cycle: (a) myosin binds to an actin filament, (b) myosin undergoes a discrete physical deformation, called the power stroke, thus generating a unitary force (F_{uni}) on the actin filament, (c) F_{uni} is maintained until inorganic phosphate (P_i) and adenosine-diphosphate (ADP) are released and adenosine-5′-triphosphate (ATP) is bound to the myosin thus initiating the detachment of myosin from actin, and (d) while detached from actin, myosin hydrolyzes the ATP thus providing the energy to recover the pre-force conformation of the myosin molecule (Geeves and Holmes 1999; Spudich 1994; Steffen and Sleep 2004; Stein et al. 1981). Myosin is now ready for another cycle. Many actin filaments exist in parallel in a muscle, and many myosin molecules act together, although not synchronously, on any one actin filament. The result is a seemingly continuous force generated by a muscle due to the many discrete forces summed over space and time.

Several properties of the myosin force-producing cycle have been studied at the level of the single molecule. Interestingly, the magnitude of F_{uni} for striated-muscle myosin does not vary much from myosin to myosin (Tyska and Warshaw 2002). On the other hand, the time duration over which myosin is bound to actin, referred to t_{on}, is inversely dependent upon ATP concentration and also varies greatly among different isoforms, species and patterns of post-translational modifications (Tyska and Warshaw 2002). The myosin t_{on} appears to bear the most significant influence on muscle performance (Spudich 1994; Tyska and Warshaw 2002). As an example of how t_{on} affects force production, consider that the force average over time generated by a single myosin molecule (F_{ave}) must be proportional to F_{uni} and the ratio $t_{on}/(t_{on}+t_{off})$, i.e., $F_{ave}=F_{uni}\times t_{on}/(t_{on}+t_{off})$, where t_{off} is the time duration when myosin is detached from actin (Fig. 1). The ratio $t_{on}/(t_{on}+t_{off})$ describes the fraction of time over which myosin is producing force and is often referred as the duty ratio. Changes in the duty ratio will result in changes in F_{ave}. For example, when the sum of t_{on} and t_{off} (t_{cycle}) does not change, a shorter t_{on} reduces F_{ave} and a longer t_{on} enhances F_{ave}, as illustrated in Fig. 1. If we now imagine a muscle containing N independent myosin molecules, then the total force (F_{total}) can be written as $F_{total}=\mathbf{N}\times F_{uni}\times t_{on}/(t_{on}+t_{off})$. Muscle performance furthermore depends upon the velocity of muscle shortening, which is inversely proportional to t_{on} (Tyska and Warshaw 2002). Muscles found throughout the animal kingdom tend to adapt myosin t_{on}, t_{off} and **N** to suit the force and velocity requirements of the muscle performance (Spudich 1994; Tyska and Warshaw 2002).

Measuring myosin t_{on} has emerged as an important component of any investigation into the molecular mechanisms that underlie muscle performance and dysfunction. Experiments in the laser trap probing the molecular performance of cardiac myosin

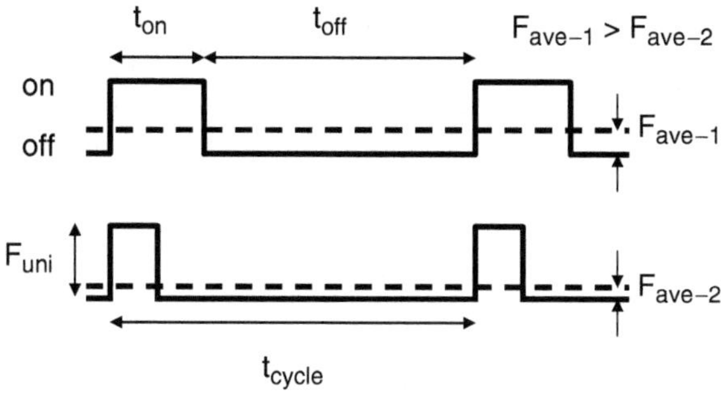

Fig. 1 Two diagrams of force production and time durations at the molecular level. The time duration over which myosin is attached to actin is indicated by as the 'on' position and referred to as time-on, t_{on}. The time duration when myosin is detached is indicated as the 'off' position and referred to as time-off, t_{off}. The time average force (F_{ave}) of an acto-myosin crossbridge is proportional to the unitary force (F_{uni}) produced by the myosin molecule during t_{on} and the crossbridge duty ratio, $F_{ave} = F_{uni} \times t_{on}/(t_{on} + t_{off})$. When t_{on} is shortened relative to t_{cycle}, the duty ratio and F_{ave} are reduced. When t_{on} is prolonged relative to t_{cycle}, the duty cycle and F_{ave} are enhanced

heavy chain (MHC), for example, have shown that myosin t_{on} is significantly shorter in the α-MHC isoform compared to β-MHC (Palmiter et al. 1999). This result is important because, when α-MHC is inadequately expressed or absent as occurs in human heart failure, one consequence is an inability for the heart to perform effectively at high frequencies (Herron and McDonald 2002; Suzuki et al. 2009). As another example, some point mutations in MHC lead to shorter t_{on} compared to that of non-mutant myosin leading to greater velocities of shortening and ultimately to a hypertrophic cardiomyopathy (Debold et al. 2007; Palmiter et al. 2000; Tyska et al. 2000; Yamashita et al. 2000). The mal-adaptation of myosin t_{on} to the required performance of the heart appears to be a significant detrimental factor in some cardiomyopathies and heart failure.

In this paper, we describe the technique of small length perturbation analysis, which provides a measure of myosin t_{on} in muscle strips dissected to <150 μm diameter cross-section. This method permits the preservation of the myofilament lattice structure found in vivo, thus providing an appropriate structural context in which t_{on} is measured, and obviates the isolation of single myosin molecules as required for the laser trap. We will also explore here how data that arise during small length perturbation analysis may be interpreted beyond the measurement of myosin t_{on}.

2 Small Sinusoidal Length Perturbation Analysis

Visco-elastic properties of muscle can be measured at the macroscopic level using small amplitude sinusoidal length perturbation analysis (Abbott and Steiger 1977; Cheung and Gray 1983; Davis and Rodgers 1995; Kawai and Brandt 1980;

Thomas and Thornhill 1995; Thorson and White 1969; White and Donaldson 1975; Zhao and Kawai 1993). Figure 2a depicts a sinusoidal length perturbation, which ensures that the force response is linear, and two example sinusoidal force responses, which lead or lag the length perturbation. Each force response can be constructed as one sinusoidal component that is in-phase with the perturbation and one cosine component that is 90° out-of-phase with the perturbation. The elastic and viscous moduli refer to the amplitudes of the in-phase and 90° out-of-phase responses, respectively, relative to the amplitude of the strain. Those force responses which lead the length perturbation result in a positive viscous modulus, while those that lag result in a negative viscous modulus (Fig. 2b). Taken together as respectively

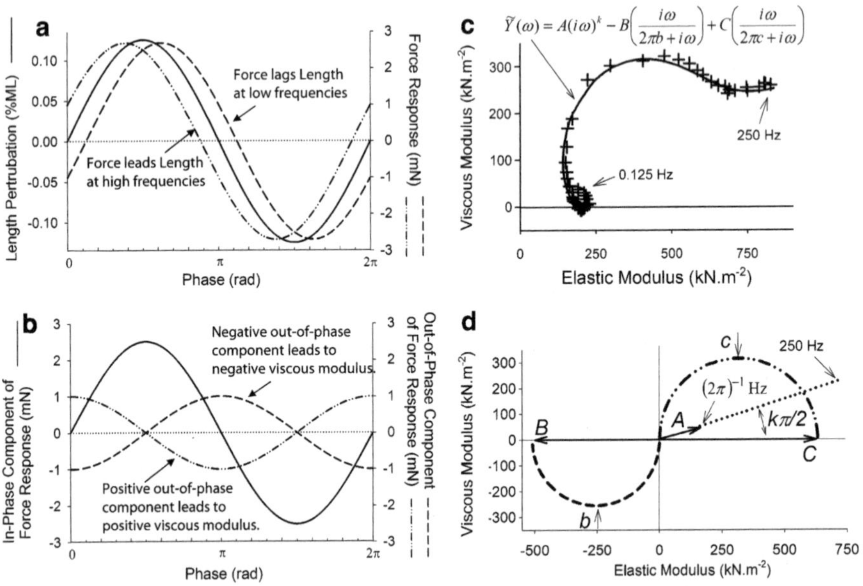

Fig. 2 Sinusoid analysis and resultant Nyquist diagram. (**a**) The application of a small amplitude sinusoidal length perturbation, i.e., 0.125% of muscle length (ML), to a muscle strip will result in a recorded force response that either leads or lags the length perturbation in time. (**b**) The force response is made up of two components, one in-phase with the length (*solid line*) and one out-of-phase (*dashed lines*) with the length. A force response that leads the length contains an out-of-phase component that is positive (shown here with amplitude +1 at time zero), which corresponds to a positive viscous modulus as observed at high frequencies. A lagging force response contains an out-of-phase component that is negative (shown here with amplitude –1), which corresponds to a negative viscous modulus as observed at low frequencies. (**c**) A plot of the viscous modulus vs. elastic modulus, i.e., Nyquist diagram, recorded from human cardiac muscles demonstrates a looped relationship between the two moduli. A negative viscous modulus occurs in the range of 1–3 Hz, and a positive viscous modulus occurs at frequencies greater than ~3 Hz. The complex modulus often conforms to the mathematical model, $\tilde{Y}(\omega)$, containing the so-called A-, B- and C-processes. (**d**) The processes describe the angular orientation of the data within the diagram (A-process), the negative viscous modulus at low frequencies (B-process) and the semi-circular relationship between the elastic and viscous moduli at higher frequencies (C-process)

the real and imaginary parts of a complex number, the elastic and viscous moduli make up the complex modulus, $\tilde{Y}(\omega)$, which describes in the frequency domain the linear response of the muscle force to a length perturbation.

In practice the relative frequency characteristics of the elastic and viscous moduli, plotted as a Nyquist diagram in Fig. 2c, demonstrate a complex, yet systematically looped relationship which arises only when the muscle is activated and therefore must reflect temporal characteristics of the myosin force-producing cycle (Campbell et al. 2004; Kawai and Brandt 1980; Machin 1964; Maughan et al. 1998; Palmer et al. 2007; Pringle 1978). A mathematical expression, (1), describing the complex modulus has been used effectively to fit these data.

$$\tilde{Y}(\omega) = A(i\omega)^k - B\left(\frac{i\omega}{2\pi b + i\omega}\right) + C\left(\frac{i\omega}{2\pi c + i\omega}\right) \quad (1)$$

where the parameters A, B and C represent the magnitudes of three terms characterized by the frequency parameters k, $2\pi b$ and $2\pi c$. The three terms of (1) correspond to three observed characteristics, often referred to as the A-, B- and C-processes, illustrated in Fig. 2c and d. When combined, these processes give rise to (a) the angular orientation of the data in the Nyquist diagram (A-process), (b) the negative viscous modulus observed at low frequencies (B-process) and (c) the semi-circular relationship between the viscous and elastic moduli observed at higher frequencies (C-process). It's important to note that the frequency characteristic $2\pi b$ is always observed to be lower than $2\pi c$. The expression in (1), however, has evolved through empirical means (Kawai and Brandt 1980; Maughan et al. 1998), and it has been long recognized that an unambiguous physiological interpretation of the parameters would be useful.

Several mathematical models have been developed attempting to describe the frequency characteristics of the complex modulus and to provide physiological meaning to the parameters in (1) (Abbott and Steiger 1977; Campbell et al. 2004; Cheung and Gray 1983; Davis and Rodgers 1995; Kawai and Brandt 1980; Thomas and Thornhill 1995; Thorson and White 1969; White and Donaldson 1975). In one previously published study from our laboratory (Palmer et al. 2007), an analytical expression was derived and a corresponding computer model was developed to describe the mechanical consequences at the macroscopic level of many myosin-actin interactions occurring at the molecular level. Each myosin-actin interaction was characterized by (a) intermittent periods of t_{on} and t_{off} governed by independent stochastic processes described by single exponential probability functions, and (b) an elastic element engaged only during t_{on}. As illustrated in Fig. 3a, a frictional force resisting the length perturbation arises during t_{on} as the length perturbation elongates and compresses the elastic element during t_{on}. This model predicted well the C-process of sinusoidal analysis (Fig. 3b) and provided a mathematical basis for estimating the mean myosin t_{on} as $(2\pi c)^{-1}$. We have since found, for example, that our estimates of myosin t_{on} for MHC isoforms and for some myosin point mutations are in qualitative agreement with those values reported using the laser trap to measure t_{on} (Palmer et al. 2004a, 2007; Suzuki et al. 2009). We have furthermore been able

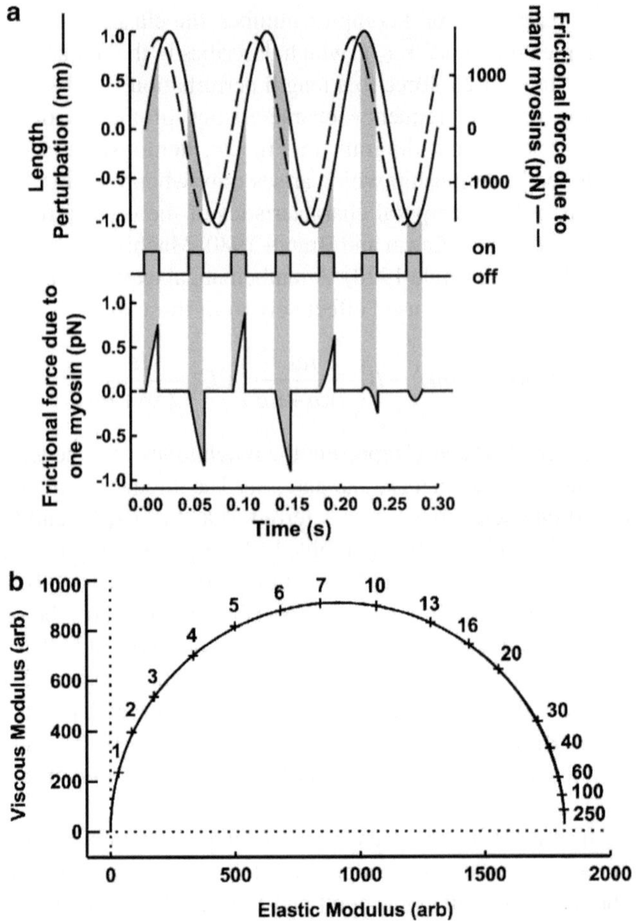

Fig. 3 The C-process arises from the frictional force due to actin-myosin interactions. (**a**) The intermittent and repeated attachment of myosin to actin results in a force that is proportional to the velocity of the length perturbation during the time of attachment, t_{on}. The net force is a frictional force that resists and leads the length perturbation. (**b**) The elastic and viscous moduli are positive due to this friction between the thick and thin filaments, and the frequency responses of the moduli conform to the mathematical expression used to describe the C-process

to use this technique to highlight how myosin t_{on} adapts to suit the function of oscillatory muscles. For example, the respective frequencies of insect flight muscle, mouse heart muscle and human heart muscle are on the order of ~100 Hz, ~10 Hz and ~1 Hz, and the corresponding myosin t_{on} for each is estimated on the order of ~0.1 ms, ~1 ms and ~10 ms.

The friction model for the C-process, however, does not explain the other frequency characteristics of the elastic and viscous moduli that appear to bear physiological importance (Maughan et al. 1998; Palmer et al. 2007; Pringle 1978; Steiger 1977). In particular, the viscous modulus of activated muscle is invariably observed to be negative in value at the specific frequency of muscle operation, e.g., in the range ~1–3 Hz for human cardiac heart as shown in Fig. 2c (Fukagawa et al. 2005). The negative viscous modulus indicates that the myosin ensemble is not absorbing mechanical work as occurs with friction, but rather producing mechanical work at that specific frequency (Maughan et al. 1998). Clearly, a more complete model of the myosin-actin system must account for this phenomenon.

Others have attempted to explain the negative viscous modulus as the result of a strain dependency on t_{on} and/or t_{off} (Abbott and Steiger 1977; Campbell et al. 2004; Cheung and Gray 1983; Thomas and Thornhill 1995; Thorson and White 1969; White and Donaldson 1975; White and Thorson 1972), however, without success in providing a mathematical representation of the negative viscous modulus. Kawai and Brandt (Kawai and Brandt 1980) offered a mathematical description of the negative viscous modulus based on a strain dependency of transition rates between biochemical states. The basis of their mathematical result was argued heuristically but not assigned mathematically. Nevertheless, a strain-dependency on t_{on} has been demonstrated at the single molecule level and should be included somehow in any model of the complex modulus (Kad et al. 2007). Below, we explore how a strain-dependence of myosin t_{on} at the single molecule level may result in the negative viscous modulus observed at the operating frequency of oscillatory muscles.

3 Analytical Results

In our previous study we used probability theory to arrive at analytical solutions describing the isometric force and frequency characteristics of the frictional force that arises from many myosin-actin interactions at the molecular level (Palmer et al. 2007). In the present study we opt not to use probability theory, but ordinary differential equations (ODEs) to explore the mechanical consequences of a strain-dependency on myosin t_{on} and a recorded muscle force. In using ODEs the mean t_{on} will be represented by its reciprocal, the myosin off-rate denoted as g by Huxley (Huxley 1957; Huxley and Simmons 1971).

We will first use ODEs to recapitulate the findings of our previous study as an illustration and validation of the use of ODEs for our main purpose here. We will then demonstrate that a strain-dependency on the myosin off-rate cannot alone result in a negative viscous modulus observed during small length perturbation analysis of oscillatory muscles. A frequency dependence on the strain-dependency on myosin off-rate does result in a negative viscous modulus like that observed as the B-process. Our analytical description of the B-process, however, relies upon a positive relationship between strain and myosin off-rate, which is contrary to observations from the laser trap (Kad et al. 2007). Nevertheless, a positive relationship between strain and myosin off-rate may arise under conditions of relatively high P_i (Baker et al. 2002; Hibberd et al. 1985), and we offer the demonstration below as the basis of a hypothesis as to the molecular mechanisms that underlie the observed negative viscous modulus in skinned muscle preparations.

3.1 ODE Solution for Isometric Force and C-Process

Huxley proposed a two-state model of myosin attaching and detaching to actin with an apparent on-rate of attachment referred to as f and an apparent off-rate of detachment referred to as g (Huxley 1957; Huxley and Simmons 1971). The ODE used to describe the rate of change of the fraction of myosin attached to actin at any time, $n(t)$, is the following:

$$\frac{dn(t)}{dt} = f\left[1 - n(t)\right] - g\,n(t) \tag{2}$$

where the term $[1 - n(t)]$ denotes the fraction of myosin detached from actin at any time. For the case when $n(t)$ does not change with time, the left hand side of (2) is zero and the steady-state solution results: $n(t) = f/(f+g)$. If each myosin attachment is now assigned a unitary force F_{uni}, then the total force of an ensemble of **N** myosin molecules is $F_{total} = \mathbf{N} \times F_{uni} \times f/(f+g)$. This particular result from (2) is reminiscent of the total force $F_{total} = \mathbf{N} \times F_{uni} \times t_{on}/(t_{on} + t_{off})$ stated in the Introduction. It should be noted that the rate constants, f and g, are first-order rate constants and are indeed equivalent to the reciprocals of mean t_{on} and mean t_{off}, respectively, when t_{on} and t_{off} are governed by independent stochastic processes described by single exponential probability functions. Under those conditions the ratios $f/(f+g)$ and $t_{on}/(t_{on}+t_{off})$ are equivalent and describe the duty ratio of the myosin crossbridge. Furthermore, the model parameter $2\pi c$ that emerges by fitting (1) to recorded data directly estimates the myosin off-rate g and therefore the mean t_{on} (Palmer et al. 2007).

Among the important contributions of Huxley's model was his use of the x-dimension to describe the length-dependence of f and g, i.e., $f = f(x)$ and $g = g(x)$. For our purposes, we will not adopt the specific $f(x)$ and $g(x)$ functions offered by Huxley. As illustrated in Fig. 4, we will assume an $f(x)$, which is reminiscent of that proposed by Hill (1974) and reflects the probability of myosin attachment to actin as being distributed along the x-dimension and symmetrical about zero displacement. For now, we will also assume $g(x) = G_0$, which is a constant and independent of the x-dimension. The function $n(t)$ now also becomes a function of the x-dimension, $n(x, t)$, which approaches zero value at $x = -\infty$ and $+\infty$.

The left hand side of (2) must now be revised according to the chain rule as follows:

$$\frac{dn(x,t)}{dt} = \frac{\partial n(x,t)}{\partial t} + v\frac{\partial n(x,t)}{\partial x} \tag{3}$$

where v is the velocity, dx/dt, of any displacement of the attached myosin relative to the thick filament, as would occur with an externally driven length perturbation. Equation (2) can now be written as follows:

$$\frac{\partial n(x,t)}{\partial t} + v\frac{\partial n(x,t)}{\partial x} = f(x)\left[1 - \int_{-\infty}^{\infty} n(x,t)dx\right] - G_0\, n(x,t) \tag{4}$$

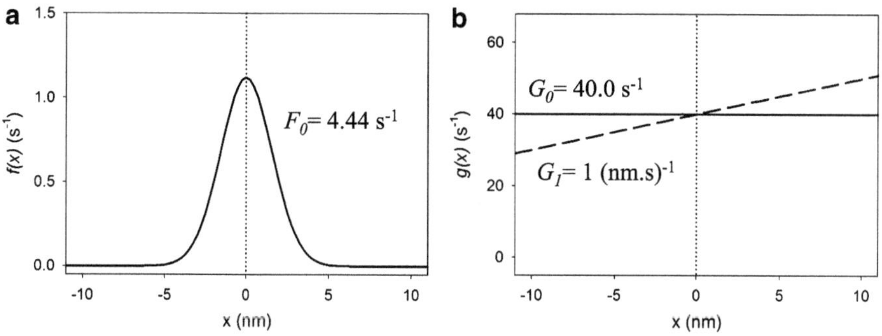

Fig. 4 The rate functions $f(x)$ and $g(x)$ used in calculations. (**a**) The function $f(x)$ describes the spatial-dependence of the probability of myosin attaching to actin along the length thin filament. The position $x=0$ indicates the central position of a myosin head and distribution about $x=0$ reflects the possible reach of the head in both the negative and positive directions. The term F_0 refers to the integral of $f(x)$ along the x-axis (see (8b)) and its value, 4.44 s^{-1}, is used to assure a duty cycle of 0.1 (see text). (**b**) The function $g(x)$ describes the spatial-dependence of the probability of myosin detaching to actin along the length thin filament. When $g(x)=G_0$, a constant, there is no spatial dependence. A monotonic strain-dependence on $g(x)$ is provided by the added term G_1x. A positive value for G_1 reflects a faster myosin off-rate, or shorter t_{on}, when myosin molecules are experiencing a positive strain, as occurs during a lengthening phase of the length perturbation

where the term $\left[1-\int_{-\infty}^{\infty} \mathrm{n}(x,t)dx\right]$ represents the fraction of myosin detached from actin at any time. It may be worthwhile to note that the definite integral over the x-axis leads to a function of time only.

Instead of solving (4) for the function n(x, t), which has two dimensions, we can choose instead to solve for the moments of the distribution n(x, t) along the x-axis, thus effectively simplifying to a one-dimension problem. Specifically, we can integrate (4) over the x-dimension, $\int_{-\infty}^{\infty} x^i(\text{Eq.4})dx$ which results in a series of one-dimensional equations. Each of these resulting equations represents an important characteristic of the distribution $\mathrm{n}(x, t)$ along the x-axis. Specifically, when $i=0$ the total fraction of myosin attached to actin is described; when $i=1$ the mean length displacement of myosin attached to actin is described; when $i=2$ the variance of the length displacement is described, etc.

The proposed integration transforms (4) into the following:

$$\int_{-\infty}^{\infty} x^i \frac{\partial \mathrm{n}(x,t)}{\partial t}dx + v\int_{-\infty}^{\infty} x^i \frac{\partial \mathrm{n}(x,t)}{\partial x}dx = \left[1-\int_{-\infty}^{\infty} \mathrm{n}(x,t)dx\right]\int_{-\infty}^{\infty} x^i f(x)dx - G_0\int_{-\infty}^{\infty} x^i\,\mathrm{n}(x,t)dx \tag{5}$$

Let's focus on evaluating the integral in the second term of (5) and perform its integration by parts.

$$\int_{-\infty}^{\infty} x^i \frac{\partial \mathrm{n}(x,t)}{\partial x} dx = \left(x^i\, \mathrm{n}(x,t)\right)\Big|_{-\infty}^{\infty} - \int_{-\infty}^{\infty} i x^{i-1} n(x,t) dx \tag{6}$$

Because the function n(x, t) goes to zero at $x=-\infty$ and $x=+\infty$, only the last term of (6) is necessary to represent the second term of (5), which can now be written as follows:

$$\int_{-\infty}^{\infty} x^i \frac{\partial \mathrm{n}(x,t)}{\partial t} dx - v \int_{-\infty}^{\infty} i x^{i-1}\, \mathrm{n}(x,t) dx =$$

$$\int_{-\infty}^{\infty} x^i f(x) dx \left[1 - \int_{-\infty}^{\infty} \mathrm{n}(x,t) dx\right] - G_0 \int_{-\infty}^{\infty} x^i\, \mathrm{n}(x,t) dx \tag{7}$$

We can also make the following definitions for $\mathrm{N}_i(t)$ and F_i:

$$N_i(t) = \int_{-\infty}^{\infty} x^i\, \mathrm{n}(x,t) dx \tag{8a}$$

and

$$F_i = \int_{-\infty}^{\infty} x^i f(x) dx \tag{8b}$$

Equation (7) then becomes the following:

$$\frac{d\mathrm{N}_i(t)}{dt} - vi\,\mathrm{N}_{i-1}(t) = F_i\left[1 - \mathrm{N}_0(t)\right] - G_0\, \mathrm{N}_i(t) \tag{9}$$

The above equation, (9), represents a series of equations for all values of i. We will use the cases $i=0$ and $i=1$ to describe the force recorded from an ensemble of myosin. Specifically, force due to the myosin force-producing cycle is $\mathbf{N}\times \mathrm{F}_{uni} \times \mathrm{N}_0(t)$, and frictional force that arises due to strain on the myosin with a stiffness k_{stiff} is $\mathbf{N}\times \mathrm{k}_{stiff} \times \mathrm{N}_1(t)$ (Palmer et al. 2007).

Referring to (8a), the function $\mathrm{N}_0(t)$ represents the fraction of myosin molecules bound to actin at any time. Equation (9) then reduces to the following:

$$\frac{d\mathrm{N}_0(t)}{dt} = F_0\left[1 - \mathrm{N}_0(t)\right] - G_0\, \mathrm{N}_0(t) \tag{10}$$

which has the same form and steady-state solution as (2), namely:

$$\mathrm{N}_0(t) = F_0 / \left(F_0 + G_0\right) \tag{11}$$

If each myosin of **N** molecules bears a unitary force F_{uni}, then $\mathrm{F}_{total} = \mathbf{N}\times \mathrm{F}_{uni} \times F_0/(F_0+G_0)$, which is the same result found from (2).

Equations (10) and (11) do not offer anything new compared to (2) and its solution, but our arriving at (10) does offer an important check as to the validity of (9). Equation (10) also provides the interpretation of f in the two-state model of (2) as the integration of any spatially-dependent $f(x)$ over the entire x-dimension.

Accordingly, the parameter g of (2) is furthermore interpreted as a constant value of the off-rate without any dependence on the x-dimension.

Referring again to (8a), the function $N_1(t)$, i.e., $i=1$, represents the mean length displacement of the ensemble of myosin molecules bound to actin at any time. For $i=1$ (9) becomes the following:

$$\frac{dN_1(t)}{dt} - v\,N_0(t) = F_1[1 - N_0(t)] - G_0\,N_1(t) \tag{12}$$

It's important to note the F_1 will have a value of zero when $f(x)$ is an even function symmetric about $x=0$, which we have assumed here (Fig. 4a). We will also assume steady-state conditions and use the relationship $N_0(t)=F_0/(F_0+G_0)$ to get the following equation:

$$\frac{dN_1(t)}{dt} - v\left(\frac{F_0}{F_0+G_0}\right) = -G_0\,N_1(t) \tag{13}$$

We can now utilize the Fourier transform and recognize that v represents the time derivative of the externally driven length perturbation, which we call $L(t)$. Specifically, we will define the following:

$$\tilde{N}_1(\omega) = \frac{1}{\sqrt{2\pi}}\int_{-\infty}^{\infty} N_1(t)e^{-i\omega t}dt \tag{14a}$$

$$\tilde{L}(\omega) = \frac{1}{\sqrt{2\pi}}\int_{-\infty}^{\infty} L(t)e^{-i\omega t}dt \tag{14b}$$

Equation (13) then becomes

$$i\omega\tilde{N}_1(\omega) - i\omega\tilde{L}(\omega)\left(\frac{F_0}{F_0+G_0}\right) = -G_0\tilde{N}_1(\omega) \tag{15}$$

The solution is

$$\tilde{N}_1(\omega) = \left(\frac{F_0}{F_0+G_0}\right)\frac{i\omega}{G_0+i\omega}\tilde{L}(\omega) \tag{16}$$

Equation (16) represents in frequency space the mean length displacement, i.e., the mean strain, of actin-attached myosin when an external perturbation has been applied to one filament relative to the other. If we assign a stiffness coefficient, k_{stiff}, to each myosin, then we would have a description of the frictional force that arises when an external length perturbation is applied while myosin intermittently and repeatedly attaches to actin. Equation (16) has the same form as the C-process of (1) and suggests, as previously (Palmer et al. 2007), that the displacement of the myosin due to the thin and thick filaments sliding past each other results in that portion of the complex modulus represented as the C-process. It should be noted that this interpretation of the C-process does not require a strain dependence on the myosin off-rate.

3.2 Simple Strain Dependency on Myosin Off-Rate

We will demonstrate in this section that a simple monotonic strain dependency on $g(x)$ cannot account for the B-process observed in small amplitude sinusoidal length perturbation analysis. As mentioned above, the use of small amplitude length perturbation analysis is restricted to systems assumed to be linear. If we were to consider (9) for the case $i=2$ or greater, the second term of the resulting equation would be non-linear. Furthermore, a result for higher order $N_i(t)$ is not related to force. For example, $N_2(t)$ describes the amount of potential energy stored in the elastic elements of the acto-myosin crossbridge. We will therefore not consider the cases $i=2$ or greater, but we will retain focus on the cases $i=0$ and $i=1$. We can still ask whether a monotonic strain dependency on the myosin off-rate can result in the negative viscous modulus at low frequencies as observed during sinusoidal analysis. The measure of a negative viscous modulus can occur only if the recorded force measurement lags in time behind the length perturbation as illustrated in Fig. 2a and b. Our fundamental understanding of friction suggests that this is not possible as a direct result of friction, i.e., friction gives rise to a force proportional to the velocity of a length perturbation and therefore leads the perturbation as suggested by the positive sign in the description of $\tilde{N}_1(\omega)$ in (16).

Considering that others have demonstrated a monotonic strain-dependency on g (Kad et al. 2007), it would be reasonable to consider how a strain dependency on g may affect the number of myosin attached, $N_0(t)$, and the resulting recorded force. As illustrated in Fig. 5, an enhanced off-rate during the lengthening phase of a perturbation would shorten t_{on} and depress F_{ave}. A reduced off-rate during the shortening phase would prolong t_{on} and enhance F_{ave}. Conceivably, a monotonic strain-dependency on g could then result in a force signal lagging the perturbation like that shown in Fig. 5. As the result for $N_1(t)$ describes the mean strain of actin-attached myosin, we will consider below whether $N_1(t)$ given in (16) may be used

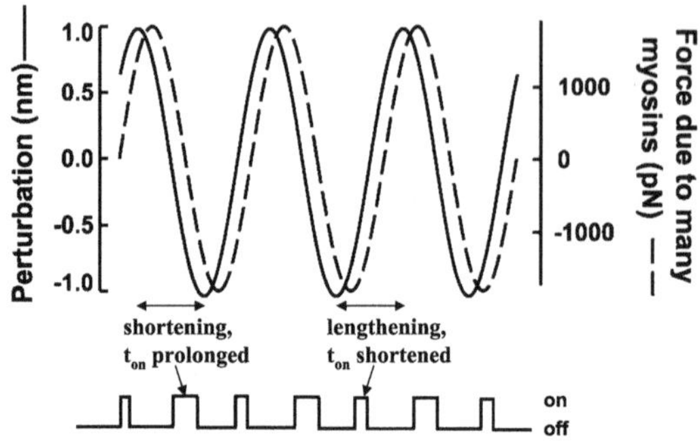

Fig. 5 A force response lagging the length perturbation may arise if t_{on} is prolonged during a shortening phase of a length perturbation and alternately shortened during a lengthening phase. The changes in F_{ave} due to changes in t_{on} would result in a lower recorded force during lengthening and higher recorded force during shortening, i.e. lagging the length perturbation. The result would be a negative viscous modulus

to affect g and in turn $N_0(t)$ to produce a force lagging the length perturbation and therefore a negative viscous modulus.

If g were given a monotonic strain dependence like that shown in Fig. 3b, i.e., $g(x) = G_0 + G_1 x$ in (4), then (10) would have the following form:

$$\frac{dN_0(t)}{dt} = F_0[1 - N_0(t)] - G_0 N_0(t) - G_1 N_1(t) \tag{17}$$

If we now use the solution of (16) to represent the last term of (17) and ignore any transient response from equilibrium, we have the following result for the steady-state response of the fraction of actin-bound myosin due to the length perturbation L(t).

$$\tilde{N}_0(\omega) = -G_1\left(\frac{F_0}{F_0 + G_0}\right)\left(\frac{i\omega}{G_0 + i\omega}\right)\left(\frac{1}{F_0 + G_0 + i\omega}\right)\tilde{L}(\omega) \tag{18}$$

Equation (18) describes in frequency space the change in the number of attached myosin due to an externally applied length perturbation. Again, if each myosin of **N** molecules bears an F_{uni}, we could use (18) to describe the total force due to the myosin force-producing cycle.

The form and sign of (18) are reminiscent of the B-process term of (1). However, the two rates embedded in the denominator terms of (18), namely G_0 and $F_0 + G_0$, are respectively equal to or greater than the myosin off-rate, g, which would be detected as $2\pi c$. Recall that the rate constant associated with the B-process, $2\pi b$, is consistently observed to be lower than $2\pi c$. We conclude that the steady-state response described in (18), and therefore a simple monotonic strain dependency on $g(x)$ in a two state model, cannot account for the B-process observed in small amplitude sinusoidal length perturbation analysis.

3.3 A Frequency Dependency on the Strain-Dependence of Myosin Off-Rate

We will demonstrate in this section that a frequency dependency on the effects of strain on myosin off-rate can result in the B-process. It would appear that a strain sensitivity on myosin off-rate would be a reflection of the mechanical attributes of the myosin molecule, i.e., affecting either the myosin interface with actin and/or the myosin affinity for ATP (Kad et al. 2007). For example, if the myosin head were the most mechanically compliant portion of the myosin molecule, then the myosin head would experience the majority of the strain that had been applied during the laser trap experiment. With this in mind, we would hypothesize that the visco-elastic characteristics of the myosin molecule, which could affect the myosin interface with actin and/or the myosin affinity for ATP, bear a characteristic rate constant that is lower than the off-rate of myosin. In other words, the intra-molecular strain of myosin must possess a frequency dependency that gives rise to significant physical

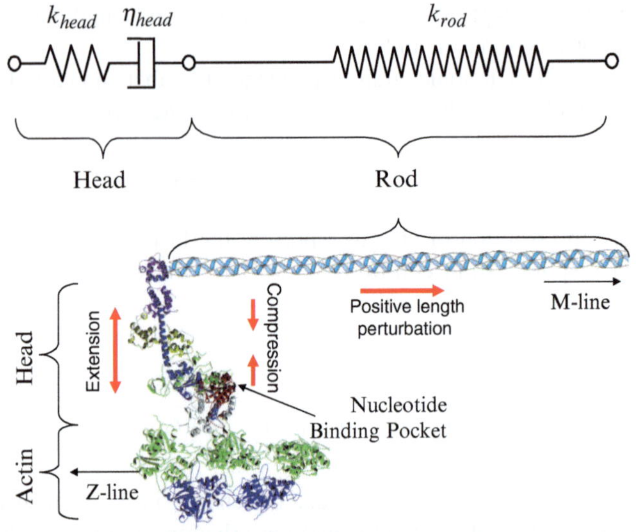

Fig. 6 A Maxwell model of the myosin head including lever arm in series with an elastic element representing the myosin rod. The nucleotide binding pocket is shown in *red*. The structure of the myosin head and lever arm shown here depicts the myosin post-power stroke state. The force generated by the power stroke and a positive length perturbation results in an extension of the myosin along the Z-line side of the myosin head and a compression of the myosin along the M-line side. A negative length perturbation would result in less extension and possibly compression along the Z-line side of the myosin head and less compression and possibly extension along the M-line side. The physical distortion of the elastic element, k_{head}, of the myosin head, as would occur during a length perturbation, is proposed as the specific strain that affects myosin off-rate, $g(x)$, via the effects on P_i and ADP release and/or P_i and ATP binding. Structure adapted from Geeves and Holmes (1999)

distortion of myosin at frequencies of perturbation lower than the off-rate of myosin and not at higher frequencies. Such would be the case if the visco-elastic characteristics of myosin head (including the lever arm) were reasonably modeled as a spring and dashpot in series, i.e., a Maxwell model, and the myosin rod were modeled as a spring in series (Fig. 6). The mean physical distortion of the spring element in the myosin head, call it $\tilde{D}(\omega)$, would be proportional to the compliance of the elastic element divided by the total compliance of the mechanical system and multiplied by the mean strain applied to the system:

$$\tilde{D}(\omega) = \frac{\left(\frac{k_{rod}}{k_{rod}+k_{head}}\right)i\omega}{\left(k_{head}/\eta_{head}\right)\left(\frac{k_{rod}}{k_{rod}+k_{head}}\right)+i\omega}\tilde{N}_1(\omega) \tag{19}$$

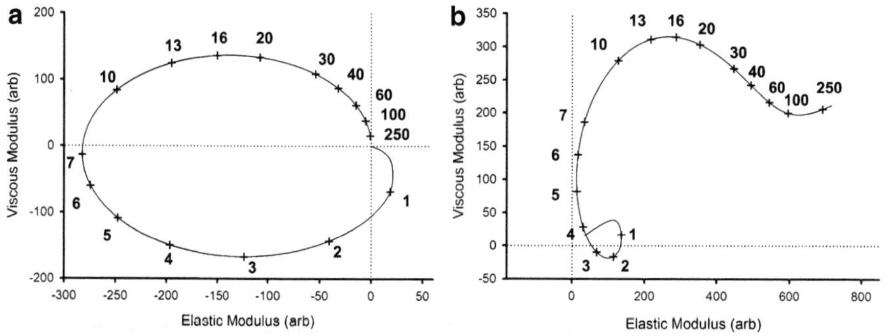

Fig. 7 Example result of frequency-dependence and strain-dependence on myosin off-rate. (**a**) Equation (20) predicts a B-process within a recorded complex modulus that includes a negative elastic modulus and a negative viscous modulus at low frequencies according to the parameter values described in the text. (**b**) When illustrative examples of the A- and C-processes are added, the Nyquist diagram qualitatively resembles that of actual recorded data like that shown in Fig. 2c

where k_{head} = the stiffness of the elastic element, η_{head} = the viscosity of the dashpot and k_{rod} = the stiffness of elastic element of the myosin rod in series with the Maxwell element.

If we use this distortion $\tilde{D}(\omega)$ instead of $\tilde{N}_1(\omega)$ to influence $\tilde{N}_0(\omega)$ in (17), then we have a steady-state solution given as:

$$\tilde{N}_0(\omega) = -G_1\left(\frac{F_0}{F_0+G_0}\right)\left(\frac{\xi i\omega}{R\xi+i\omega}\right)\left(\frac{i\omega}{G_0+i\omega}\right)\left(\frac{1}{F_0+G_0+i\omega}\right)\tilde{L}(\omega) \quad (20)$$

where $\xi = k_{rod}/(k_{rod}+k_{head})$ and $R = (k_{head}/\eta_{head})$. Note that the value of ξ is always less than 1. The rate constant represented by $R\xi$ describes the characteristic rate of the myosin head deformation due to a length perturbation. This rate constant depends on the mechanical attributes of the myosin head and may well be lower than the myosin off-rate, G_0, which would be detected as $2\pi c$. It should be noted that the form of (20) does not strictly comply with the analytical expression for the B-process provided in (1). Nevertheless, if we assume the values $F_0 = 4.44\ \mathrm{s}^{-1}$ and $G_0 = 40.0\ \mathrm{s}^{-1}$, which represent a duty cycle = 0.1 (Harris and Warshaw 1993) and t_{on} = 25 ms similar to that in human cardiac muscle, $G_1 = 1\ (\mathrm{nm\ s})^{-1}$, $\xi = 0.8$ and $R = 5\ \mathrm{s}^{-1}$, we find that (20) predicts negative values for the elastic and viscous moduli at low frequencies (Fig. 7a) as would be expected from a B-process shown in Fig. 2d. The result of (20) also predicts a positive viscous modulus at higher frequencies, which has not been modeled previously or included in (1) but is also not contrary to observation. The addition of the A- and C-process as given in (16) results in a loop in the Nyquist diagram (Fig. 7b) that is qualitatively similar in shape to that observed in practice and shown in Fig. 2c.

4 Discussion

This paper demonstrates that a frequency dependent, positively monotonic strain-dependency on the myosin off-rate can result in negative values for the elastic and viscous moduli, which could account for the observed B-process of sinusoidal analysis of oscillatory muscles. The frequency dependency utilized here was modeled as the result of the visco-elastic properties of the myosin molecule. Although the actual physical distortion of myosin that occurs with an imposed length perturbation is not known, the frequency characteristic of myosin strain may well be less than that of the C-process, $2\pi c$, and could therefore underlie the frequency characteristics of the B-process. A similar concept of visco-elastic properties leading to changes in acto-myosin kinetics has been demonstrated for smooth muscle myosin (Kad et al. 2007) and has been forwarded to explain the processivity of myosin-V (Veigel et al. 2005).

We showed in our development to (20) that, if the myosin off-rate is positively related to the physical distortion of the visco-elastic elements of myosin, then negative values for the elastic and viscous moduli would arise. The form of the model presented here also offers an explanation how mutations in the myosin rod, like those that lead to cardiomyopathies (Debold et al. 2007; Palmiter et al. 2000), can affect myosin off-rate. According to our results, a mutation that results in a stiffer myosin rod would raise the characteristic frequency of distortion in the myosin head (Miller et al. 2009). A more compliant rod would lead to a lower characteristic frequency.

There is, however, one significant caveat to the present model that must be addressed. The B-process developed here relies upon the myosin off-rate being positively related to the positive strain of myosin, i.e., a physical elongation of myosin. Studies of the strain dependency of the myosin off-rate using the laser trap have demonstrated a negative relationship between the myosin off-rate and the imposed load or strain, i.e., off-rate is enhanced with a negative or assistive load and is reduced with a positive or resistive load (Harris and Warshaw 1993; Kad et al. 2007). We reconcile these experimental findings with our analysis by suggesting that the negative relationship between myosin off-rate and strain observed in the laser trap is restricted to those myosin t_{on} which reflect only an ATP-dependent force-producing cycle without thin filament regulatory proteins. According to Kad et al. (2007) the positive strain of the myosin head due to a resistive load inhibits the release of ADP and/or the binding of ATP, thus prolonging the myosin t_{on}. If we recognize that myosin t_{on} may also be governed via a P_i-dependent cycle (Hibberd et al. 1985), which is much shorter in duration than that of the ATP-dependent cycle (Baker et al. 2002), then we have the possibility of a positive relationship between myosin off-rate and strain. Considering that P_i release and rebinding occur at different sites of the myosin head relative to the nucleotide binding pocket where ADP is released and ATP is bound (Geeves and Holmes 1999), it is conceivable that the P_i-dependent cycle bears a positive relationship between myosin off-rate and strain. Indeed, we find in practice that P_i concentration will enhance the amplitude of the

B-process (Palmer et al. 2004b; Swank et al. 2006). In the scheme that we propose here, a rise in the P_i concentration would effectively raise the myosin off-rate, G_0, and the sensitivity of the strain-dependency, G_1.

We suggest then that a positive strain-dependency on the myosin off-rate could arise if both ATP- and P_i-dependent cycles are considered. We furthermore hypothesize that (20) underlies the observed B-process of sinusoidal analysis of oscillatory muscles in a manner that relies upon the release rates of P_i and ADP and the binding rates of P_i and ATP. We believe the hypothesis raised in (20) is important because it suggests a molecular mechanism by which myosin can regulate temporal characteristics of force production necessary for the effective and efficient function of oscillatory muscles like insect flight muscle and cardiac muscle (Maughan et al. 1998). We expect to proceed with these investigations with analytical solutions using probability theory and a computer model that will test the validity of (20) based on data available in the literature and the assumptions outlined in the present work.

Acknowledgements This paper is dedicated in honor of Dr. Mark R. Hilty. This work was supported by a grant from the NIH P01-HL59408.

References

Abbott RH, Steiger GJ (1977) Temperature and amplitude dependence of tension transients in glycerinated skeletal and insect fibrillar muscle. J Physiol 266:13–42

Baker JE, Brosseau C, Joel PB, Warshaw DM (2002) The biochemical kinetics underlying actin movement generated by one and many skeletal muscle myosin molecules. Biophys J 82:2134–2147

Campbell KB, Chandra M, Kirkpatrick RD, Slinker BK, Hunter WC (2004) Interpreting cardiac muscle force-length dynamics using a novel functional model. Am J Physiol Heart Circ Physiol 286:H1535–H1545

Cheung AS, Gray BF (1983) Muscle tension response to sinusoidal length perturbation: a theoretical study. J Muscle Res Cell Motil 4:615–623

Davis JS, Rodgers ME (1995) Indirect coupling of phosphate release to de novo tension generation during muscle contraction. Proc Natl Acad Sci USA 92:10482–10486

Debold EP, Schmitt JP, Patlak JB, Beck SE, Moore JR, Seidman JG, Seidman C, Warshaw DM (2007) Hypertrophic and dilated cardiomyopathy mutations differentially affect the molecular force generation of mouse alpha-cardiac myosin in the laser trap assay. Am J Physiol Heart Circ Physiol 293:H284–H291

Finer JT, Simmons RM, Spudich JA (1994) Single myosin molecule mechanics: piconewton forces and nanometre steps. Nature 368:113–119

Fukagawa NK, Palmer BM, Barnes WD, Leavitt BJ, Ittleman FP, Lewinter MM, Maughan DW (2005) Acto-myosin crossbridge kinetics in humans with coronary artery disease: influence of sex and diabetes mellitus. J Mol Cell Cardiol 39:743–753

Geeves MA, Holmes KC (1999) Structural mechanism of muscle contraction. Annu Rev Biochem 68:687–728

Harris DE, Warshaw DM (1993) Smooth and skeletal muscle myosin both exhibit low duty cycles at zero load in vitro. J Biol Chem 268:14764–14768

Herron TJ, McDonald KS (2002) Small amounts of alpha-myosin heavy chain isoform expression significantly increase power output of rat cardiac myo-cyte fragments. Circ Res 90:1150–1152

Hibberd MG, Dantzig JA, Trentham DR, Goldman YE (1985) Phosphate re-lease and force generation in skeletal muscle fibers. Science 228:1317–1319

Hill TL (1974) Theoretical formalism for the sliding filament model of contraction of striated muscle. Part I. Prog Biophys Mol Biol 28:267–340

Huxley AF (1957) Muscle structure and theories of contraction. Prog Biophys Biophys Chem 7:255–318

Huxley AF, Simmons RM (1971) Proposed mechanism of force generation in striated muscle. Nature 233:533–538

Kad NM, Patlak JB, Fagnant PM, Trybus KM, Warshaw DM (2007) Mutation of a conserved glycine in the SH1-SH2 helix affects the load-dependent kinetics of myosin. Biophys J 92:1623–1631

Kawai M, Brandt PW (1980) Sinusoidal analysis: a high resolution method for correlating biochemical reactions with physiological processes in activated skeletal muscles of rabbit, frog and crayfish. J Muscle Res Cell Motil 1:279–303

Machin KE (1964) Feedback theory and its application to biological systems. Symp Soc Exp Biol 18:421–445

Maughan D, Moore J, Vigoreaux J, Barnes B, Mulieri LA (1998) Work production and work absorption in muscle strips from vertebrate cardiac and insect flight muscle fibers. Adv Exp Med Biol 453:471–480

Miller MS, Dambacher CM, Knowles AF, Braddock JM, Farman GP, Irving TC, Swank DM, Bernstein SI, Maughan DW (2009) Alternative S2 hinge regions of the myosin rod affect myofibrillar structure and myosin kinetics. Biophys J 96:4132–4143

Palmer BM, Fishbaugher DE, Schmitt JP, Wang Y, Alpert NR, Seidman CE, Seidman JG, VanBuren P, Maughan DW (2004a) Differential cross-bridge kinetics of FHC myosin mutations R403Q and R453C in heterozygous mouse myocardium. Am J Physiol Heart Circ Physiol 287:H91–H99

Palmer BM, Noguchi T, Wang Y, Heim JR, Alpert NR, Burgon PG, Seidman CE, Seidman JG, Maughan DW, LeWinter MM (2004b) Effect of cardiac myosin binding protein-C on mechanoenergetics in mouse myocardium. Circ Res 94:1615–1622

Palmer BM, Suzuki T, Wang Y, Barnes WD, Miller MS, Maughan DW (2007) Two-state model of acto-myosin attachment-detachment predicts C-process of sinusoidal analysis. Biophys J 93:760–769

Palmiter KA, Tyska MJ, Dupuis DE, Alpert NR, Warshaw DM (1999) Kinetic differences at the single molecule level account for the functional diversity of rabbit cardiac myosin isoforms. J Physiol 519:669–678

Palmiter KA, Tyska MJ, Haeberle JR, Alpert NR, Fananapazir L, Warshaw DM (2000) R403Q and L908V mutant beta-cardiac myosin from patients with familial hypertrophic cardiomyopathy exhibit enhanced mechanical performance at the single molecule level. J Muscle Res Cell Motil 21:609–620

Pringle JW (1978) The Croonian Lecture, 1977. Stretch activation of muscle: function and mechanism. Proc R Soc Lond B Biol Sci 201:107–130

Spudich JA (1994) How molecular motors work. Nature 372:515–518

Steffen W, Sleep J (2004) Repriming the actomyosin crossbridge cycle. Proc Natl Acad Sci USA 101:12904–12909

Steiger GJ (1977) Tension transients in extracted rabbit heart muscle preparations. J Mol Cell Cardiol 9:671–685

Stein LA, Chock PB, Eisenberg E (1981) Mechanism of the actomyosin ATPase: effect of actin on the ATP hydrolysis step. Proc Natl Acad Sci USA 78:1346–1350

Suzuki T, Palmer BM, James J, Wang Y, Chen Z, Van Buren P, Maughan DW, Robbins J, LeWinter MM (2009) Effects of cardiac myosin isoform variation on myofilament function and crossbridge kinetics in transgenic rabbits. Circ Heart Fail 2(4):334–341

Swank DM, Vishnudas VK, Maughan DW (2006) An exceptionally fast actomyosin reaction powers insect flight muscle. Proc Natl Acad Sci USA 103:17543–17547

Thomas N, Thornhill RA (1995) A theory of tension fluctuations due to muscle cross-bridges. Proc Biol Sci 259:235–242

Thorson J, White DC (1969) Distributed representations for actin-myosin interaction in the oscillatory contraction of muscle. Biophys J 9:360–390

Tyska MJ, Warshaw DM (2002) The myosin power stroke. Cell Motil Cytoskeleton 51:1–15

Tyska MJ, Hayes E, Giewat M, Seidman CE, Seidman JG, Warshaw DM (2000) Single-molecule mechanics of R403Q cardiac myosin isolated from the mouse model of familial hypertrophic cardiomyopathy. Circ Res 86:737–744

Veigel C, Schmitz S, Wang F, Sellers JR (2005) Load-dependent kinetics of myosin-V can explain its high processivity. Nat Cell Biol 7:861–869

White DC, Donaldson MM (1975) Mechanical and biochemical cycles in muscle contraction. Ciba Found Symp 341–353

White DC, Thorson J (1972) Phosphate starvation and the nonlinear dynamics of insect fibrillar flight muscle. J Gen Physiol 60:307–336

Yamashita H, Tyska MJ, Warshaw DM, Lowey S, Trybus KM (2000) Functional consequences of mutations in the smooth muscle myosin heavy chain at sites implicated in familial hypertrophic cardiomyopathy. J Biol Chem 275:28045–28052

Zhao Y, Kawai M (1993) The effect of the lattice spacing change on cross-bridge kinetics in chemically skinned rabbit psoas muscle fibers. II. Elementary steps affected by the spacing change. Biophys J 64:197–210

Electron Microscopic Visualization of the Cross-Bridge Movement Coupled with ATP Hydrolysis in Muscle Thick Filaments in Aqueous Solution, Reminiscences and Future Prospects

Haruo Sugi

Abstract Although it has been well established that muscle contraction results from cyclic attachment–detachment between the cross-bridges extending from the thick filaments and the sites on the thin filaments, the movement of the cross-bridges coupled with ATP hydrolysis still remains to be a matter for debate and speculation. The most straightforward way to elucidate this mystery is to record individual cross-bridge movement in response to ATP. Using a gas environmental chamber (EC, or hydration chamber), with which biological specimens retaining their physiological function can be observed under an electron microscope, my coworkers and I succeeded in recording the ATP-induced individual cross-bridge movement in two different kinds of synthetic thick filaments in 1997 and 2008. In the synthetic bipolar filaments consisting of rabbit skeletal muscle myosin, the amplitude of cross-bridge movement exhibits a peak at 5–7.5 nm, and the direction of cross-bridge movement is away from, but not towards, the filament bare region in the absence of thin filaments. After exhaustion of ATP, the cross-bridges return towards their initial position, indicating that the initial cross-bridge state may be analogous to that after completion of power stroke. These results constitute the first visualization of the cross-bridge recovery stroke, indicating that the EC is a powerful tool to open new horizons in the research fields of life sciences.

Keywords Myosin • Cross-bridges • Electron microscope • Thick filaments

1 Introduction

It has long been a dream of investigators in the field of life sciences to observe living microorganisms under an electron microscope. In order to realize this dream, the gas environmental chamber (EC, or hydration chamber) has been developed in a

H. Sugi (✉)
Department of Physiology, School of Medicine, Teikyo University, Itabashi-ku, Tokyo 173-8605, Japan
e-mail: sugi@med.teikyo-u.ac.jp

D.E. Rassier (ed.), *Muscle Biophysics: From Molecules to Cells*, Advances in Experimental Medicine and Biology 682,
DOI 10.1007/978-1-4419-6366-6_5,

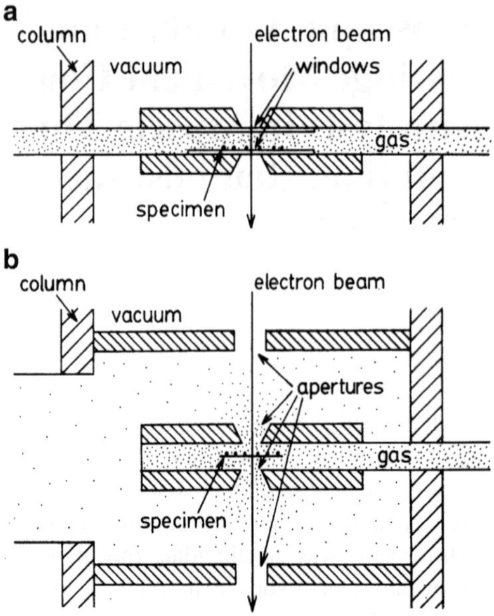

Fig. 1 Film-sealed EC (**a**) and aperture-limited EC (**b**) (Fukushima 1988)

number of laboratories, in which biological specimens are placed in a small chamber in an electron microscope, and kept wet by circulating gas (water vapor) through the chamber (for an extensive review, see Butler and Hale 1981). Two different types of EC have hitherto been developed; one is film-sealed EC, which is insulated from high vacuum of electron microscope by means of sealing film at its upper and lower windows to pass electron beam (Fig. 1a), while the other is aperture-limited EC, which has small apertures to pass electron beam, but is not insulated from high vacuum of electron microscope. In the latter type of EC, gas going out through its apertures is sucked out with a pumping device (Fig. 1b). Attempts to observe living microorganisms with the EC attached to an electron microscope, however, proved to be impossible since their physiological function was readily impaired by electron beam. Meanwhile, biological macromolecules such as proteins, lipids and nucleic acids are expected to be much more resistant against electron beam irradiation.

In 1982, I had the idea that an appropriate use of EC would make it possible to study dynamic structural changes of contractile proteins in hydrated state. Fortunately just at that time, I had an opportunity to meet Professor Akira Fukami in the Department of Applied Physics, Nihon University. He had just succeeded in preparing a carbon sealing film suitable for the film-sealed type EC, and was looking for coworkers in the field of life sciences. Professor Fukami generously allowed me to use the carbon sealing film for my research work on muscle proteins. Furthermore, President Kazuo Ito of Japan Electron Optics Laboratory (JEOL Ltd) kindly helped me in constructing the whole EC system attached to a 200 kV electron microscope in my laboratory, so that I could start experiments with EC in 1985. My main coworker was Dr. Tsuyoshi Akimoto, who had been trained to use the EC in Fukami's laboratory.

In this article, I will first explain the experimental apparatus and the outline of the methods in some detail, to make general readers to understand our novel experimental methods. Next, I will describe the progress of research work in a chronological order, so that the readers may understand what I had in mind at various stages of research work, as well as difficulties encountered. Finally, I will discuss the attachment–detachment cycle between the cross-bridge and the thin filament based on the results obtained from the synthetic bipolar filaments, with a conclusion that our results constitute the first visualization of the cross-bridge recovery stroke.

2 The EC System

2.1 Carbon Sealing Film

The most important element of the film-sealed type EC is the carbon sealing film, developed in Fukami's laboratory. In principle, the spatial resolution and contrast of electron micrographs taken by the EC increases with decreasing thickness of sealing film. Preliminary experiments in Fukami's laboratory indicated that, to obtain a special resolution <1 nm, the thickness of the sealing film should be 15–20 nm. On the other hand, resistibility of a sealing film against pressure difference decreases sharply with increasing area of sealing film; the thickness of sealing film covering an aperture of 50 μm diameter should be ~100 nm to bear a reasonable amount of pressure difference. Since it is practically difficult to make a hole <50 μm in a metal block constituting the EC, Fukami and Adachi (1965) developed a plastic microgrid made from a high-molecular organic compound (cellulose acetobutylate). Figure 2 shows photomicrographs of microgrids. Microgrids with holes of small diameters (a), and with holes of nonuniform diameters (b) are unsuitable for observation of the specimen. Microgrids with fairly uniform holes of 5–8 μm diameter (c) are suitable for observation of the specimen and has a reasonable resistivity against pressure difference, providing ideal sealing films for high-resolution electron microscopy of hydrated biological specimens (Fukushima 1988).

2.2 The EC

As shown diagrammatically in Fig. 3, the EC, developed in Fukami's laboratory, is a cylindrical compartment (diameter, 3.5 mm; depth, 0.8 mm) with upper and lower window frames (copper grids) to pass electron beam. Each window frame has nine apertures, each having a diameter of 0.1 mm. The apertures are further covered by the carbon sealing film (thickness, 15–20 nm). The specimen placed on the lower sealing film is kept in a thin layer of aqueous solution (experimental solution) by constantly circulating the air saturated with water vapor (26–28°C) through the EC. To obtain clear specimen images, the internal pressure of the EC is made 60–80 Torr. The flow rate of water vapor (0.1–0.2 l/min) can be adjusted in such a way that the

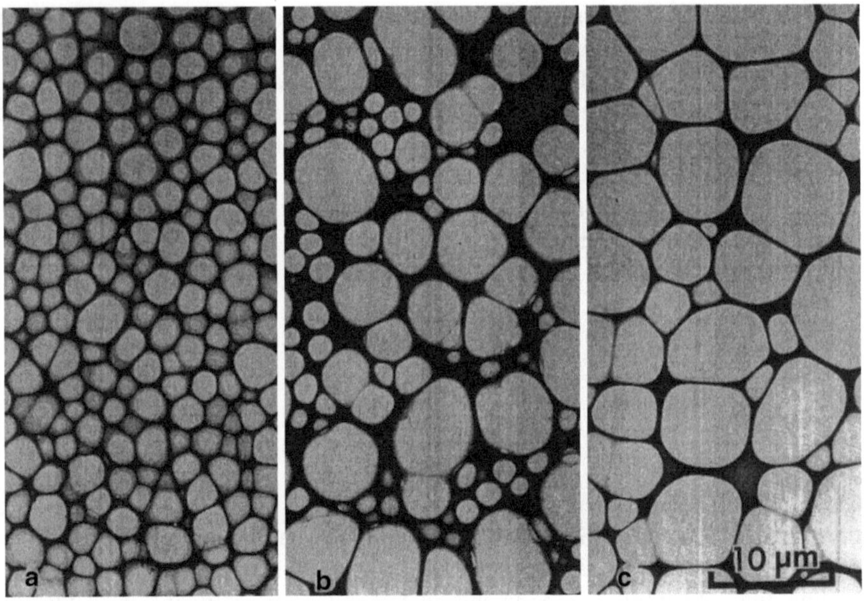

Fig. 2 Photomicrographs of microgrids with holes of small diameters (**a**), with holes of nonuniform diameters (**b**) and with fairly uniform holes of 5–8 μm diameters (**c**) (Fukushima 1988)

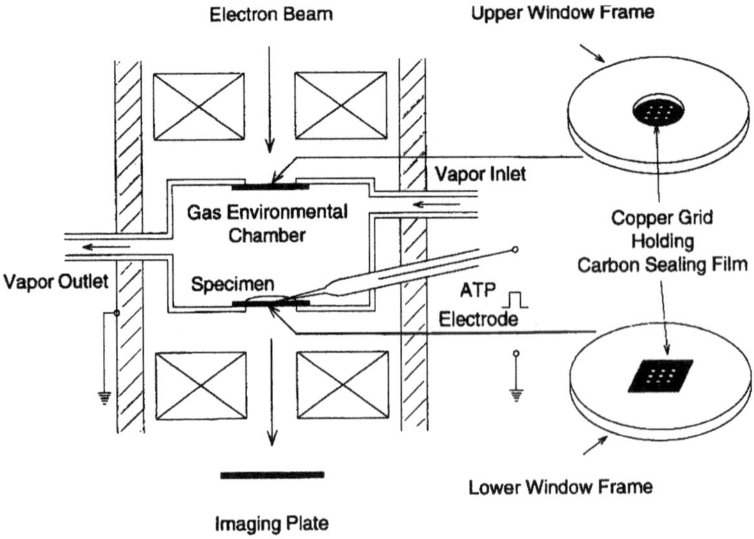

Fig. 3 Diagram of the EC. The upper and lower windows (copper grids with nine apertures) are covered with carbon sealing films held on copper grids with nine apertures. The EC is circulated with water vapor to keep the specimen placed on the lower carbon film in wet state. The EC contains an ATP-containing electrode to apply ATP to the specimen. Specimen images are recorded on the imaging plate (IP) (Sugi et al. 1997)

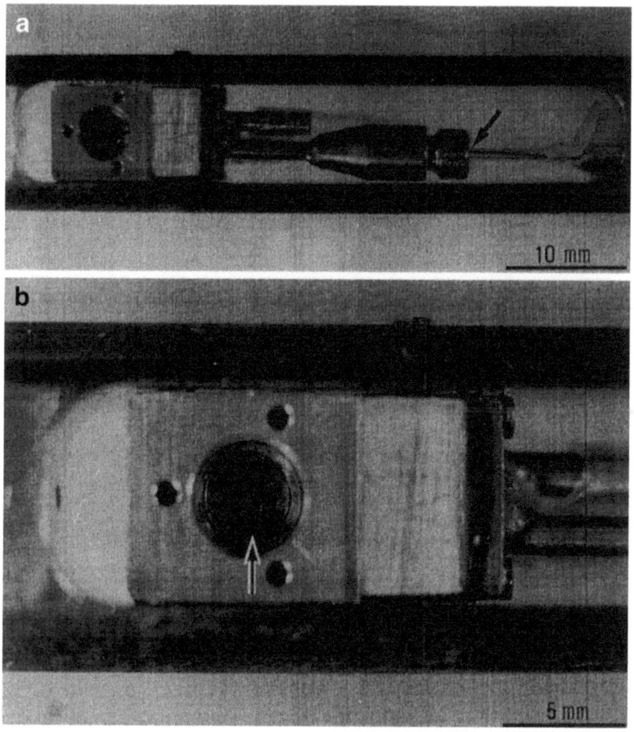

Fig. 4 Photographs of the EC. *Arrows* in (**a** and **b**) indicate the part to which the ATP electrode is connected, and the interior of the EC where the specimen is placed, respectively

thin layer of experimental solution around the specimen is in equilibrium with the vapor pressure in the EC. Details of the water vapor circulation device have been described elsewhere (Fukushima et al. 1985; Fukami et al. 1991). In my laboratory, a glass capillary microelectrode filled with ATP solution was included in the EC. Figure 4 show photographs of the EC.

3 Observation and Recording of Specimen Image

3.1 The Critical Incident Electron Dose to Impair Function of Contractile Proteins

The critical incident electron dose to impair physiological function of muscle contractile proteins was determined in Fukami's laboratory (Suda et al. 1992). They prepared hydrated myofibrils isolated from crayfish skeletal muscle for electron microscopic observation (magnification, 2,500×), and activated them to shorten by

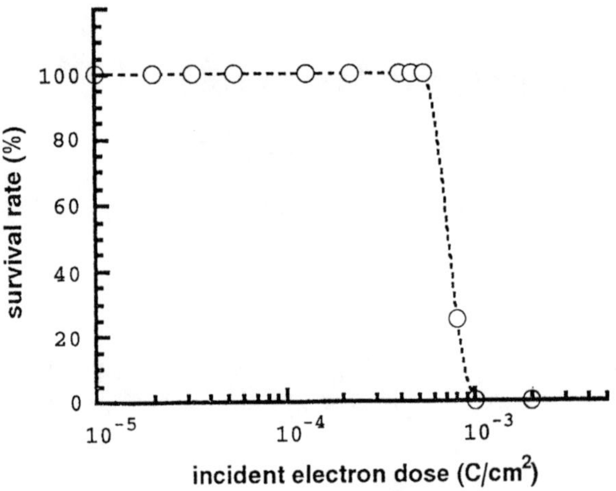

Fig. 5 Relation between the total incident electron dose and the survival rate of muscle myofibrils, expressed as percentage of myofibrils shortened in response to ATP (Suda et al. 1992)

injecting ATP-containing solution into the EC. The results are shown in Fig. 5. With total incident electron dose $<5\times10^{-4}$ C/cm^2, all the myofibrils shorten in response to applied ATP. If, however, the total incident electron dose was further increased, the ATP-induced shortening of myofibrils disappeared in a nearly all-or-none manner, though the myofibrils exhibited no appreciable change in appearance.

The above critical incident electron dose to impair physiological function of contractile proteins was also confirmed by us with respect to both the ATP-induced cross-bridge movement and the ATPase activity of the filaments. Based on these results, electron microscopic observation and recording of the specimen were made with a total incident electron dose $<10^{-4}$ C/cm^2, being well below the critical dose to impair function of contractile proteins. For this purpose, the specimen was observed with extremely weak electron beam intensities (at the fluorescent screen) $<5\times10^{-13}$ A/cm^2, so that observation and focusing of the specimen required enormous technical skill and patience. The actual beam intensity through the specimen with a magnification of 10,000× was $5\times10^{-13}\times(10{,}000)^2=5\times10^{-5}$ A/cm^2. As soon as the specimen was brought into focus, electron beam was stopped until the time of recording.

3.2 *Recording of Specimen Image*

The specimen image (magnification, 10,000×) was recorded on an imaging plate (IP) system (PIX system, JEOL). The IP is 10.2×7.7 cm in size, and had sensitivity ~60 times that of X-ray film. The exposure time was 0.1 s with a beam intensity of $1\text{–}2\times10^{-12}$ A/cm^2. In the experiments with myosin–paramyosin hybrid filaments (Sugi et al. 1997), the number of pixels in the IP was ~3,000,000 ($2{,}045\times1{,}536$). With a magnification of 10,000×, the spatial resolution was ~5 nm.

In the experiments with bipolar thick filaments (Sugi et al. 2008), the pixel number in the IP increased to ~12,000,000 (4,090×3,072), and the special resolution of the IP was ~2.5 nm under the same magnification. Due to limitation of the total incident electron dose, recording of the specimen image can be only made three to four times. The IP system had been developed in Fuji Photofilm, Co. and became commercially available for TEM use from JEOL shortly before our start of experiments.

3.3 Application of ATP and ADP

Before starting experiments, I made some revisions on the Fukami-type EC. The most important revision was inclusion of an ATP-containing glass capillary microelectrode (see Fig. 3). In Fukami's laboratory, application of ATP to the specimen was made by simply injecting ATP-containing solution into the EC, and this procedure inevitably caused marked displacement of the specimen due to flow of solution. To eliminate displacement of specimen during ATP application, we decided to apply ATP iontophoretically by passing current through the electrode to move negatively charged ATP ions out of the electrode; the ATP released from the electrode reached the specimen by diffusion in ATP-free experimental solution. A current pulse (intensity, 10 nA; duration, 1 s) from an electronic stimulator was applied to the electrode containing ATP (or ADP) through a current-clamp circuit (Oiwa et al. 1993). The total amount of ATP released was estimated to be ~10–14 mol (Oiwa et al. 1991). The time required for the released ATP to reach the specimen by diffusion was estimated to be <30 s by video recording ATP-induced shortening of myofibrils mounted in the EC under a light microscope. Hexokinase (50 units/ml) and D-glucose (2 mM) were added to the experimental solution to eliminate contamination of ATP.

3.4 Position Marking of Individual Cross-Bridges

To mark the position of individual cross-bridges in the filaments under unstained condition, colloidal gold particles (diameter, 15 or 20 nm; coated with protein A; EY laboratories) were attached to the cross-bridges, using a site-directed antibody (IgG) to the junctional peptide between 50- and 20-kDa segments of myosin heavy chain (Sutoh et al. 1989). The attachment of the antibody has already been well characterized by Sutoh et al. (1989); it mostly attaches to only one of the two myosin cross-bridges at the region ~16 nm distant from the cross-bridge-myosin rod junction. Technical details to position mark the cross-bridges have been described previously (Sugi et al. 1997). It was important to label the cross-bridges sparsely, so that each gold particle was reasonably distant from the neighboring ones; otherwise, changes in position of individual cross-bridges in response to ATP could not be distinguished from those of neighboring ones.

3.5 Data Analysis

Under a magnification of 10,000×, the pixel size in the IP was 5 × 5 nm in the 1997 experiments, and 2.5 × 2.5 nm in the 2008 experiments, reflecting an improvement in spatial resolution of the IP. The average number of electrons reaching each pixel during the exposure time was about 30 in the 1997 experiments, and 7–8 in the 2008 experiments. Each IP record was divided into a number of subframes, and each subframe was observed on the monitor screen. Gold particles suitable for analysis were selected after an appropriate binning procedure, i.e., the procedure to determine each particle configuration consisting of pixels with electron counts above a chosen level. Particle shapes were not markedly altered by the level of binning.

The center of mass position of each gold particle in the filaments was determined with an image processor (Nexus Qube System, Nexus Tokyo) in the experiments with hybrid filaments (Sugi et al. 1997), and with an ordinary personal computer in the experiments with bipolar filaments (Sugi et al. 2008), reflecting an enormous improvement in the performance of personal computers. The center of mass position for each selected particle was determined as the coordinates (two significant figures) within a single pixel where the center of mass position was located, and these coordinates, representing the position of the particle (and therefore the position of the cross-bridge), were compared between two different IP records. The absolute coordinates common to the two IP records were obtained based on the position of natural markers (bright spots on the carbon sealing film, see lower part of Fig. 6a). The distance (D) between the two center of mass positions (with the coordinates x1, y1 and x2, y2, respectively) was calculated as $D = (x1 - x2)^2 + (y1 - y2)^2$.

4 Experiments with Myosin–Paramyosin Hybrid Filaments

4.1 Reasons to Use the Hybrid Filaments

At the start of our work, we first tried to use native thick filaments isolated from rabbit skeletal muscle fibers, but gave up to use them because (1) they tended to curl and aggregate, (2) small filament diameters (<20 nm) made it difficult to clearly distinguish the filaments from the background, and (3) due to small filament diameters, we felt anxious about the possibility that most cross-bridges, extending laterally from the filaments together with antibodies and gold particles, might attach to the carbon sealing film to result in a marked restriction of their possible ATP-induced movement. We next tried to use synthetic thick filaments with much larger diameters, but again gave up using such filaments because of their marked tendency to aggregate. Finally, we decided to use hybrid synthetic filaments, in which rabbit skeletal muscle myosin was bound around the surface of extremely long and thick paramyosin filaments (diameter, 50–200 nm; length, 10–30 μm). The synthetic myosin–paramyosin core complex filaments are analogous in structure to native

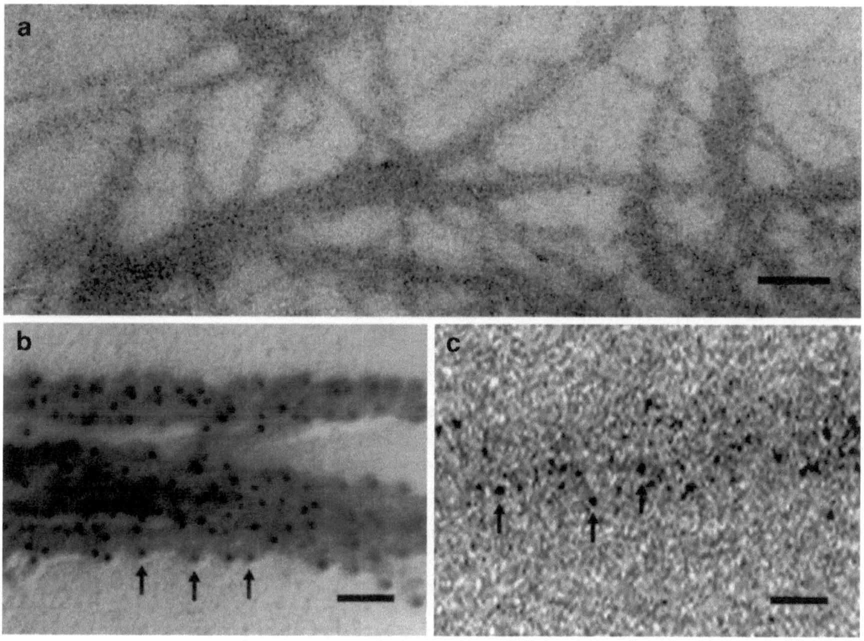

Fig. 6 Images of synthetic hybrid thick filaments. (**a**) IP record of the filaments with a number of gold particles. Bar, 500 nm. (**b**) Conventional electron micrograph of the filaments after negative staining and rotary shadowing. Bar, 100 nm. (**c**) Enlarged IP record showing part of a filament with gold particles attached. Bar, 100 nm. *Arrows* in (**b** and **c**) indicate gold particles (Sugi et al. 1997)

thick filaments in the anterior byssus retractor muscle (ABRM) of a bivalve mollusc, *Mytilus edulis*. Since antibodies for ABRM myosin were not available, we replaced ABRM myosin by rabbit skeletal muscle myosin (Nonomura 1974; Sugi et al. 1997).

The hybrid filaments had the following advantages: (1) they were stiff and nearly straight in shape, and showed no tendency to aggregate; (2) their profile was clearly distinguished from the background due to their large diameters; and (3) as myosin molecules were bound around the paramyosin core in parallel rows (Squire 1981), most cross-bridges were located at the upper side of the filaments, and free from possible constraints arising from their attachment to the carbon film. The filaments exhibited an Mg-ATPase activity (~0.13 s^{-1}, 28°C), which was not appreciably affected after mixing with the antibody.

4.2 Stability of the Cross-Bridge Position in the Absence of ATP

Figure 6a shows a typical IP record of the hybrid filaments. The gold particles, sparsely attached to the cross-bridges, are clearly seen as discrete dark spots on the filaments, which are also clearly distinguished from the background. Figure 6b is a conventional electron micrograph of the filaments after negative staining and rotary shadowing. The filaments were actually covered by the cross-bridges with gold particles attached to them. An enlarged IP record showing part of the filament (Fig. 6c) indicated that the gold particles are somewhat variable both in size and in shape, reflecting electron statistics. To examine whether the particle (and therefore the cross-bridge) positions are stable or fluctuate with time, we compared the particle positions between two IP records of the same filaments taken at an interval of 3–5 min. Inspection of pairs of records by eye, we immediately became aware that position of each particle did not change appreciably. This was confirmed by precise determination of center of mass position for each particle as coordinates within a single pixel, and by comparing the coordinates representing the particle (cross-bridge) position between the two records.

An example of the results is shown in Fig. 7a, in which a circle of 15 nm diameter (the same as the particle diameter) is drawn around each center of mass position of particles.

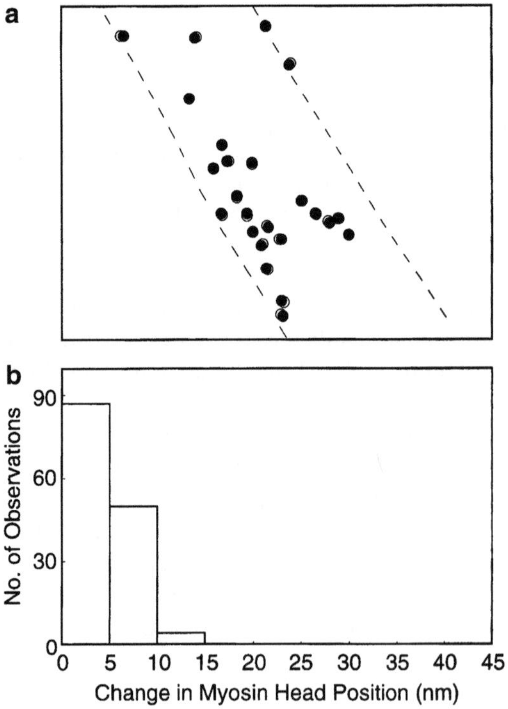

Fig. 7 Stability of time-averaged cross-bridge mean position in the absence of ATP. (**a**) Comparison of cross-bridge positions between two IP records of the same filament. *Filled* and *open circles* (diameter, 15 μm) are drawn around the center of mass position of each particle in the first and the second records, respectively. *Broken lines* show the contour of the filament on which the particles are located. (**b**) Histogram showing distribution of distance between the two center of mass positions of the same particles in the first and the second records (Sugi et al. 1997)

It was found that the position for each particle, i.e. the position for each cross-bridge, remained almost unchanged with time, if the limit of spatial resolution determined by the pixel size (5×5 nm) is taken into consideration. Figure 7b is a histogram showing distribution of the distance (D) between the two positions of the same particle. Among 141 particles examined on three different pairs of IP records, 87 particles showed no significant position changes ($D<5$ nm), and 50 particles showed only small position changes (5 nm$<D<$10 nm). These results indicate that, though the each cross-bridge is expected to fluctuate due to thermal agitation, its time-averaged mean position does not change appreciably with time in the ATP-free experimental solution.

4.3 ATP-Induced Cross-Bridge Movement

The stability of the time averaged cross-bridge position in the absence of ATP was very encouraging for me, since it indicated that ATP-induced change in position of individual cross-bridges could be determined without considering thermal fluctuation. When we started to apply ATP to the filaments, we encountered serious difficulties in passing current through the ATP electrode; first, the tip of the ATP electrode was frequently made apart from the thin layer of experimental solution containing the filaments, due to mechanical vibrations during attachment of the EC to the electron microscope, and secondly, reduction of internal pressure in the EC tended to produce minute bubbles in the ATP electrode, which tended to move to tapered end of the electrode to block current to release ATP. To be free from these difficulties, we attempted to introduce technique of laser-flash photolysis of caged ATP to the EC system. Unfortunately, we gave up this planning because the carbon sealing film was readily broken by laser light. Thus, we had to overcome the above difficulties inherent to the ATP electrode with technical skill and patience. Up to the present time, the percentage of success to pass current to the ATP electrode, i.e. to apply ATP to the filaments, is still <50%, and this makes the progress of research work very slow.

Based on the calculated diffusion time of ATP from the ATP electrode to the filaments (<30 s), we took the first IP record at 2–3 min before the ATP application, and took the second record at 60 s after ATP application. When we finally succeeded in passing current through the ATP electrode, we were excited to find that the position of each particle was completely different between the first and the second records. To compare particle (cross-bridge) positions precisely and unambiguously, we focused attention only on particles that were approximately round-shaped and reasonably distant from neighboring particles. As already mentioned, we compared particle positions relative to the position of natural markers on the carbon film.

The results obtained are summarized in Fig. 8. The amplitude of ATP-induced cross-bridge movement was unexpectedly large. As schematically illustrated in Fig. 8a and b, the center of mass position of each particle was found to move by up to 30 nm in the direction parallel to the filament long axis after ATP application, indicating the marked ATP-induced movement of individual cross-bridges.

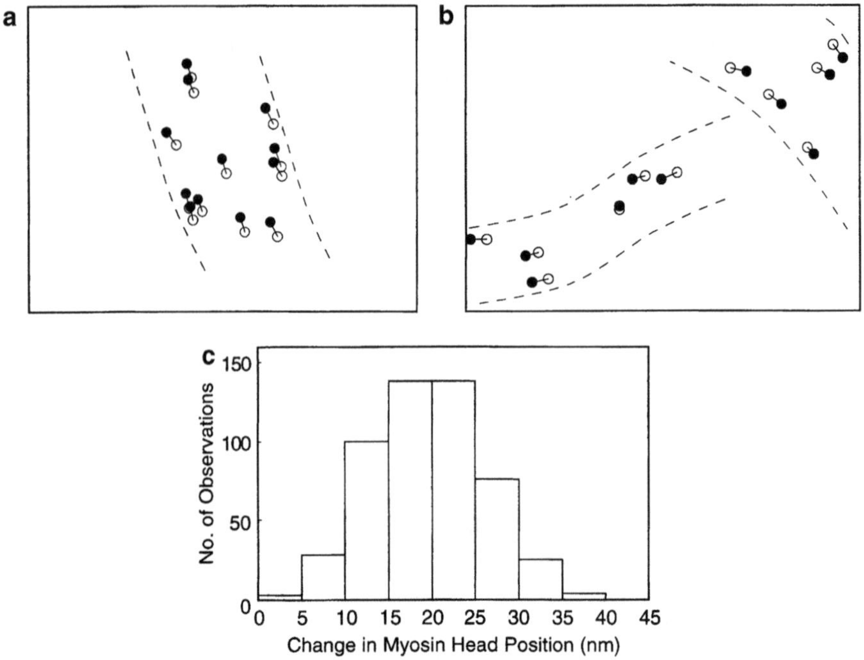

Fig. 8 ATP-induced cross-bridge movement. (**a** and **b**) Examples showing changes in the center of mass positions of the same particles after ATP application. *Filled* and *open circles* (diameter, 15 μm) are also drawn around the center of mass positions of the same particles in two records taken before and after ATP application, respectively. (**c**) Histogram showing distribution of distance between the two center of mass positions of the same particles in the two records, representing the amplitude of ATP-induced cross-bridge movement (Sugi et al. 1997)

Figure 8c is a histogram showing distribution of the amplitude of cross-bridge movement, constructed from 512 measurements on three different pairs of records. The histogram exhibited a peak at ~20 nm. The average amplitude of cross-bridge movement was 19.6 ± 0.3 nm (mean ± SD, $n = 512$). The occasional presence of particles whose position did not change appreciably by ATP application indicated that the filaments were fixed in position.

As the cross-bridges moved only in one direction because we failed to observe the thick filament bare region, it was necessary to exclude the possibility that the cross-bridge movement was associated with any drift of the filaments. This was made by determining the contour of the filament from the contrast-enhanced filament image after appropriate binning and smoothing procedures, to reach a conclusion that the filament drift, if any, may not seriously affect the results obtained (Sugi et al. 1997).

The ATP-induced cross-bridge movement was not observed in the filaments whose ATPase activity had been completely eliminated by N-ethylmaleimide. We also applied ADP iontophoretically to the filaments, and observed no appreciable cross-bridge movement in response to ADP.

4.4 Interpretation of the Results Obtained

When ATP released from the ATP electrode reaches the filaments, it may reacts rapidly with the cross-bridge (M) to form complex M•ADP•Pi, having an average lifetime of >10 s due to its slow Pi release (Lymn and Taylor 1971). Though some cross-bridges would repeat ATPase reaction cycle more than once, the majority of the cross-bridges in the IP records may be in the state of M•ADP•Pi, indicating that the cross-bridge movement is coupled with reaction, M + ATP → M•ADP•Pi. It is generally believed that, during attachment–detachment cycle between the cross-bridges and the thin filaments, the cross-bridge power stroke producing force and motion takes place when they attach to the thin filaments, while the cross-bridge recovery stroke takes place when they are detached from the thin filaments. Since our experimental system does not contain the thin filaments, the ATP-induced cross-bridge movement may correspond to the cross-bridge recovery stoke.

If this explanation is correct, the direction of cross-bridge movement should be away from the thick filament bare region, across which the cross-bridge polarity is reversed. Unfortunately, we could not prove this point by recording movement of the cross-bridges at both sides of the filament bare region, probably because of the following reasons: (1) as the synthetic filaments used were extremely long, the probability to observe the filament bare region within a limited electron microscopic field of view was very small; and (2) it seemed also possible that the synthetic filaments were mostly monopolar in structure, and did not have bare region. Thus, to our great pity, the direction of cross-bridge movement with respect to the filament bare region remained to be an open question until our experiments with bipolar thick filaments were performed about 10 years later.

The unexpectedly large amplitude of cross-bridge movement (average, ~20 nm) may not readily fit into the contraction models on the basis of three-dimensional filament-lattice structures in vertebrate skeletal muscle. In the synthetic filaments used, however, myosin molecules are attached around the thick paramyosin filaments in parallel rows (Squire 1981), and it seems possible that, in the above condition, the subfragennt-2 (S-2) region of myosin molecules might change their length by reversible transition between helical and random coil structures (Lovell et al. 1988). Evidence for the S-2 shortening coupled with ATP hydrolysis in vertebrate skeletal muscle has been presented by Sugi et al. (1992). Although the cross-bridge cannot move over 20 nm while attached to the thin filament due to structural constraints, the tendency of the S-2 to shorten would serve as a mechanism to contribute force production, as suggested by Harrington et al. (1988).

Our success in recording the ATP-induced cross-bridges electron in the synthetic filaments was highly evaluated by Professor Andrew Huxley, and he kindly communicated our paper to PNAS in 1997. In general, however, our work did not draw much attention of muscle researchers, since the hybrid filaments used by us differed too far in structure from that in vertebrate skeletal muscle. Of course, we were going to proceed to start experiments using bipolar thick filaments prepared from vertebrate skeletal muscle. Unfortunately, however, we had to stop research work with the EC due to a number of political and financial problems.

5 Experiments with Bipolar Thick Filaments Consisting of Rabbit Skeletal Muscle Myosin

5.1 Resumption of Experiments at JEOL Ltd

Shortly before my retirement from Teikyo University in 2004, President Terukazu Etoh of JEOL Ltd very generously offered me to provide facilities to carry out research work with the EC system in the JEOL campus located in the suburbs of Tokyo. I accepted this offer with great pleasure, and all experimental equipments in my laboratory were transferred to the JEOL campus. At that time, however, I had two serious problems to actually resume research work. First, my past coworkers in my laboratory had already moved to other institutions, and I could no longer expect them to work together. Secondly, a severe age discrimination characteristic of Japan virtually prevented retired professors like me from applying scientific grants.

Fortunately, Mr Hidehiro Ito, business manager in JEOL, kindly introduced me Dr. Hideki Minoda, Associate Professor in Applied Physics, Tokyo University of Agriculture and Technology, who had been interested in performing experimental work with the EC after reading our PNAS paper (Sugi et al. 1997). As Dr. Minoda's laboratory is geographically close to JEOL campus, I could have a very enthusiastic and able coworker together with his post graduate students. Furthermore, Professor Suguru Tanokura in the Department of Agriculture and Life Sciences, University of Tokyo kindly supplied thick filament samples for our work. Thus, the first problem was completely solved with an enormous success. Meanwhile, the second problem was only partly solved by generous financial support from investigators interested in our work.

Here, I would like to take an opportunity to thank the late Professor Akira Fukami, who developed the carbon sealing film and the EC system. His pupils in Nihon University include two presidents of JEOL, Mr Terukazu Etoh and his successor Dr. Yoshiyasu Harada, and without Fukami's influence on JEOL people my rather miraculous resumption of research work could not be realized (Fig. 9). To my great pity, Professor Fukami passed away shortly after my resumption of research work at JEOL.

5.2 Problems with Bipolar Thick Filaments

Since our primary purpose was to record ATP-induced cross-bridge movement at both sides of the filament bare region, we used synthetic bipolar thick filaments consisting of myosin–myosin rod mixture, prepared from rabbit psoas muscle. Myosin was prepared by the method of Perry (1955), while myosin rod was obtained by chymotryptic digestion of myosin by the method of Margossian and Lowey (1982). Myosin and myosin rod were mixed (molar ratio, 1:1), and the myosin–myosin rod mixture was slowly polymerized by dialysis against a solution of low ionic strength (KCl concentration was reduced from 500 to 120 mM) to

Fig. 9 Photograph taken at the time when I resumed experiments (Etoh, T., H. Ito, A. Fukami, H. Sugi)

obtain bipolar thick filaments. The bipolar filaments were 1.5–3 μm in length and 50–200 nm in diameter at the center. The advantages of using the synthetic bipolar filaments are: (1) they are stiff and tend to form nearly straight rods, though filaments with diameters <100 nm were wavy in shape; (2) their large diameter makes it possible to distinguish the filament profile from the background; and (3) the cross-bridges extending from the upper side of the filaments can move freely without possible constraints arising from attachment to the carbon film. In order that each colloidal gold particle is reasonably separated from neighboring ones, the molar ratio between the cross-bridges and the antibody was carefully chosen to be 1:1.4–1.6. The ATPase activity of the filaments was 0.14–0.16 s^{-1} at 28°C, and was not appreciably affected by mixing with antibody.

The synthetic bipolar filaments had a marked tendency to aggregate, as has been mentioned previously, and this was a great trouble to carry out experiments. After a number of trials, we noticed that, outside bulky filament aggregates, there are groups of filaments whose bipolar shape could be seen clearly. Although component filaments in such filament groups also showed tendency to overlap each other, we could occasionally find a single bipolar filaments without overlap with other filaments or with partial overlap with other filaments. Figure 10a–c are examples of the spindle-shaped bipolar filaments with a number of colloidal gold particles attached to individual cross-bridges. As shown in Fig. 10c, the image of gold particles consisted of 20–50 dark pixels with a wide range of gradation. The particle image differed from that in our previous work (Sugi et al. 1997) because (1) the particle diameter increased from 15 to 20 nm, and (2) the pixel size changed from 5×5 to 2.5×2.5 nm. We selected particles with approximately round-shaped configuration to determine their center of mass positions.

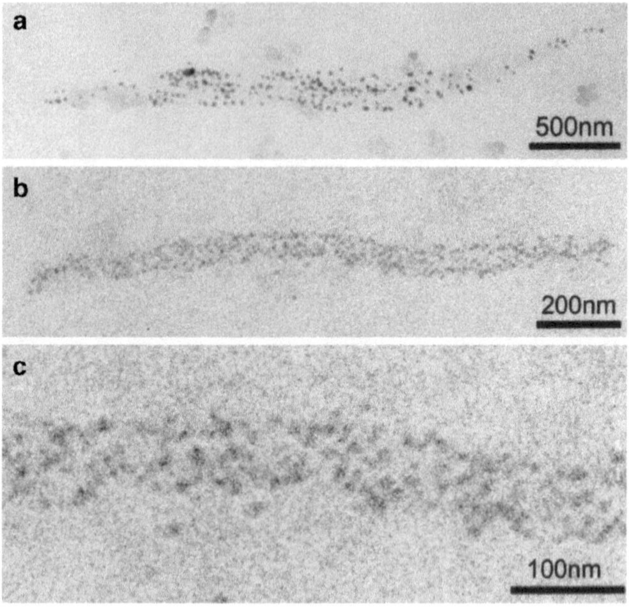

Fig. 10 Typical IP records of single bipolar thick filaments with a number of gold particles attached to individual cross-bridges. (**a** and **b**) The whole spindle-shaped thick filaments with tapered ends. (**c**) Enlarged IP record showing part of the thick filament in (**a**). Each gold particle image consists of 20–50 dark pixels (Sugi et al. 2008)

5.3 Stability of the Cross-Bridge Position in the Absence of ATP

We also examined whether the particle (and therefore the cross-bridge) positions were stable or changed with time in the absence of ATP, by comparing the center of mass positions of the same particle between the two IP records of the same filament, taken at intervals of 5–10 min. An example of the results is shown in Fig. 11a, in which a circle of 20 nm diameter is drawn around the center of mass position of each selected particle. As has been the case in our previous work (Sugi et al. 1997), the position of each particle, i.e. the position of each cross-bridge, remained almost unchanged with time, if the limit of spatial resolution determined by the pixel size (2.5×2.5nm) was taken into consideration. Figure 11b is a histogram showing distribution of the distance between the two positions of the same particle. Among 120 particles examined on three different pairs of IP records, 93 particles exhibited no significant change in position ($D < 2.5$ nm), and 27 particles exhibited only small position changes ($2.5\ \mathrm{nm} < D > 5$ nm). These results indicate, as with our previous study on the hybrid filaments (Sugi et al. 1997), that (1) the filaments are firmly fixed in position to the carbon sealing film, and (2) despite thermal fluctuation, the cross-bridge mean position, time averaged over 0.1 s, remains almost unchanged with

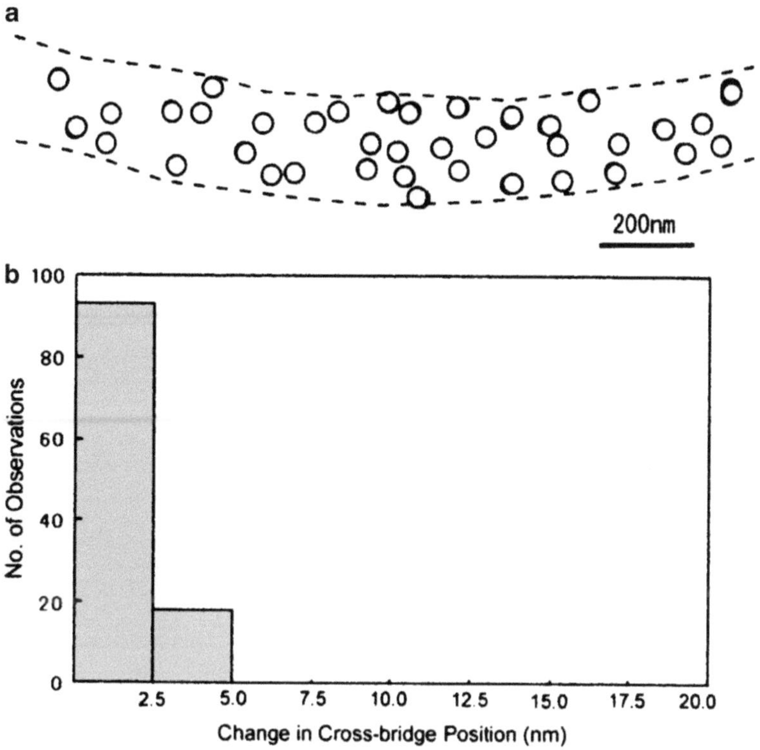

Fig. 11 Stability of time-averaged cross-bridge mean position in the absence of ATP (**a**) Comparison of the cross-bridge position between the two IP records of the same filament. *Open* and *filled circles* (diameter, 20 nm) are drawn around the center of mass position of each gold particle in the first and the second records, respectively. In this and subsequent figures, *broken lines* indicate contour of the filament. Note that *filled circles* are barely visible because of the nearly complete overlap of *open* and *filled circles*. (**b**) Histogram showing distribution of distance between the two center of mass positions of the same particles in the first and the second records (Sugi et al. 2008)

time in the absence of ATP. The stability of time-averaged cross-bridge position observed in the two different filaments suggests that it is a common property in all kinds of muscle thick filament. The stability of time-averaged cross-bridge mean position is consistent with the Huxley contraction model (Huxley 1957), in which each cross-bridge fluctuates around a definite equilibrium position.

5.4 Amplitude of the ATP-Induced Cross-Bridge Movement

Based on the stability of the cross-bridge position with time, we examined the cross-bridge movement in response to ATP, by comparing two IP records of the same filaments, taken at 3–4 min before and 40–60 s after ATP application.

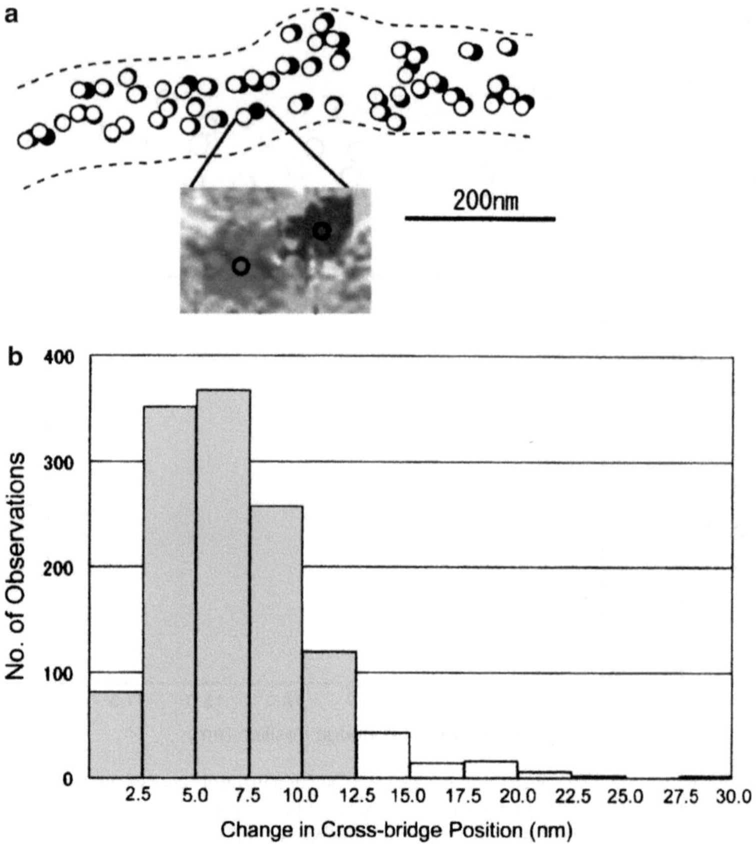

Fig. 12 ATP-induced cross-bridge movement. (**a**) Comparison of the cross-bridge position between the two IP records, taken before and after ATP application. *Open* and *filled circles* (diameter, 20 nm) are drawn around the center of mass positions of the same particles before and after ATP application, respectively. (*Inset*) an example of superimposed IP records showing the change in position of the same particle, which is *colored dark grey* (before ATP application) and *light grey* (after ATP application). The center of mass position for each particle image is located in the *circle* in the particle image. (**b**) Histogram showing distribution of the amplitude of ATP-induced cross-bridge movement, determined from the ATP-induced changes in the center of mass position of each gold particle (Sugi et al. 2008)

Since it was not easy to focus part of a filament including the bare region, we first examined ATP-induced unidirectional cross-bridge movement at one side of the bare region. In response to ATP, the position of particles, i.e. the position of individual cross-bridges, were observed to move nearly parallel to the filament long axis as shown in Fig. 12a.

The amplitude of ATP-induced cross-bridge movement was measured not only on the filaments completely separated or largely separated from adjacent filaments over a distance >1 μm, but also on the filaments clumped with adjacent filaments, so that the whole profile of individual filaments was not clear. Figure 12b is a histogram showing distribution of the amplitude of ATP-induced cross-bridge movement, constructed from 1,285 measurements on eight different pairs of IP records obtained from eight different filaments. The histogram exhibited a peak at 5–7.5 nm. The average amplitude of ATP-induced cross-bridge movement (excluding values <2.5 nm) was 6.5 ± 3.7 nm (mean ± SD, n = 1,210).

In the present experimental condition, the particles located on both upper and lower side of the filaments were equally in focus in the microscopic field. While the cross-bridges located at the upper side of the filaments may move freely in response to ATP, the cross-bridges at the lower side of the filaments would attach to the carbon film and their movement would be inhibited or markedly reduced in amplitude. This possibility is consistent with the frequent presence of particles that did not move appreciably or move with small amplitude <5 nm (see Fig. 12a). If this explanation is correct, the mean amplitude of ATP-induced movement of the cross-bridge that can move freely would be >7.5 nm. As with the previous study (Sugi et al. 1997), the ATP-induced cross-bridge movement was eliminated by N-ethylmaleimide, and the cross-bridges did not move in response to ADP as with our previous study (Sugi et al. 1997), again indicating that the cross-bridge movement is associated with reaction between the cross-bridges and ATP.

5.5 *Reversal in the Direction of ATP-Induced Cross-Bridge Movement Across the Filament Bare Region*

After a number of painstaking efforts, we finally succeeded in recording the ATP-induced cross-bridge movement at both sides of the bipolar filament bare region, across which the cross-bridge polarity was reversed. Examples are shown in Fig. 13. Similar results were obtained in several other pairs of IP records. In my long period of research work over 50 years, I was most excited with the finding that, in response to ATP, the cross-bridges moved away from, but not towards, the thick filament bare region, since it has long been my dream to visualize the cross-bridge movement coupled with their physiological function. As the present experiments are made in the absence of thin filaments, the observed cross-bridge movement is consistent with the general view that it corresponds with the cross-bridge recovery stroke, coupled with reaction, $M + ATP \rightarrow M{\bullet}ADP{\bullet}Pi$. This point will be discussed in detail later. This interpretation is supported by the result that ADP produced no appreciable cross-bridge movement.

Judging from the minimum distance between the two particles moving in opposite directions, the bare region width of the synthetic filaments was <100 nm in most cases, being much shorter than that of natural thick filament in vertebrate skeletal muscle. This may be explained as being caused by incomplete lateral alignment of myosin rods constituting the thick filament bare region.

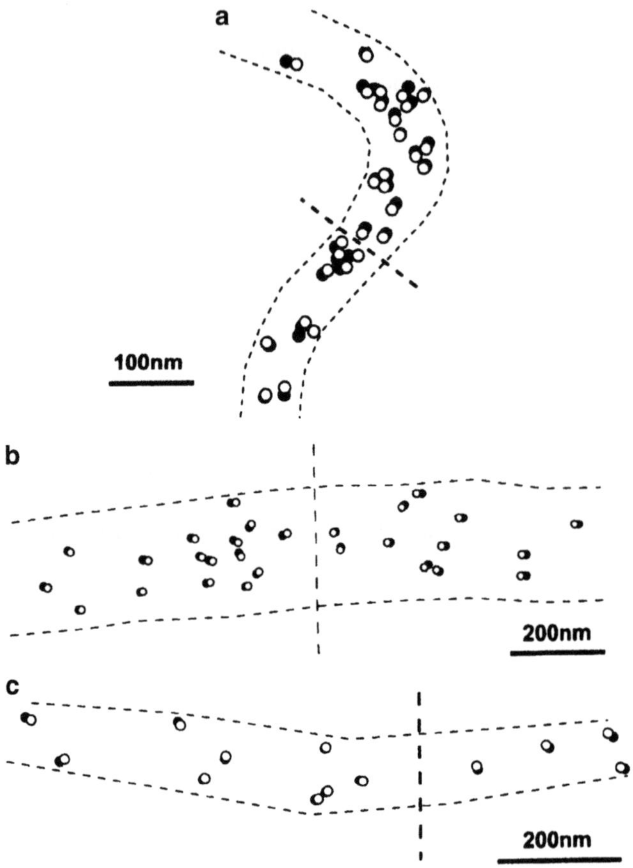

Fig. 13 Examples of IP records showing the ATP-induced cross-bridge movement at both sides of thick filament bare region, across which the cross-bridge polarity is reversed. *Open* and *filled circles* (diameter, 20 μm) are drawn around the center of mass positions of the same particles in the IP records, taken before and after ATP application, respectively. Note that the cross-bridges move away from the bare region. Approximate location of the bare region is indicated by *broken lines* across the center of the filament. The filament is thin and bent in (**a**), while the filament is thick and nearly straight in (**b** and **c**) (Sugi et al. 2008)

5.6 Reversibility of the ATP-Induced Cross-Bridge Movement

To ascertain whether the cross-bridges return to their initial position after exhaustion of applied ATP, experiments were further made in which the images of the same filaments were recorded three times in the following sequence: (1) before ATP application, (2) 40–60 s after ATP application, i.e. during the formation of M•ADP•Pi, and (3) 5–6 min after ATP application, i.e. after exhaustion of applied ATP. Since the experimental solution contained hexokinase and D-glucose serving as scavenger for ATP (Oiwa et al. 1993), all ATP molecules released from the ATP electrode were completely exhausted before the time of the third record. Therefore, the first and third records were taken in the absence of ATP, while the second record was taken when most cross-bridges were in the state of M•ADP•Pi.

Figure 14 shows examples of the sequential changes in position of nine different pixels (2.5×2.5 nm) where the center of mass positions of the different particles are located. Therefore, these pixel positions in the first (grey), second (black), and third

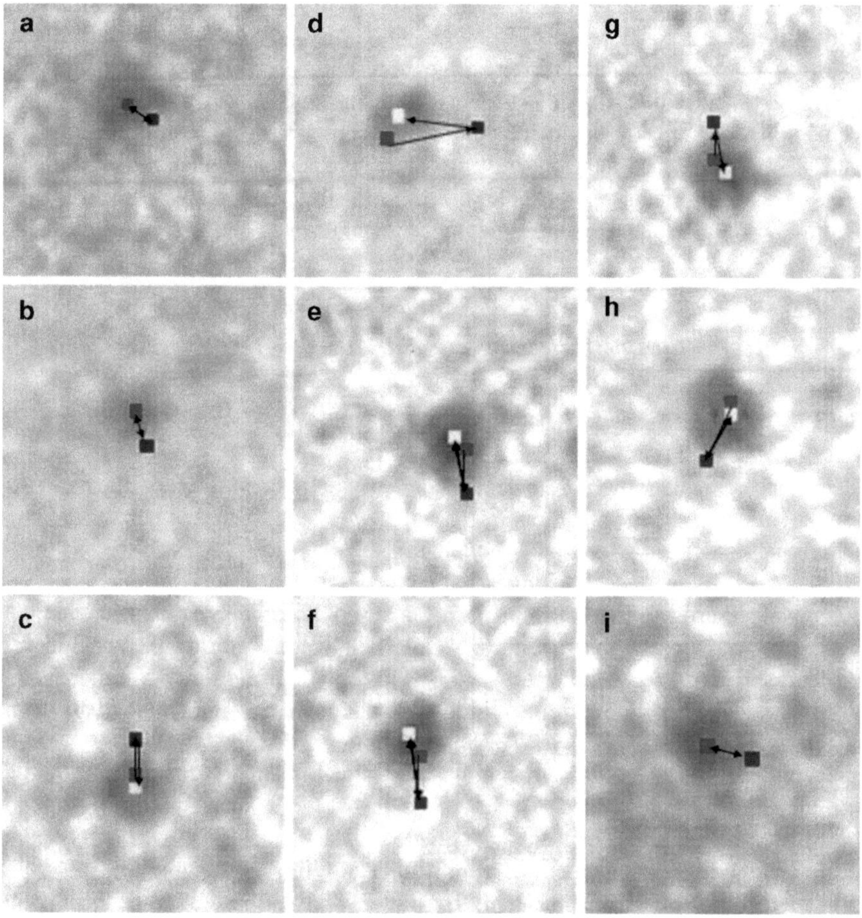

Fig. 14 Examples showing sequential changes in position of nine different pixels (each 2.5×2.5 nm) where the center of mass positions of corresponding nine different particles are located. In each frame, the pixel positions were recorded three times, i.e., before ATP application, during ATP application, and after exhaustion of applied ATP. The changes in position of the pixels in the first (*dark grey*), second (*light grey*), and third (*white*) records indicate changes in the cross-bridge position before ATP application, during ATP application, and after exhaustion of ATP. Direction of each cross-bridge movement is indicated by *arrow*. The angle between the filament long axis, on which the particle was located, and the X–Y coordinates of the IP was random. Note that the cross-bridges return toward their initial position after exhaustion of ATP. The cross-bridge returned almost exactly to its initial position in (**a**, **b** and **i**), and close to its initial position in (**c**–**h**) (Sugi et al. 2008)

(white) records can be taken to represent sequential position changes of nine different individual cross-bridges in the three IP records.

It can be seen that the cross-bridges first move in response to ATP, as associated with reaction, M + ATP → M• ATP → M• ADP• Pi, and then return toward their initial position after complete exhaustion of ATP, i.e. detachment of Pi and ADP from M. When the amplitude of ATP-induced cross-bridge movement is small (~5 nm), the cross-bridges are observed to return almost exactly to their initial position, as indicated by the complete overlap of the pixel position in the first and the third records, so that white pixels are entirely covered by grey pixels (Fig. 14a, b). With larger amplitudes of cross-bridge movement, the cross-bridges are mostly observed to return to the position close to their initial position, as indicated by the small distance between grey and white pixels (<5 nm) (Fig. 14c–h), although the complete overlap of black and white pixels is also seen occasionally (Fig. 14i). These results indicate that the ATP-induced cross-bridge movement is reversible; the cross-bridges, in the state of M• ADP• Pi, return toward their initial position after detachment of Pi and ADP from them.

5.7 *Interpretation of the Results Obtained*

5.7.1 Relation to Previous Studies

Using the EC, with which biological specimens can be kept to retain their physiological function in the electron microscope, we have succeeded in visualizing individual cross-bridge movement coupled with ATP hydrolysis in the synthetic bipolar thick filaments consisting of rabbit skeletal muscle myosin. The observed ATP-induced cross-bridge movement in the absence of the thin filament seems to be related to the early observation of ATP-dependent cross-bridge arrangement in insect thick filaments (Clarke et al. 1986). Up to the present time, ATP-dependent order-disorder transition of cross-bridge arrangement has been studied in detail (Padron and Craig 1989; Xu et al. 1999; Zoghbi et al. 2004).

In our experiments with the EC, only a very small fraction of the cross-bridges in the thick filaments are labeled to record movements of individual cross-bridges unambiguously, and therefore bears no direct relation to these studies, in which information about cross-bridge arrangement (largely azimuthal) are sampled from all the cross-bridges in the experimental material. The present results have been obtained from changes in the time-averaged cross-bridge position, in which any cross-bridge flexibility is included, and also bears no direct relation to the X-ray studies on large and random cross-bridge displacements in the whole muscle, in which all the cross-bridges are sampled over a long time (Poulsen and Lowy 1983; Lowy and Poulsen 1990).

5.7.2 Direct Evidence for the Cross-Bridge Recovery Stroke

Figure 15 is a diagram showing a most plausible attachment–detachment cycle between the cross-bridges and the thin filament. It is generally believed that a

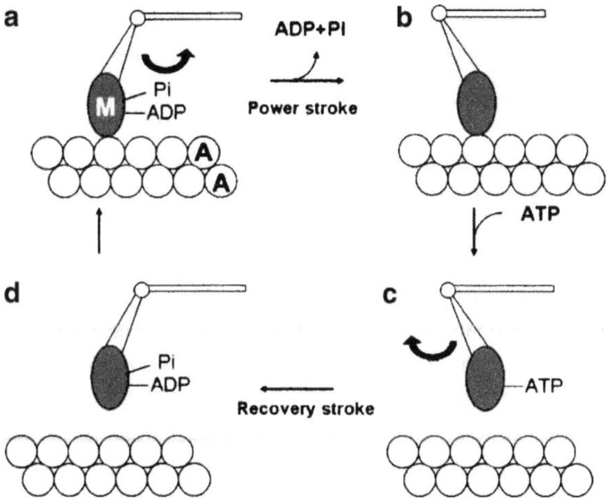

Fig. 15 Schematic diagram of attachment–detachment cycle between the cross-bridge (M) extending from the thick filament and actin (A) in the thin filament. M in the form of M • ADP • Pi attaches to actin to exert a power stroke, associated with release of Pi and ADP (from **a** to **b**). After the end of the power stroke, M remains attached to A, taking its post-power stroke configuration (**b**). Upon binding with ATP, M detaches from A to exert a recovery stroke, associated with reaction M • ATP → M • ADP • Pi (from **c** to **d**). M • ADP • Pi again attaches to A to again exert a power stroke (from **d** to **a**). M is assumed to attach rigidly to A, while its power and recovery strokes are assumed to result from swinging of the lever arm around the hinge

cross-bridge (M), in the form of M• ADP• Pi, attaches actin (A) in the thin filament to exert a power stroke, associated with release of Pi and ADP, so that at the end of power stroke, M forms rigor linkage with A. Then, M detaches from A upon binding with ATP to exert a recovery stroke, associated with reaction, M• ATP → M• ADP• Pi. Since the thin filament is absent in our experiments, it seems likely that the ATP-induced cross-bridge movement corresponds to the recovery stroke in the attachment–detachment cycle between M and A (corresponding to the cross-bridge states from c to d in Fig. 15). This view is supported by our finding that the direction of ATP-induced cross-bridge movement is away from the filament bare region, i.e. the direction opposite to that of power stroke (Fig. 15).

In order that a cross-bridges repeats attachment–detachment cycle with the thin filament, the recovery stroke should be the same in amplitude as, but opposite in direction to, the power stroke. As a matter of fact, the cross-bridges, that had moved for a distance in response to ATP, were found to return to their initial position after exhaustion of ATP (Fig. 14), providing additional evidence that the ATP-induced cross-bridge movement actually corresponds to the recovery stroke. This implies that, when a cross-bridges return to their initial position after exhaustion of ATP, i.e. after detachment of Pi and ADP from M, they undergo conformational changes analogous or similar to those of the power stroke (corresponding to the cross-bridge states from a to b in Fig. 15). These results strongly suggest that, in the absence of ATP, the cross-bridges take configurations analogous or similar to those at the end of the power stroke.

In the lever arm hypothesis of muscle contraction (Uyeda et al. 1996; Geeves and Holmes 1999; Fisher et al. 2005; Huxley et al. 2006), the cross-bridge catalytic domain (M) is believed to attach rigidly to actin (A) before the cross-bridge movement is produced by tilting of the lever arm (Fig. 15), that in turn implies that the lever arm should tumble around its hinge until M comes to an appropriate position to attach A. The reversibility of the ATP-induced cross-bridge movement (Fig. 14) suggests that, even in the absence of the thin filament, the lever can make M take its appropriate position to attach A without being guided by the thin filament. It seems possible that the tumbling of M caused by hinge region flexibility always takes place around a certain neutral point, as has been supposed concerning thermal motion of the cross-bridges in the Huxley contraction model (Huxley 1957). This again implies that the cross-bridge configuration in the absence of ATP is analogous or similar to that at the end of the power stroke. Of course, our results give no information about the actual attachment angle of M to A in the thin filament. Much more experimental work is needed.

5.7.3 Variation in the Amplitude of Cross-Bridge Movement

The amplitude distribution of ATP-induced cross-bridge movement exhibited a peak at 5–7.5 nm. This value is consistent with the general framework of sliding filament mechanism in muscle contraction (e.g., Huxley 1957). Meanwhile, a small proportion of the cross-bridges showed amplitudes >12.5 nm (Fig. 12b). In the hexagonal myofilament lattice structure, the stroke of cross-bridges attached to actin is obviously limited to ~10 nm because of structural constraints. On this basis, the observed cross-bridge strokes >12.5 nm might bear no direct relation to physiological muscle contraction. Although the ATP-induced cross-bridge movement of ~20 nm has been recorded on the myosin–paramyosin core complex filaments (Sugi et al. 1997), the arrangement of myosin molecules in the complex filaments is entirely different from that in the bipolar filaments used in the present study (Squire 1981). It seems, however, possible that, in the synthetic bipolar filaments used in the present work, a small proportion of the component myosin molecules take special configurations so that they exhibit shortening of subfragment-2 region coupled with reaction, M•ATP→M•ADP•Pi. As has been suggested by Harrington et al. (1988), the tendency of the S-2 region to shorten might work as an additional mechanism for force production in muscle.

To summarize, the present wok constitutes the first direct evidence for the cross-bridge recovery stroke, which has been achieved by visualizing the ATP-induced individual cross-bridge movement under a high magnification of electron microscope.

6 Future Prospects

The experiments described in this article have proved that the EC is an extremely powerful tool in elucidating mysteries not only in the research field of muscle contraction, but also in other research fields of life sciences. At present, the EC is the only means to record movements of biomolecules, retaining their physiological function, under a high magnification of electron microscope. The fundamental problem in the use of EC for investigating movements (or structural changes) of biomolecules related to their physiological function is electron beam damage to the specimen; to avoid it, the total incident electron dose should be limited below 10^{-4} C/cm^2 (Suda et al. 1992). Due to this absolute limitation of total incident electron dose, it has been difficult for us to take IP records of the filaments more than three times, because more than 80% of the critical electron dose is used up to find and focus filaments suitable for experiments under relatively low electron microscopic magnifications. If a device can be developed, which makes it possible to find and focus suitable specimens rapidly (for example, by use of a computer memorizing the shape of suitable specimens), then it would be possible to take a number of IP records on one and the same specimen under various experimental conditions. Furthermore, it would become possible to record an animation visualizing actual movement of biomolecules related to their function. We have already started discussing this possibility with a research group in Japan, who has already succeeded in developing a novel ultra high-speed video system.

Another limitation concerning the use of the EC is electron microscopic magnification. The present spatial resolution (pixel size, 2.5×2.5 nm under a magnification of 10,000×) is suitable for recording elementary steps of various molecular motors, but insufficient for recording structural changes in ionic channels, transporters and receptors. While I am skeptical about future improvement with respect to spatial resolution of the IP system, it is possible to amplify minute structural changes in such organelles. We can find such an example in the work of Kinosita et al. (2000). They succeeded in visualizing the ATP-dependent rotary motion of F0-F1 ATPase located in the mitochondrial membrane under a light microscope by attaching a fluorescent rod-shaped F-actin filament to each rotary motor. The ingenious method can be applied to other kinds of organelles. Finally, I heartily hope that the EC system will be widely used by investigators to open new horizons in the research field of life sciences.

Acknowledgements I wish to express my sincere thanks to the late Professor Akira Fukami for allowing me to perform experiments using the EC system developed in his laboratory, and for making my resumption of experiments in JEOL campus. My thanks are also due to President Kazuo Ito, President Terukazu Eto and President Yoshiyasu Harada of JEOL, Co., who generously providing facilities to continue experiments with the EC. I am also indebted to Professor Hugh E. Huxley for his interest and evaluation of our work, Mr Hidehiro Ito for introducing my coworkers to me, Dr. Kurio Fukushima for kind help at the start of experiments in the JEOL campus, and Drs. Kuniaki Nagayama, Kazuhiko Kinosita and Eisaku Katayama for financial support. The names of my coworkers are listed below with thanks ; Drs. Tsuyoshi Akimoto, Kazuo Sutoh, Shigeru Chaen, Noboru Oishi, Suechika Suzuki in the work published in 1997, and Drs. Hiroki Minoda, Yuhri Inayoshi, Fumiaki Yumoto, Takuya Miyakawa, Yumiko, Masaru Tanokura, Tsuyoshi Akimoto, Takakazu Kobayashi, Shigeru Chaen and Seiryo Sugiura in the work published in 2008.

References

Butler EP, Hale KF (eds) (1981) Dynamic experiments in the electron microscope. Chapters 5 and 6, In: Practical Methods in Electron Microscopy, Vol. 9. North-Holland, Amsterdam

Clarke ML, Hofman W, Wray JS (1986) ATP binding and crossbridge structure in muscle. J Mol Biol 191:581–585

Fisher S, Bjorn W, Horak D, Holmes KC, Smith JC (2005) Structural mechanism of the recovery stroke in the myosin molecular motor. Proc Natl Acad Sci U S A 102:6873–6878

Fukami A, Adachi K (1965) A new method of preparation of a self-perforated micro plastic grid and its applications. J Electron Microsc (Tokyo) 14:112–118

Fukami A, Fukushima K, Kohyama N (1991) Observation technique for wet clay minerals using film-sealed environmental cell equipment attached to high-resolution electron microscope. In: Bennett RH, Bryant WR, Hulbert MH (eds) Microstructure of fine-grained sediments from mud to shale, Springer, Heidelberg, 321–331

Fukushima K (1988) Application of the gas environmental chamber for electron microscopy. Ph.D Thesis (Nagoya University, Nagoya) (in Japanese)

Fukushima K, Ishikawa A, Fukami A (1985) Injection of liquid into environmental cell for in situ observations. J Electron Microsc 34:47–51

Geeves MA, Holmes KC (1999) Structural mechanism of muscle contraction. Annu Rev Biochem 68:687–728

Harrington WF, Ueno H, Davis JS (1988) Helix-coil melting in rigor and activated cross-bridges of skeletal muscle. In: Sugi H, Pollack GH (eds) Molecular mechanism of muscle contraction. Plenum, New York

Huxley AF (1957) Muscle structure and theories of contraction. Prog Biophys Biophys Chem 7:255–318

Huxley HE, Reconditi M, Stewart A, Irving T (2006) X-ray interference studies of cross-bridge action in muscle contraction: evidence from quick releases. J Mol Biol 363:743–761

Kinosita K Jr, Yasuda R, Noji H, Adachi K (2000) A rotary molecular motor that can work at near 100% efficiency. Philos Trans R Soc Lond B Biol Sci 335:473–489

Lovell S, Karr T, Harrington WF (1988) Suppression of contractile force in muscle fibers by antibody to myosin subfragment 2. Proc Natl Acad Sci U S A 85:1849–1853

Lowy J, Poulsen FR (1990) Studies of the diffuse x-ray scattering from contracting frog skeletal muscles. Biophys J 57:977–985

Lymn RW, Taylor RW (1971) Mechanism of adenosine triphosphate hydrolysis by actomyosin. Biochemistry 10:4617–4627

Margossian SS, Lowey S (1982) Hybridization and reconstitution of thick filament structure. Methods Enzymol 85:20–55

Nonomura Y (1974) Fine structure of the thick filament in molluscan catch muscle. J Mol Biol 88:445–455

Oiwa K, Kawakami T, Sugi H (1993) Unitary distance of actin-myosin sliding studied using an in vitro force-movement assay system combined with ATP iontophoresis. J Biochem 114:28–32

Padron R, Craig R (1989) Disorder induced in nonoverlap myosin cross-bridges by loss of adenosine triphosphate. Biophys J 56:927–933

Perry SV (1955) Myosin adenosine triphosphatase. Methods Enzymol 2:582–588

Poulsen FR, Lowy J (1983) Small angle x-ray scattering from myosin heads in relaxed and rigor frog skeletal muscles. Nature 303:146–152

Squire JM (1981) The structural basis of muscle contraction. Plenum, New York

Suda H, Ishikawa A, Fukami A (1992) Evaluation of the critical electron dose on the contractile activity of hydrated muscle fibers in the film-sealed environmental cell. J Electron Microsc 41:223–229

Sugi H, Kobayashi T, Gross T, Noguchi K, Karr T (1992) Contraction characteristics and ATPase activity of skeletal muscle fibers in the presence of antibody to myosin subfragment 2. Proc Natl Acad Sci U S A 89:6134–6137

Sugi H, Akimoto T, Sutoh K, Chaen S, Oishi N, Suzuki S (1997) Dynamic electron microscopy of ATP-induced myosin head movement in living muscle thick filaments. Proc Natl Acad Sci U S A 94:4378–4382

Sugi H, Minoda H, Inayoshi Y, Yumoto F, Miyakawa T, Miyauchi Y, Tanokura S, Akimoto T, Kobayashi T, Chaen S, Sugiura S (2008) Direct demonstration of the cross-bridge recovery stroke in muscle thick filaments in aqueous solution by using the hydration chamber. Proc Natl Acad Sci U S A 105:17396–17401

Sutoh K, Tokunaga M, Wakabayashi T (1989) Electron microscopic mapping of myosin head with site-directed antibodies. J Mol Biol 206:357–363

Uyeda TQP, Abramson PD, Spudich JA (1996) The neck region of the myosin motor domain acts as a lever arm to generate movement. Proc Natl Acad Sci U S A 93:4459–4464

Xu S, Gu J, Rhodes T, Belknap B, Rosenbaum G, Offer G, White H, Yu LC (1999) The M.ADP.P(i) state is required for helical order in the thick filaments of skeletal muscle. Biophys J 77:2665–2676

Zoghbi ME, Woodhead JL, Craig R, Padron R (2004) Helical order in tarantula thick filaments requires the "closed" conformation of the myosin head. J Mol Biol 342:1223–1236

Role of Titin in Skeletal Muscle Function and Disease

Coen A.C. Ottenheijm and Henk Granzier

Abstract This review covers recent developments in the titin field. Most recent reviews have discussed titin's role in cardiac function: here we will mainly focus on skeletal muscle, and discuss recent advances in the understanding of titin's role in skeletal muscle function and disease.

Keywords Striated muscle • Sarcomere • PEVK • Titin isoforms • Single molecule

1 Introduction

More than a quarter century after the sliding filament model of the sarcomere was proposed the existence of an additional sarcomeric filament was discovered, consisting of the elastic protein titin (also known as connectin) (Wang et al. 1979; Maruyama 1976). Titin is the largest protein known to date (Mr of a single polypeptide, 3–4 MDa) (Labeit and Kolmerer 1995) and is the third most abundant muscle protein (after myosin and actin) (Labeit et al. 1997). Thus, although titin was discovered relatively late, it is a major constituent of muscle. Since its initial discovery important advances have been made regarding the layout of the titin filament in the sarcomere. It is now clear that a single titin polypeptide spans the full distance from the Z-disk (N-terminus) to M-line region (C-terminus) of the sarcomere, and that the segment of titin in the I-band of the sarcomere is elastic. This elastic segment extends as the sarcomere is stretched, giving rise to the titin-based passive force. Recent reviews have discussed titin's role in cardiac function (Granzier and Labeit 2004, 2007; LeWinter et al. 2007) and here we will mainly discuss recent advances in the understanding of titin's role in skeletal muscle function.

H. Granzier (✉)
Department of Physiology, University of Arizona, Tucson, AZ 85724, USA
e-mail: granzier@email.arizona.edu

D.E. Rassier (ed.), *Muscle Biophysics: From Molecules to Cells*,
Advances in Experimental Medicine and Biology 682,
DOI 10.1007/978-1-4419-6366-6_6,

2 Brief Overview of Titin's Layout

The past two decades important advances have been made regarding titin's layout in the sarcomere, and here we will present a brief overview of the established concepts (for a more detailed discussion we refer to Linke (2008) and Granzier and Labeit (2004)). The N-terminal ~80 kDa of titin spans the entire Z-line region (Gregorio et al. 1998), and titin filaments from opposite sarcomeres therefore overlap within the Z-lines. Between 0.8 and 1.5 MDa of titin are located within the I-band, depending on the particular isoform (see below). The central section of I-band titin – corresponding to its elastic region – is formed by tandemly arranged Ig repeats which are contained in two separate blocks, the proximal and the distal tandem Ig segments. Both of the tandem Ig groups are separated by up to ~2,200 residues that are unusually rich in P (proline), E (glutamate), V (valine), and K (lysine). This elastic segment from the central I-band titin is referred to as the "PEVK domain" (Labeit and Kolmerer 1995). N2A (skeletal and cardiac muscle) and N2B (cardiac muscle only) segments function as additional spring elements. The C-terminal ~2 MDa of titin is located in the A-band; this region of titin is rendered inextensible by its tight association with the myosin thick filament, and is composed of regular patterns of Ig-like and fibronectin type III repeats (Labeit and Kolmerer 1995). The M-line region of titin corresponds to ~200 kDa of the final C-terminal region of the protein, and contains at the periphery a kinase domain (Obermann et al. 1996). As in Z-lines, where titin filaments from opposite sarcomeres fully overlap, titin filaments from opposite half-sarcomeres fully overlap within the M-line region (Obermann et al. 1997). The overlapping ends are interconnected by their binding to other sarcomeric components. Thus, titin filaments with opposite polarity overlap in both the Z-lines and M-lines, forming a contiguous system along myofibrils.

3 Modulation of Titin's Elasticity Through Isoform Diversity

Sarcomeric proteins are typically encoded by multiple genes that are expressed in both developmental and tissue specific patterns. In contrast, there is only a single titin gene and multiple splice pathways in the I-band encoding region of the titin gene give rise to isoforms with distinct spring compositions (Freiburg et al. 2000). In humans the titin gene contains 363 exons coding for a protein that is maximally ~4.2 MDa in size (Bang et al. 2001): alternative splicing of the titin gene has shown to render isoforms that range from ~3.0 MDa to ~3.9 MDa. Exon 49 (containing the N2B sequence) is excluded in skeletal muscle titins, but is present in all cardiac titin isoforms (Bang et al. 2001; Freiburg et al. 2000). Splicing of exons 49/50 to exon 219 results in the "small" 2,970-kDa cardiac isoform known as N2B titin (so named because it contains the N2B element) (Freiburg et al. 2000). A second class of cardiac isoforms contains, in addition to exon 49 (N2B element), also exons 102–109 (coding for the N2A element), the so-called N2BA titin isoform (name reflects presence of both N2B and N2A elements). N2BA titins have a longer PEVK segment and contain additional Ig domains. Titin isoforms in skeletal muscle express the longest PEVK segment, with human soleus expressing a 2,174 residue PEVK segment vs. a 180 residue segment in cardiac N2B titin (Labeit and Kolmerer 1995; Freiburg et al. 2000).

Using a titin exon microarray that allows all 363 titin exons of the human titin gene to be monitored simultaneously, we studied titin in fetal, neonatal, and adult myocardium (Lahmers et al. 2004). The titin microarray revealed fetal cardiac titins that contain additional spring elements both in the tandem Ig and PEVK region of the molecule (many of these elements have not been found in adult titins). The fetal isoform predominates in fetal and neonatal myocardium and gradually disappears during postnatal development with a time course that varies in different species. Consistent with the presence of additional spring elements in fetal cardiac titin, passive myocardium is less stiff in the neonate than in the adult: the regulation of titin's spring composition in fetal and neonatal myocardium may allow adjustment of diastolic filling behavior during development of the heart. Changes in titin isoform expression in skeletal muscle during development was first suggested by work that revealed an increase in resting tension of differentiating skeletal muscle fibers of the rat (Mutungi et al. 2003). Recently, we have found that in contrast to cardiac tissue development, where large and small isoforms that remain of the same size are co-expressed, but whose expression ratio shifts towards that of the adult, (Lahmers et al. 2004; Krüger et al. 2006, 2008; Warren et al. 2004), neonatal skeletal muscle expresses a single titin isoform that gradually increases its mobility on gels during development (Ottenheijm et al. 2009). This finding implies that during skeletal muscle development, many titin isoforms can be expressed that are close in mobility and that gradually shift from large to small. Depending on species and muscle type, titin isform size ranges from 3.92 kDa in neonates to 3.24 kDa and adult mouse skeletal muscles revealed that the decrease in titin isoform size during development is largely caused by restructuring of titin's extensible I-band region, in particular the PEVK segment. Apparently, the number of splice pathways is not restricted to a few, but a large number of pathways exist in the PEVK region of the titin gene, thereby rendering titin isoforms that can differ several hundred kDa in size. An interesting finding was that during skeletal muscle development PPAK repeats appear to be spliced out of the titin molecule thereby increasing the proportion of E-rich motifs in the PEVK segment. The functional significance of this remains to be established, but considering that Ca^{2+} binding to E-rich motifs in the PEVK segment increases its bending rigidity (Labeit et al. 2003) it could be that a higher proportion of E-rich motifs in the PEVK segment elevates the Ca^{2+}-responsiveness of titin-based passive stiffness in adult muscle fibers. Thus, we propose that alternative splicing of titin during development renders stiffer titin isoforms with elevated Ca^{2+} responsiveness and that this improves fine motor control during muscle during development. Furthermore, it was found that the PEVK domain displays a spatially hierarchical arrangement of local elasticity: the N-terminal region is the most rigid and the C-terminal region is the most flexible (Nagy et al. 2005). This led Nagy et al. (2005) to propose that PEVK-bound Ca^{2+}

might dissociate gradually from titin's PEVK region driven by its magnitude of extension, thereby providing the sarcomere means to monitor its current position across a wide range of sarcomere lengths. Thus, addition or excision of PEVK exons might not only affect passive stiffness by altering the PEVK's contour length but presumably also through changes in its hierarchical extensibility and Ca^{2+} dissociation properties.

In cardiac tissue of large mammals coexpression of N2B and N2BA isoforms occurs at the level of the half-sarcomere. Each half-thick filament binds six titin molecules (Liversage et al. 2001) and this value appears to be constant despite widely varying coexpression ratios (Cazorla et al. 2000). Thus, varying the coexpression ratio of isoforms while keeping the stoichiometry constant allows development of graded passive force levels in between those of isoform-pure muscle cells (Trombitás et al. 2001). However, unlike cardiac muscle, the vast majority of skeletal muscles from both developing and adult mice and rabbits appear to express a single titin isoform (unpublished observations, and Prado et al. (2005)), which can vary in size substantially. Thus, it seems that skeletal muscles tune the sarcomere's passive stiffness by expressing a single titin isoform of appropriate size.

4 Molecular Basis of Titin's Elasticity

The molecular basis of titin's elasticity has been studied in single-molecule experiments using laser tweezers and atomic force microscopy (AFM). This has shown that the titin molecule behaves as a wormlike-chain (WLC) entropic spring (Kellermayer et al. 1997; Rief et al. 1997; Tskhovrebova et al. 1997). The WLC model describes the molecule as a deformable continuum of persistence length (PL) (measure of bending rigidity, or minimal distance along chain for orientations to be uncorrelated) and contour length (CL). In this model, force increases in a nonlinear fashion with fractional extension (end-to-end length, divided by CL), and is inversely proportional to PL. Thus, if the flexibility of a chain is high, PL is short, and a relatively high force is needed to extend the chain. The PL values of titin are on the order of 0.5–10 nm (see also below) which is much below that of other cytoskeleton filaments (microtubules ~1 mm, microfilaments ~1 μm (Tskhovrebova et al. 1997)). Thus, by virtue of its high flexibility a given fractional extension of titin will result in a relatively high force.

Tandem Ig segments. Early work suggested that titin's elasticity results from the reversible unfolding of Ig domains (Erickson 1994). However, follow-up studies revealed that the unfolding force of Ig domains varies with stretch speed, and ranges from ~150–300 pN for stretch rates from ~1–1,000 nm/s (Li and Fernandez 2003; Li et al. 2002; Watanabe et al. 2002a). These results led to the realization that Ig domains unfold with a probability that increases with increasing force and passing time. The unfolding forces of single molecules in vitro are high compared to physiological forces encountered in muscle (estimated at ~0–10 pN/titin molecule

(Watanabe et al. 2002b)), questioning whether unfolding takes place in vivo. Whether unfolding takes place in vivo was experimentally tested by stretching and holding relaxed soleus muscle fibers for extended durations and measuring the end-to-end length of tandem Ig segments by immuno-electronmicroscopy. Monte Carlo simulations were also performed of the extensible region of a single titin molecule in a soleus sarcomere with unfolding/refolding kinetics of Ig domains similar to those obtained in single-molecule experiments. The simulations predict that stretching and then holding the molecule results within ~30 min in unfolding of 70 Ig domains and an ensuing large increase in the length of the tandem Ig segments, at the expense of the PEVK segment that will shorten. However, IEM on fibers that had been held at a constant length for various durations (up to 64 h) failed to reveal the predicted length increase of tandem Ig segments. Thus, it appears that Ig unfolding in the sarcomere is less likely than the single molecule experiments suggest (for details, see (Trombitás et al. 2003)). Unfolding in muscle may be undesirable because refolding only takes place during subsequent release to low force (Kellermayer et al. 1997) giving rise to hysteresis and energy loss when muscle is cyclically stretch-released (see also discussion below). Hence, it is likely that extension of the tandem Ig segments in short to intermediate long sarcomeres results from unbending of sequences that link Ig domains. Indeed, at a force of 1 pN (this at the lower end of the physiological force range) tandem Ig segments extend to ~50% of their contour length (assuming a PL value of ~15 nm), as indicated by WLC model simulations (Trombitás et al. 2003).

PEVK segment. A major source of titin extensibility in intermediate to long sarcomeres is the PEVK segment (Linke et al. 1998; Trombitás et al. 1998a, b). IEM studies have measured the extension of the PEVK segment in sarcomeres of skeletal muscle stretched to different sarcomere lengths (Trombitás et al. 1998a, b). In sarcomeres of highly stretched human soleus fibers, the PEVK region was found to extend to a length of ~750 nm, a value that is close to the 826 nm contour length of the PEVK segment, assuming that the PEVK is fully unfolded [maximal residue spacing in an unfolded polypeptide (0.38 nm) times the number of residues in the soleus PEVK (2174, (Labeit and Kolmerer 1995))]. Furthermore, the extensible behavior of the PEVK segment measured at a wide range of sarcomere lengths can be simulated well by assuming WLC behavior without unfolding/refolding transitions. These observations led to the notion that the PEVK segment behaves as an unfolded polypeptide (random coil) at all sarcomere lengths, consistent with the preponderance of proline residues and charge clusters along the PEVK sequence that are likely to prevent the formation of stable structures (Labeit and Kolmerer 1995). Single-molecule experiments have been carried out with AFM on constructs that represent the PEVK segment of cardiac titin (N2B isoform). These experiments showed that stretching the PEVK segment from a short length to close to its unfolded contour length gives rise to a force–extension curve that does not reveal abrupt force fluctuations (unlike tandem Ig segments), but that is relatively smooth and that can be fit well with a single WLC equation. The featureless appearance of the force–extension curve indicates the absence of major structural transitions and suggests that the mechanical behavior of the PEVK segment is largely dominated by unfolded polypeptide (Li et al. 2001; Watanabe et al. 2002b). The man PL that was obtained was 0.91 nm (Li et al. 2001). A subsequent AFM study on a skeletal muscle PEVK construct with flanking Ig domains revealed a PL of ~1.3 nm (Labeit et al. 2003). Nagy et al. (2005) studied recombinant subdomains of human soleus muscle: an N-terminal, a middle, and a C-terminal subdomain and found PL values of ~1.2, ~1.0, and ~0.65 nm, respectively (resulting in hierarchical extensibility of the PEVK segment, as discussed above). In a recent AFM study, Sarkar et al. (2005) measured the PL value of three different PEVK exons and found of all exons similar PL values of ~0.9–1.0 nm. Finally, using laser-tweezers, Leake et al. (2004) studied the native PEVK segment of skeletal muscle titin as well as a skeletal muscle PEVK construct. Values obtained were ~1.4 and ~2.2 nm for native titin and the construct, respectively. The mean PL value of all studies conducted so far is ~1.2 nm. This relatively low mean PL value for the PEVK segment indicates that high forces are required to stretch it and, thus, at low passive force levels (short sarcomeres) PEVK extension is limited (at a force of 1 pN, the extension is 20% of the PEVK contour length, several-fold less than obtained for tandem Ig segments; see above).

For a more detailed discussion of N2B mechanics we refer to (Helmes et al. 1999; Linke et al. 1999; Trombitás et al. 1999; Li et al. 2002; Watanabe et al. 2002b; Leake et al. 2006; Zhu et al. 2009).

In Vivo behavior. Titin's extensible region in the sarcomere can be modeled as serially linked WLCs with different contour and persistence lengths. The tandem Ig segments (see above) have a PL value of ~10 nm and their CL in the various isoforms can be calculated from the combined number of Ig domains contained in the tandem Ig segments and assuming an average domain spacing (domain plus the fully extended linker sequence) of ~5 nm (Trombitás et al. 2003). As for the PEVK region, its elasticity is likely derived mainly from straightening of random coil sequences and its average PL is ~1.2 nm. The CL of the PEVK region in different isoforms may be obtained from the number of amino acid residues multiplied by the maximal residue spacing in an unfolded polypeptide (0.38 nm). Using the PL and CL of the tandem Ig, and PEVK spring elements and then linking them in a serially linked WLC model allows calculation of the extension of the individual spring elements as well as the force of a single titin molecule in sarcomeres of different length (this is based on the principle of equivalence of forces within a serially linked mechanical system, as detailed elsewhere (Watanabe et al. 2002a). Although there is some uncertainty in the exact PL values and, in addition, PL values of different regions of the PEVK may vary somewhat (see above), the results are mainly determined by the relative PL values of the spring elements, with tandem Ig segments having a PL approximately tenfold higher than the PEVK. The thus obtained force-SL relations of different titin isoforms indicate that the passive force level that titin generates at a given sarcomere length is greatly influenced by differential splicing of its spring elements. For instance, titin molecules from human psoas muscle are predicted to generate greater than threefold higher passive tension upon stretch than titin molecules from soleus muscle.

5 Stretch-Release Force Hysteresis of Titin Molecules

Passive muscle cells are not purely elastic but instead exhibit viscoelasticity, as revealed by passive force hysteresis in stretch-release cycles (for a give sarcomere length, the force measured during stretch is greater than during release) and by force relaxation when the passive myocyte is held at a stretched length (Helmes et al. 1999; Minajeva et al. 2001; Kellermayer et al. 1997, 2001; Leake et al. 2004). The mechanisms that underlie titin's viscoelasticity may involve (1) dynamic interaction between titin and actin (Yamasaki et al. 2001; Kulke et al. 2001), (2) dynamic crosslinks in the PEVK region (either intramolecular or between adjacent molecules) (Kellermayer et al. 2001; Trombitás et al. 2003), and (3) unfolding of Ig domains (Minajeva et al. 2001). None of these sources, however, appear to be the sole explanation for viscoelasticity. First, hysteresis is still present in thin filament-extracted myocytes (Granzier et al. 1997); second, single molecule force-extension curves of the PEVK display (at relatively high force) little hysteresis (Watanabe et al. 2002b; Li et al. 2002); third, there is no direct evidence for Ig domain unfolding at physiological force levels (Trombitás et al. 2003). Thus, despite its physiological relevance the exact molecular basis of titin's viscoelasticity remains to be established. Our working hypothesis is that a number of mechanisms contribute, including intra- and intermolecular events that involve the PEVK segment. It is especially important to study passive force hysteresis and titin's sensitivity to mechanical fatigue in striated muscles that undergo repetitive contractions, such as the myocardium, but also the diaphragm muscle which is continuous active during life (see also discussion below) and can be subject to increased loading during pulmonary or cardiac diseases (Ottenheijm et al. 2007; Mancini et al. 1992). Future research using mouse models in which extensible segments of titin are selectively removed might elucidate the nature of titin's force hysteresis.

6 Titin-Based Protein Complexes as Stress Sensors

Recent evidence indicates that titin may function as a biomechanical sensor by virtue of the structural and signaling proteins that bind to titin. Interestingly, the binding sites are not randomly distributed along the titin filament but seem restricted to "hot spots": one in and near the Z-disk, another in the central I-band, and a third in the M-line regions.

Z-disk: The Z-disk titin-capping protein T-cap, or telethonin, binds to titin's N-terminus and links titin to signaling and structural molecules (Mues et al. 1998; Gregorio et al. 1998). T-cap interacts with cytoplasmic domains of two membrane associated proteins, the K channel subunit minK/isk in the T-tubules (Furukawa et al. 2001) and small ankyrin-1 (sANK-1), a transmembrane sarcoplasmic reticulum (SR) protein. The minK/isk interaction may function to modulate K channel function in response to stretch while the sANK-1 interaction may serve to fix the position of the SR around the Z-disk. An interaction between titin's near Z-disk domains and obscurin (Young et al. 2001) suggests an additional role for titin in relation to organization of the SR since obscurin in turn interacts with ankyrin isoform 1.5 (Young et al. 2001; Russell et al. 2002). The latter appears to link the SR to the sarcomere and regulate ryanodine receptor distribution in the SR (Bagnato et al. 2003). These Z-disk region interactions suggest a general role for titin in positioning SR and T-tubule systems in close proximity to the I-band and perhaps ensuring that these systems move with the Z-disk during contraction and relaxation, preventing excessive strains on the membranes. Thus, during contraction-relaxation cycles titin maintains the structural organization of the sarcomere (by keeping the A-band in a central location) as well as the organization of the SR and T-tubular systems.

I-band: The N2A element in skeletal muscle titin interacts with calpain protease P94 (also known as calpain 3), and may modulate the function of P94 in protein degradation (Ono et al. 2004). For instance, in vitro studies revealed that p94-N2A interactions suppress p94 autolysis and protected titin from proteolysis (Hayashi et al. 2008). In addition to p94, calpain 1 has been proposed to bind to titin's N2-line presumably in a calcium dependent manner (Raynaud et al. 2005). Additional titin Ig segments bind to three homologous muscle ankyrin repeat proteins (MARPs), cardiac ankyrin repeat protein (CARP), ankrd2 (or Arpp) and diabetes ankyrin repeat protein (DARP) (Miller et al. 2003). One or more of these ankyrin repeat proteins are induced by cardiac injury, skeletal muscle denervation and stretch, and metabolic challenge (Nakada et al. 2003; Kemp et al. 2000; Ikeda et al. 2009). Thus, these ankyrin repeat proteins may be a component of muscle stress responses in general. Importantly, recent work with MARP knock-out models revealed that when MARPs are absent, skeletal muscle has reduced passive stiffness and expresses larger titin isoforms (Barash et al. 2007), suggesting that MARPs could play a role in the changes during development in passive mechanical behavior of muscle. T-cap also interacts with muscle LIM protein (MLP). MLP has been suggested to be an important mediator of mechanical stress in cardiac tissue (Knöll et al. 2002), but its role in skeletal muscle remains unclear. A MLP-KO model shows severe cardiac hypertrophy, myofibrillar disarray, and a less compliant heart (Ehler et al. 2001; Omens et al. 2002). Perhaps as a result of this dramatic cardiac muscle phenotype, the effect of loss of MLP in skeletal muscle has been masked. Recent studies (Barash et al. 2005) revealed that MLP might play a subtle role in the maintenance of normal muscle characteristics and in the early events of the recovery process of skeletal muscle to injury, serving both structural and gene-regulatory roles. The I-band ligands of titin and their associated binding partners are also found in the nucleus, where they function in transcriptional and cell cycle regulation. It is intriguing to consider that this dual localization may link titin's cytoplasmic stretch-based sensing to gene expression via transcriptional regulation. Along these lines, a naturally occurring "muscular dystrophy with myositis" mouse has been found to contain an 83 amino acid deletion mutation in the

N2A element along with marked changes in the expression of MLP, CARP and MURF-1 (a titin-ligand M-line protein) (Witt et al. 2004). Thus, this deletion mutation appears to affect ligands in multiple domains.

M-line: The M-line region of titin also contains a number of sites that may be involved in sensing and signaling. These include a serine/threonine kinase domain (Gautel et al. 1993). Little is known about substrates and function of the titin kinase. In vitro studies with a mutant kinase domain have provided evidence that T-cap is a titin kinase substrate in embryonic muscle (Mayans et al. 1998). Furthermore, the titin kinase has been suggested to play a role in embryonic sarcomere development, i.e. the integration of titin into thc A-band (Weinert et al. 2006) and sarcomere structure maintenance (Gotthardt et al. 2003). Finally, it has been proposed that the TK is a mechano-sensor that regulates muscle protein expression in a strain-dependent fashion (Lange et al. 2005; Puchner et al. 2008). Puchner et al. (2008) using AFM techniques revealed that mechanical strain activates ATP binding before unfolding of the structural titin domains, and that the titin kinase can thus act as a biological force sensor. Moreover, Lange et al. (2005) proposed that the titin kinase in response to the mechanical load placed on the sarcomere binds MURF-2 and together with nbr1 and p62 forms a signalosome that communicates with the nucleus and modulates protein expression and turnover. Finally, recent studies from our lab suggest that excision of titin M-line exons MEx1 and MEx2 (which encode the titin kinase) affects skeletal muscle contractility through enhancing myofilament Ca^{2+} sensitivity of force generation, which appeared to be mediated at the cross bridge level, presumably through effects on MyBP-C isoform expression (Ottenheijm et al. 2009). These data suggest a functional relation between the titin kinase and MyBP-C that is isoform-specific and that is involved in tuning skeletal muscle contractility.

In addition to the kinase, titin's M-line region contains additional binding sites for obscurin (Young et al. 2001; Fukuzawa et al. 2008) and obscurin-like 1 (Fukuzawa et al. 2008), which appear important for proper myofibril maintenance, and possibly for T-Cap (Mayans et al. 1998). Furthermore, titin's M-line region contains a binding site for the RING finger protein MURF-1 (muscle specific ring finger protein-1) (Centner et al. 2001; Mrosek et al. 2007). MURF-1 in particular may be involved in multiple signaling pathways. MURF-1 is a sarcomere-associated skeletal and cardiac muscle protein that functions as an E3 ubiquitin ligase that conjugates ubiquitin to proteins destined for proteolysis (Bodine et al. 2001), suggesting that MURF-1 may play a regulatory role in degradation of muscle proteins. MURF-1 also binds to a number of other myofibrillar proteins than titin, including troponin I (Kedar et al. 2004) and myosin heavy chain (Clarke et al. 2007), which results in increased degradation and possibly modulation of contractility by MURF-1.

Thus various lines of evidence implicate diverse roles for titin in biomechanical sensing and signaling. This suggests that titin has complex and important integrative functions beyond its purely mechanical properties and constitutes an exciting area for future research.

7 Titin-Associated Skeletal Muscle Diseases

Hereditary myopathies: Because of its size and diverse functions, titin is a prime candidate for disease causing mutations. Although titin's role in dilated cardiomyopathies has gained much attention and DCM causing mutation have been identified (Gerull et al. 2002), data on titin mutations causing skeletal myopaythies are relatively scarce. The skeletal muscular dystrophy TMD (tibial muscular dystrophy) is a genetic skeletal muscle disease of dominant inheritance; sequencing of the entire gene in TMD families revealed a mutation in M-line exon 363 (MEx6) (Hackman et al. 2002). This point mutation causes an amino acid change in the ultimate c-terminal domain M10, without interrupting the reading frame. M10 is next to a p94 (or calpain 3) binding site (see above), and secondary p94 deficiency was shown in these patients (Haravuori et al. 2001). Two other TMD-causing point-mutations in MEx6 were found in one Belgian and two unrelated French families (Hackman et al. 2002; Van den Bergh et al. 2003). Apparently, the clinically more severe mutations in the Finnish and French families severely reduce titin-obscurin binding whereas this binding capacity in the less-severe Belgian mutation is not detectably reduced. This might suggest an important role for obscurin in the pathophysiology of these myopathies. A mutation in exon Mex1 in the kinase domain of titin causes hereditary myopathy with early respiratory failure (HMREF) (Lange et al. 2005). It was shown that this mutation disrupts formation of the MURF-2-nbr1-p62 pathway and impairs muscle gene expression (see above). Recently, three novel dominant truncating mutations in the last two exons of titin, MEx5 and MEx6, were found in Spanish and French families with distal myopathy phenotypes (Hackman et al. 2008).

It is important to highlight that the titin mutations identified thus far are in regions of the titin gene that are expressed in all striated muscles but that not all muscles are affected identically. For example, exon 363 is expressed in all muscles, but the TMD mutation in this exon affects selectively only the extensor muscles of the anterior leg compartment, whereas adjacent muscles are spared (Hackman et al. 2002). Similarly, Z-disk mutations identified in DCM are expressed in all titin isoforms but there is no clinically detectable phenotype in skeletal muscle (Gerull et al. 2002). The mechanisms that underlie these muscle type-specific phenotypes are unknown.

Acquired titin diseases: Spasticity: Measurements on single skeletal muscle fibers from patients with cerebral palsy indicate that fibers taken from subjects with spasticity have a passive elastic modulus of the stress–strain relationship that is almost double that measured in normal fibers, indicating remodeling of titin (Friden and Lieber 2003). Similar findings were made in spinal cord-injured patients with spasticity (Olsson et al. 2006). Thus, the increased muscle stiffness in spastic patients arises not only from extracellular matrix remodeling, but also from structural and functional adaptations inside muscle cells, with titin as a prime candidate.

Chronic Obstructive Pulmonay Disease (COPD): Dysfunction of the respiratory muscles frequently occurs in patients with severe COPD (Ottenheijm et al. 2007), and appears to be an important contributor to mortality in these patients. Recently, it was demonstrated that, in addition to reduced active tension, diaphragm fibers from patients with COPD generate less passive tension upon fiber-stretch compared to fibers from non-COPD patients (Ottenheijm et al. 2006; Moore et al. 2006). Titin gene transcript studies revealed increased expression of exons coding for the PEVK region in COPD diaphragm, resulting in a larger extensible titin segment (Ottenheijm et al. 2006), suggesting that alternative splicing of the titin gene reduces the passive tension generated by diaphragm fibers from COPD patients (see also Fig. 1). It is tempting to speculate that the reduced stiffness of titin in the

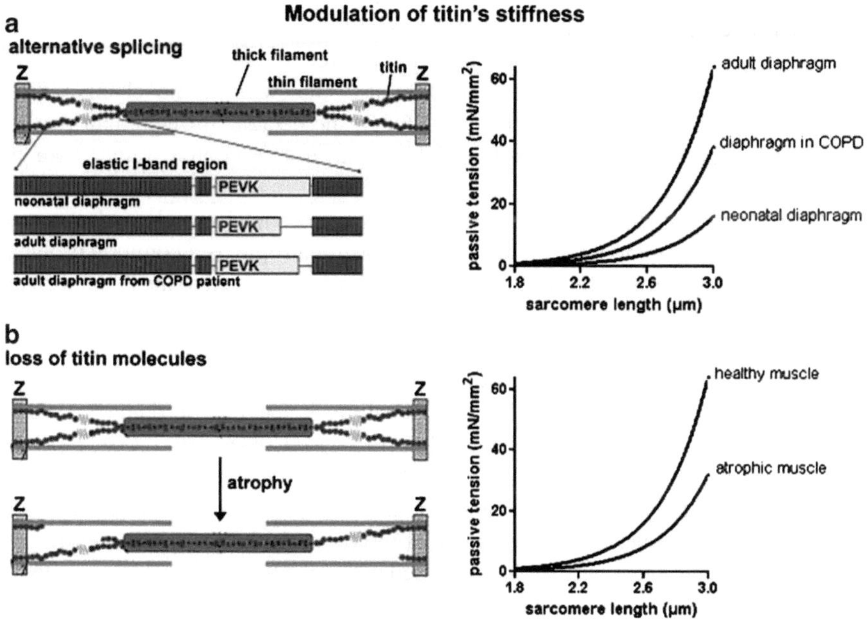

Fig. 1 Schematic of two distinct mechanisms for modulating titin-based skeletal muscle stiffness. (**a**) Modulation of passive tension through titin isoform diversity. Alternative splicing of the titin gene results in isoforms that range greatly in size. For instance, during skeletal muscle development in mice titin isoform size has been shown to decrease, with the magnitude of decrease depending on muscle type and ranging from ~100 to ~500 kDa (Ottenheijm et al. 2009). Titin microarray analysis revealed that the decrease in titin isoform size during development is largely caused by restructuring of titin's extensible I-band region, in particular the PEVK segment. Recent studies suggested that titin isoform expression in the diaphragm muscle in patients with COPD switches back to a neonatal program: titin gene transcript studies revealed increased expression of exons coding for the PEVK region in COPD diaphragm, resulting in a larger extensible titin segment and reduced passive tension generation upon diaphragm fiber-stretch (Ottenheijm et al. 2006). (**b**) A second mechanism by which titin-based passive stiffness can be reduced is through loss of titin molecules. For simplicity, only two titin molecules per half-sarcomere are depicted (the true number is approximately six). Recent work (Udaka et al. 2008) has shown that during disuse-atrophy titin passive stiffness is reduced through this mechanism

diaphragm of patients with mild to moderate COPD reduces titin's kinase activity, thereby inducing loss of signaling between titin and the nucleus, resulting in decreased transcription of muscle-specific genes. This could affect the production of contractile proteins, and contribute to the loss of myosin in the diaphragm of these patients, especially because diaphragm protein degradation appears to be elevated. Evaluating titin kinase activity and the distribution of sarcomeric and nuclear components of this signaling pathway in COPD diaphragm could test this hypothesis. It might also be speculated that the increased compliance of titin in the diaphragm of COPD patients results in less hysteresis and thus less energy loss, which might be a beneficial adaptation that offsets the increased frequency of contraction in these patients. Thus, the elongated titin molecule might reduce energy loss and compensate for the increased activity the diaphragm. Changes in diaphragm passive properties also occur in an animal model for chronic heart failure (van Hees et al. 2010); passive tension generation upon stretch was significantly reduced (>35%) in diaphragm fibers from heart failure rats, whereas no changes were observed in soleus fibers. In contrast to COPD, the decreased diaphragm passive stiffness in heart failure was explained by reduced titin content rather than by changes in titin size.

Disuse atrophy: In response to decreased usage, skeletal muscle undergoes an adaptive reductive remodeling. This so-called disuse atrophy is a common problem for patients with chronic diseases, individuals who are confined to the bed, mechanically ventilated septic patients in the intensive care unit who develop rapid disuse atrophy of the inspiratory muscles (especially the diaphragm), and for the elderly. Recent work revealed that titin is reduced preferentially following long-term disuse (Toursel et al. 2002; Udaka et al. 2008), resulting in sarcomeric damage (Udaka et al. 2008) and reduction in passive tension (Fig. 1b). Udaka et al. (2008) reported decreased Ca^{2+} sensitivity of active force following 6 weeks of hindlimb immobilization in the soleus muscle of the rat. Structural analysis revealed marked changes in the disused sarcomere, with expansion of interfilament lattice spacing leading to a reduction in Ca^{2+} sensitivity. How titin-based passive tension affects Ca^{2+} sensitivity of force generation in skeletal muscle is not completely understood. Several recent lines of evidence indicate that the lateral separation of thick and thin filaments is regulated by titin as a lattice stabilizer via its passive force, thereby affecting the likelihood of myosin attaching to the thin filament (Cazorla et al. 2001; Fukuda et al. 2003). Therefore, long-term disuse may result in the formation of myofibrils with expanded lattice spacing, due to preferential titin loss and ensuing defects in lattice stabilization. The underlying mechanisms for preferential loss of titin during skeletal muscle unloading are unknown. However, given the increased titin degradation products during long-term muscle disuse (Udaka et al. 2008), enhanced titin degradation seems to play a role, possibly through pathways that involve muscle specific ring finger protein 1 (MURF-1), an ubiquitin-ligase that binds to titin's M-line region. Alternatively, preferential loss of titin might reduce the load on the thick filament and thereby the activation of the titin kinase in the M-line region. Previous work has shown that this may lead to reduced muscle protein synthesis (Lange et al. 2005).

8 Summary

Progress in understanding titin's functions has been rapid during the last several years. Titin-based stiffness adjustments have been reported to occur in various skeletal muscle diseases and, furthermore, several disease-causing titin mutations have been discovered. Titin plays an important role in passive and active skeletal muscle contractility, and contributes to muscle weakness in titin-associated skeletal muscle diseases. Recent work also suggests that titin functions as a sensor of mechanical load that triggers adaptations in response to disease.

References

Bagnato P, Barone V, Giacomello E, Rossi D, Sorrentino V (2003) Binding of an ankyrin-1 isoform to obscurin suggests a molecular link between the sarcoplasmic reticulum and myofibrils in striated muscles. J Cell Biol 160(2):245–253

Bang ML, Centner T, Fornoff F, Geach AJ, Gotthardt M, McNabb M, Witt CC, Labeit D, Gregorio CC, Granzier H, Labeit S (2001) The complete gene sequence of titin, expression of an unusual approximately 700-kDa titin isoform, and its interaction with obscurin identify a novel Z-line to I-band linking system. Circ Res 89(11):1065–1072

Barash IA, Mathew L, Lahey M, Greaser ML, Lieber RL (2005) Muscle LIM protein plays both structural and functional roles in skeletal muscle. Am J Physiol Cell Physiol 289(5):C1312–C1320

Barash IA, Bang ML, Mathew L, Greaser ML, Chen J, Lieber RL (2007) Structural and regulatory roles of muscle ankyrin repeat protein family in skeletal muscle. Am J Physiol Cell Physiol 293(1):C218–C227

Bodine SC, Latres E, Baumhueter S, Lai VK, Nunez L, Clarke BA, Poueymirou WT, Panaro FJ, Na E, Dharmarajan K, Pan ZQ, Valenzuela DM, DeChiara TM, Stitt TN, Yancopoulos GD, Glass DJ (2001) Identification of ubiquitin ligases required for skeletal muscle atrophy. Science 294(5547):1704–1708

Cazorla O, Freiburg A, Helmes M, Centner T, McNabb M, Wu Y, Trombitas K, Labeit S, Granzier H (2000) Differential expression of cardiac titin isoforms and modulation of cellular stiffness. Circ Res 86(1):59–67

Cazorla O, Wu Y, Irving TC, Granzier H (2001) Titin-based modulation of calcium sensitivity of active tension in mouse skinned cardiac myocytes. Circ Res 88(10):1028–1035

Centner T, Yano J, Kimura E, McElhinny AS, Pelin K, Witt CC, Bang ML, Trombitas K, Granzier H, Gregorio CC, Sorimachi H, Labeit S (2001) Identification of muscle specific ring finger proteins as potential regulators of the titin kinase domain. J Mol Biol 306(4):717–726

Clarke BA, Drujan D, Willis MS, Murphy LO, Corpina RA, Burova E, Rakhilin SV, Stitt TN, Patterson C, Latres E, Glass DJ (2007) The E3 Ligase MuRF1 degrades myosin heavy chain protein in dexamethasone-treated skeletal muscle. Cell Metab 6 (5):376–385

Coen AC, Ottenheijm, Anna M. Knottnerus, Danielle Buck, Xiuju Luo, Kevin Greer, Adam Hoying, Siegfried Labeit, and Henk Granzier (2009) Tuning passive mechanics through differential splicing of titin During Skeletal Muscle Development. J Biol 97(8):2277–2286

Ehler E, Horowits R, Zuppinger C, Price RL, Perriard E, Leu M, Caroni P, Sussman M, Eppenberger HM, Perriard JC (2001) Alterations at the intercalated disk associated with the absence of muscle LIM protein. J Cell Biol 153(4):763–772

Erickson HP (1994) Reversible unfolding of fibronectin type III and immunoglobulin domains provides the structural basis for stretch and elasticity of titin and fibronectin. Proc Natl Acad Sci U S A 91(21):10114–10118

Freiburg A, Trombitas K, Hell W, Cazorla O, Fougerousse F, Centner T, Kolmerer B, Witt C, Beckmann JS, Gregorio CC, Granzier H, Labeit S. (2000) Series of exon-skipping events in the elastic spring region of titin as the structural basis for myofibrillar elastic diversity. Circ Res 86(11):1114–1121

Friden J, Lieber RL (2003) Spastic muscle cells are shorter and stiffer than normal cells. Muscle Nerve 27(2):157–164

Fukuda N, Wu Y, Farman G, Irving TC, Granzier H. (2003) Titin isoform variance and length dependence of activation in skinned bovine cardiac muscle. J Physiol 553. 1:147–154

Fukuzawa A, Lange S, Holt M, Vihola A, Carmignac V, Ferreiro A, Udd B, Gautel M (2008) Interactions with titin and myomesin target obscurin and obscurin-like 1 to the M-band: implications for hereditary myopathies. J Cell Sci 121:1841–1851

Furukawa T, Ono Y, Tsuchiya H, Katayama Y, Bang ML, Labeit D, Labeit S, Inagaki N, Gregorio CC (2001) Specific interaction of the potassium channel beta-subunit minK with the sarcomeric protein T-cap suggests a T-tubule-myofibril linking system. J Mol Biol. 313(4):775–784

Gautel M, Leonard K, Labeit S (1993) Phosphorylation of KSP motifs in the C-terminal region of titin in differentiating myoblasts. EMBO J 12(10):3827–3834

Gerull B, Gramlich M, Atherton J, McNabb M, Trombitás K, Sasse-Klaassen S, Seidman JG, Seidman C, Granzier H, Labeit S, Frenneaux M, Thierfelder L (2002) Mutations of TTN, encoding the giant muscle filament titin, cause familial dilated cardiomyopathy. Nat Genet 30(2):201–204

Gotthardt M, Hammer RE, Hübner N, Monti J, Witt CC, McNabb M, Richardson JA, Granzier H, Labeit S, Herz J (2003) Conditional expression of mutant M-line titins results in cardiomyopathy with altered sarcomere structure. J Biol Chem 278(8):6059–6065

Granzier H, Labeit S (2004) The giant protein titin: a major player in myocardial mechanics, signaling, and disease. Circ Res 94(3):284–295

Granzier H, Labeit S (2007) Structure-function relations of the giant elastic protein titin in striated and smooth muscle cells. Muscle & Nerve 36(6):740–755

Granzier H, Kellermayer M, Helmes M, Trombitás K (1997) Titin elasticity and mechanism of passive force development in rat cardiac myocytes probed by thin-filament extraction. Biophys J 73(4):2043–2053

Gregorio CC, Trombitás K, Centner T, Kolmerer B, Stier G, Kunke K, Suzuki K, Obermayr F, Herrmann B, Granzier H, Sorimachi H, Labeit S (1998) The NH2 terminus of titin spans the Z-disc: its interaction with a novel 19-kD ligand (T-cap) is required for sarcomeric integrity. J Cell Biol 143(4):1013–1027

Hackman P, Vihola A, Haravuori H, Marchand S, Sarparanta J, De Seze J, Labeit S, Witt C, Peltonen L, Richard I, Udd B (2002) Tibial muscular dystrophy is a titinopathy caused by mutations in TTN, the gene encoding the giant skeletal-muscle protein titin. Am J Hum Genet 71(3):492–500

Hackman P, Marchand S, Sarparanta J, Vihola A, Pénisson-Besnier I, Eymard B, Pardal-Fernández JM, Hammouda EL-H, Richard I, Illa I, Udd B (2008) Truncating mutations in C-terminal titin may cause more severe tibial muscular dystrophy (TMD). Neuromuscul Disord 18(12):922–928

Haravuori H, Vihola A, Straub V, Auranen M, Richard I, Marchand S, Voit T, Labeit S, Somer H, Peltonen L, Beckmann JS, Udd B (2001) Secondary calpain3 deficiency in 2q-linked muscular dystrophy: titin is the candidate gene. Neurology 56(7):869–877

Hayashi C, Ono Y, Doi N, Kitamura F, Tagami M, Mineki R, Arai T, Taguchi H, Yanagida M, Hirner S, Labeit D, Labeit S, Sorimachi H (2008) Multiple molecular interactions implicate the connectin/titin N2A region as a modulating scaffold for p94/calpain 3 activity in skeletal muscle. J Biol Chem 283(21):14801–14814

Helmes M, Trombitás K, Centner T, Kellermayer M, Labeit S, Linke WA, Granzier H (1999) Mechanically driven contour-length adjustment in rat cardiac titin's unique N2B sequence: titin is an adjustable spring. Circ Res 84(11):1339–1352

Ikeda K, Yamamoto R, Wirschell M, Yagi T, Bower R, Porter ME, Sale WS, Kamiya R (2009) A novel ankyrin-repeat protein interacts with the regulatory proteins of inner arm dynein f (I1) of Chlamydomonas reinhardtii. Cell Motil Cytoskeleton 66(8):448–56

Kedar V, McDonough H, Arya R, Li HH, Rockman HA, Patterson C (2004) Muscle-specific RING finger 1 is a bona fide ubiquitin ligase that degrades cardiac troponin I. Proc Natl Acad Sci U S A 101(52):18135–18140

Kellermayer MS, Smith SB, Granzier HL, Bustamante C (1997) Folding-unfolding transitions in single titin molecules characterized with laser tweezers. Science 276 (5315):1112–1116

Kellermayer MS, Smith SB, Bustamante C, Granzier HL (2001) Mechanical fatigue in repetitively stretched single molecules of titin. Biophys J 80(2):852–863

Kemp TJ, Sadusky TJ, Saltisi F, Carey N, Moss J, Yang SY, Sassoon DA, Goldspink G, Coulton GR (2000) Identification of Ankrd2, a novel skeletal muscle gene coding for a stretch-responsive ankyrin-repeat protein. Genomics 66(3):229–241

Knöll R, Hoshijima M, Hoffman HM, Person V, Lorenzen-Schmidt I, Bang ML, Hayashi T, Shiga N, Yasukawa H, Schaper W, McKenna W, Yokoyama M, Schork NJ, Omens JH, McCulloch AD, Kimura A, Gregorio CC, Poller W, Schaper J, Schultheiss HP, Chien KR. (2002) The cardiac mechanical stretch sensor machinery involves a Z disc complex that is defective in a subset of human dilated cardiomyopathy. Cell 111(7):943–955

Krüger M, Kohl T, Linke WA (2006) Developmental changes in passive stiffness and myofilament Ca^{2+} sensitivity due to titin and troponin-I isoform switching are not critically triggered by birth. Am J Physiol Heart Circ Physiol 291(2):H496–H506

Krüger M, Sachse C, Zimmermann WH, Eschenhagen T, Klede S, Linke WA (2008) Thyroid hormone regulates developmental titin isoform transitions via the phosphatidylinositol-3-kinase/AKT pathway. Circ Res 102(4):439–447

Kulke M, Fujita-Becker S, Rostkova E, Neagoe C, Labeit D, Manstein DJ, Gautel M, Linke WA (2001) Interaction between PEVK-titin and actin filaments: origin of a viscous force component in cardiac myofibrils. Circ Res 89(10):874–881

Labeit S, Kolmerer B (1995) Titins: giant proteins in charge of muscle ultrastructure and elasticity. Science 270(5234):293–296

Labeit S, Kolmerer B, Linke WA (1997) The giant protein titin. Emerging roles in physiology and pathophysiology. Circ Res 80(2):290–294

Labeit D, Watanabe K, Witt C, Fujita H, Wu Y, Lahmers S, Funck T, Labeit S, Granzier H (2003) Calcium-dependent molecular spring elements in the giant protein titin. Proc Natl Acad Sci USA 100(23):13716–13721

Lahmers S, Wu Y, Call DR, Labeit S, Granzier H (2004) Developmental control of titin isoform expression and passive stiffness in fetal and neonatal myocardium. Circ Res 94(4):505–513

Lange S, Xiang F, Yakovenko A, Vihola A, Hackman P, Rostkova E, Kristensen J, Brandmeier B, Franzen G, Hedberg B, Gunnarsson LG, Hughes SM, Marchand S, Sejersen T, Richard I, Edström L, Ehler E, Udd B, Gautel M (2005) The kinase domain of titin controls muscle gene expression and protein turnover. Science 308 (5728):1599–1603

Leake MC, Wilson D, Gautel M, Simmons RM (2004) The elasticity of single titin molecules using a two-bead optical tweezers assay. Biophys J 87(2):1112–1135

Leake MC, Grützner A, Krüger M, Linke WA (2006) Mechanical properties of cardiac titin's N2B-region by single-molecule atomic force spectroscopy. J Struct Biol 155(2):263–272

LeWinter MM, Wu Y, Labeit S, Granzier H (2007) Cardiac titin: structure, functions and role in disease. Clin Chim Acta 375(1–2):1–9

Li H, Fernandez JM (2003) Mechanical design of the first proximal Ig domain of human cardiac titin revealed by single molecule force spectroscopy. J Mol Biol 334(1):75–86

Li H, Oberhauser AF, Redick SD, Carrion-Vazquez M, Erickson HP, Fernandez JM (2001) Multiple conformations of PEVK proteins detected by single-molecule techniques. Proc Natl Acad Sci USA 98(19):10682–10686

Li H, Linke WA, Oberhauser AF, Carrion-Vazquez M, Kerkvliet JG, Lu H, Marszalek PE, Fernandez JM (2002) Reverse engineering of the giant muscle protein titin. Nature 418(6901):998–1002

Linke WA (2008) Sense and stretchability: the role of titin and titin-associated proteins in myocardial stress-sensing and mechanical dysfunction. Cardiovasc Res 77(4):637–648

Linke WA, Ivemeyer M, Mundel P, Stockmeier MR, Kolmerer B (1998) Nature of PEVK-titin elasticity in skeletal muscle. Proc Natl Acad Sci U S A 95(14):8052–8057

Linke WA, Rudy DE, Centner T, Gautel M, Witt C, Labeit S, Gregorio CC (1999) I-band titin in cardiac muscle is a three-element molecular spring and is critical for maintaining thin filament structure. J Cell Biol 146(3):631–644

Liversage AD, Holmes D, Knight PJ, Tskhovrebova L, Trinick J (2001) Titin and the sarcomere symmetry paradox. J Mol Biol 305(3):401–409

Mancini DM, Henson D, LaManca J, Levine S (1992) Respiratory muscle function and dyspnea in patients with chronic congestive heart failure. Circulation 86(3):909–918

Maruyama K (1976) Connectin, an elastic protein from myofibrils. J Biochem 80(2):405–407

Mayans O, van der Ven PF, Wilm M, Mues A, Young P, Fürst DO, Wilmanns M, Gautel M (1998) Structural basis for activation of the titin kinase domain during myofibrillogenesis. Nature 395(6705):863–869

Miller MK, Bang ML, Witt CC, Labeit D, Trombitas C, Watanabe K, Granzier H, McElhinny AS, Gregorio CC, Labeit S (2003) The muscle ankyrin repeat proteins: CARP, ankrd2/Arpp and DARP as a family of titin filament-based stress response molecules. J Mol Biol 333(5):951–964

Minajeva A, Kulke M, Fernandez JM, Linke WA (2001) Unfolding of titin domains explains the viscoelastic behavior of skeletal myofibrils. Biophys J 80(3):1442–1451

Moore AJ, Stubbings A, Swallow EB, Dusmet M, Goldstraw P, Porcher R, Moxham J, Polkey MI, Ferenczi MA (2006) Passive properties of the diaphragm in COPD. J Appl Physiol 101(5):1400–1405

Mrosek M, Labeit D, Witt S, Heerklotz H, von Castelmur E, Labeit S, Mayans O (2007) Molecular determinants for the recruitment of the ubiquitin-ligase MuRF-1 onto M-line titin. FASEB J 21(7):1383–1392

Mues A, van der Ven PF, Young P, Fürst DO, Gautel M (1998) Two immunoglobulin-like domains of the Z-disc portion of titin interact in a conformation-dependent way with telethonin. FEBS Lett 428(1–2):111–114

Mutungi G, Trinick J, Ranatunga KW (2003) Resting tension characteristics in differentiating intact rat fast- and slow-twitch muscle fibers. J Appl Physiol 95(6):2241–2247

Nagy A, Grama L, Huber T, Bianco P, Trombitás K, Granzier HL, Kellermayer MS (2005) Hierarchical extensibility in the PEVK domain of skeletal-muscle titin. Biophys J 89(1):329–336

Nakada C, Oka A, Nonaka I, Sato K, Mori S, Ito H, Moriyama M (2003) Cardiac ankyrin repeat protein is preferentially induced in atrophic myofibers of congenital myopathy and spinal muscular atrophy. Pathol Int 53(10):653–658

Obermann WM, Gautel M, Steiner F, van der Ven PF, Weber K, Fürst DO (1996) The structure of the sarcomeric M band: localization of defined domains of myomesin, M-protein, and the 250-kD carboxy-terminal region of titin by immunoelectron microscopy. J Cell Biol 134(6):1441–1453

Obermann WM, Gautel M, Weber K, Fürst DO (1997) Molecular structure of the sarcomeric M band: mapping of titin and myosin binding domains in myomesin and the identification of a potential regulatory phosphorylation site in myomesin. EMBO J 16(2):211–220

Olsson MC, Krüger M, Meyer LH, Ahnlund L, Gransberg L, Linke WA, Larsson L (2006) Fibre type-specific increase in passive muscle tension in spinal cord-injured subjects with spasticity. J Physiol 577.1:339–352

Omens JH, Usyk TP, Li Z, McCulloch AD (2002) Muscle LIM protein deficiency leads to alterations in passive ventricular mechanics. Am J Physiol Heart Circ Physiol 282(2):H680–H687

Ono Y, Kakinuma K, Torii F, Irie A, Nakagawa K, Labeit S, Abe K, Suzuki K, Sorimachi H (2004) Possible regulation of the conventional calpain system by skeletal muscle-specific calpain, p94/calpain 3. J Biol Chem 279(4):2761–2771

Ottenheijm CA, Heunks LM, Hafmans T, van der Ven PF, Benoist C, Zhou H, Labeit S, Granzier HL, Dekhuijzen PN (2006) Titin and diaphragm dysfunction in chronic obstructive pulmonary disease. Am J Respir Crit Care Med 173(5):527–534

Ottenheijm CA, Heunks LM, Dekhuijzen PN (2007) Diaphragm muscle fiber dysfunction in chronic obstructive pulmonary disease: toward a pathophysiological concept. Am J Respir Crit Care Med 175(12):1233–1240

Prado LG, Makarenko I, Andresen C, Krüger M, Opitz CA, Linke WA (2005) Isoform diversity of giant proteins in relation to passive and active contractile properties of rabbit skeletal muscles. J Gen Physiol 126(5):461–480

Puchner EM, Alexandrovich A, Kho AL, Hensen U, Schäfer LV, Brandmeier B, Gräter F, Grubmüller H, Gaub HE, Gautel M (2008) Mechanoenzymatics of titin kinase. Proc Natl Acad Sci U S A 105(36):13385–13390

Raynaud F, Fernandez E, Coulis G, Aubry L, Vignon X, Bleimling N, Gautel M, Benyamin Y, Ouali A (2005) Calpain 1-titin interactions concentrate calpain 1 in the Z-band edges and in the N2-line region within the skeletal myofibril. FEBS J 272(10):2578–2590

Rief M, Gautel M, Oesterhelt F, Fernandez JM, Gaub HE (1997) Reversible unfolding of individual titin immunoglobulin domains by AFM. Science 276(5315):1109–1112

Russell MW et al. (2002) Identification, tissue expression and chromosomal localization of human Obscurin-MLCK, a member of the titin and Dbl families of myosin light chain kinases. Gene 282 (1–2):237–246

Sarkar A, Caamano S, and Fernandez JM (2005) The elasticity of individual titin PEVK exons measured by single molecule atomic force microscopy. J Biol Chem 280 (8):6261–6264

Toursel T, Stevens L, Granzier H, Mounier Y (2002) Passive tension of rat skeletal soleus muscle fibers: effects of unloading conditions. J Appl Physiol 92(4):1465–1472

Trombitás K, Greaser M, French G, Granzier H (1998a) PEVK extension of human soleus muscle titin revealed by immunolabeling with the anti-titin antibody 9D10. J Struct Biol 122(1–2):188–196

Trombitás K, Greaser M, Labeit S, Jin JP, Kellermayer M, Helmes M, Granzier H (1998b) Titin extensibility in situ: entropic elasticity of permanently folded and permanently unfolded molecular segments. J Cell Biol 140(4):853–859

Trombitás K, Freiburg A, Centner T, Labeit S, Granzier H (1999) Molecular dissection of N2B cardiac titin's extensibility. Biophys J 77(6):3189–3196

Trombitás K, Wu Y, Labeit D, Labeit S, Granzier H (2001) Cardiac titin isoforms are coexpressed in the half-sarcomere and extend independently. Am J Physiol Heart Circ Physiol 281(4):H1793–H1799

Trombitás K, Wu Y, McNabb M, Greaser M, Kellermayer MS, Labeit S, Granzier H (2003) Molecular basis of passive stress relaxation in human soleus fibers: assessment of the role of immunoglobulin-like domain unfolding. Biophys J 85(5):3142–3153

Tskhovrebova L, Trinick J, Sleep JA, Simmons RM (1997) Elasticity and unfolding of single molecules of the giant muscle protein titin. Nature 387(6630):308–312

Udaka J, Ohmori S, Terui T, Ohtsuki I, Ishiwata S, Kurihara S, Fukuda N (2008) Disuse-induced preferential loss of the giant protein titin depresses muscle performance via abnormal sarcomeric organization. J Gen Physiol 131(1):33–41

Van den Bergh PY, Bouquiaux O, Verellen C, Marchand S, Richard I, Hackman P, Udd B (2003) Tibial muscular dystrophy in a Belgian family. Ann Neurol 54(2):248–251

van Hees HW, Ottenheijm CA, Granzier HL, Dekhuijzen PN, Heunks LM (2010) Int J Cardiol. 141(3):275–283

Wang K, McClure J, Tu A (1979) Titin: major myofibrillar components of striated muscle. Proc Natl Acad Sci USA 76(8):3698–3702

Warren CM, Krzesinski PR, Campbell KS, Moss RL, Greaser ML (2004) Titin isoform changes in rat myocardium during development. Mech Dev 121(11):1301–1312

Watanabe K, Muhle-Goll C, Kellermayer MS, Labeit S, Granzier H (2002a) Different molecular mechanics displayed by titin's constitutively and differentially expressed tandem Ig segments. J Struct Biol 137(1–2):248–258

Watanabe K, Nair P, Labeit D, Kellermayer MS, Greaser M, Labeit S, Granzier H (2002b) Molecular mechanics of cardiac titin's PEVK and N2B spring elements. J Biol Chem 277(13):11549–11558

Weinert S, Bergmann N, Luo X, Erdmann B, Gotthardt M (2006) M line-deficient titin causes cardiac lethality through impaired maturation of the sarcomere. J Cell Biol 173(4):559–570

Witt CC, Ono Y, Puschmann E, McNabb M, Wu Y, Gotthardt M, Witt SH, Haak M, Labeit D, Gregorio CC, Sorimachi H, Granzier H, Labeit S (2004) Induction and myofibrillar targeting of CARP, and suppression of the Nkx2.5 pathway in the MDM mouse with impaired titin-based signaling. J Mol Biol 336(1):145–154

Yamasaki R, Berri M, Wu Y, Trombitás K, McNabb M, Kellermayer MS, Witt C, Labeit D, Labeit S, Greaser M, Granzier H (2001) Titin-actin interaction in mouse myocardium: passive tension modulation and its regulation by calcium/S100A1. Biophys J 81(4):2297–2313

Young P, Ehler E, Gautel M (2001) Obscurin, a giant sarcomeric Rho guanine nucleotide exchange factor protein involved in sarcomere assembly. J Cell Biol 154(1):123–136

Zhu Y, Bogomolovas J, Labeit S, Granzier H (2009) Single molecule force spectroscopy of the cardiac titin N2B element: effects of the molecular chaperone alphaB-crystallin with disease-causing mutations. J Biol Chem 284(20):13914–13923

Contractile Characteristics of Sarcomeres Arranged in Series or Mechanically Isolated from Myofibrils

Dilson E. Rassier and Ivan Pavlov

Abstract The mechanisms of contraction are intrinsically connected to sarcomere mechanics during muscle activation and relaxation. This chapter presents two sets of experiments performed with (1) myofibrils, in which individual sarcomeres in series can be evaluated during contractions, and (2) mechanically isolated sarcomeres. When activated at optimal length (~2.0–2.4 μm), myofibrils and sarcomeres produce similar forces. However, their dependence on length differs: sarcomeres in series in a myofibril are able to produce similar forces at distinct lengths, while isolated sarcomeres show a force–length relation that resembles that obtained in original studies performed with single muscle fibers. Although force in isolated sarcomeres is rapidly stabilized during activation, significant movements of A-band are present when the contraction is produced at optimal lengths, which leads to different dynamics in the two half-sarcomeres. A-band movements decrease linearly with increasing lengths between 2.6 and 3.6 μm. Myofibrils and sarcomeres represent reliable techniques to evaluate contractile mechanisms at the most basic level of muscle organization. However, they present different mechanical characteristics that must be taken into account when scientists evaluate mechanisms of contraction.

Keywords Sarcomeres • Myofibrils • Stability • Force–length relation • Myosin • Actin

1 Introduction

Sarcomeres are the smallest functional units of striated muscles that still retain the three-dimensional lattice, with all the main proteins involved in contraction. The study of sarcomere mechanics is of primary importance; it has been used to investigate

D.E. Rassier (✉)
Department of Kinesiology and Physical Education, McGill University,
Pine Avenue West 475, Montreal, QC, Canada H2W1S4
e-mail: dilson.rassier@mcgill.ca

D.E. Rassier (ed.), *Muscle Biophysics: From Molecules to Cells*,
Advances in Experimental Medicine and Biology 682,
DOI 10.1007/978-1-4419-6366-6_7,

the relation between force and filament overlap (Gordon et al. 1966; ter Keurs et al. 1978), cross-bridge kinetics during force development and relaxation (Reconditi et al. 2004; Edman and Flitney 1982), the length dependence of muscle activation (Martyn and Gordon 2001; Wang and Fuchs 2001), and the Frank-Starling law of the heart (Fukuda et al. 2001; de Tombe et al. 2007), to cite a few phenomena.

Historically, sarcomere mechanics has been investigated in isolated muscle fibers. In these preparations, the average sarcomere length (SL) is based on the striation spacing produced by thick and thin filaments, comprising millions of sarcomeres arranged in series and in parallel. Since measurements of individual sarcomeres are not possible in single fibers, the interpretation of sarcomere mechanics is not without limitations: muscle activation leads to substantial SL non-uniformity (ter Keurs et al. 1978; Julian and Morgan 1979a; Edman and Reggiani 1984b), resulting in contradictory interpretations. In the past years, a few laboratories initiated experiments with isolated myofibrils, which allows the evaluation of individual sarcomeres (Rassier 2008; Rassier et al. 2003; Joumaa et al. 2008a, b; Stehle et al. 2002; Shimamoto et al. 2008, 2009) and half-sarcomeres (Telley et al. 2006a, b) in series, which represents a technical achievement.

In studies with myofibrils, a complex behavior is observed, as sarcomeres may change their length continuously during activation and relaxation (Rassier 2008; Telley et al. 2006a, b). Adjacent sarcomeres influence each other considerably, suggesting an inherent inter-sarcomere coordination (Shimamoto et al. 2009). Such coordination may led to results that are difficult to interpret; when myofibrils are activated first and then stretched, sarcomeres of distinct lengths are able to share similar forces without presenting large instabilities (Joumaa et al. 2008a; Rassier et al. 2003).

While myofibrils represent an optimal preparation for evaluating the mechanics of sarcomeres arranged in series, they do not allow the investigation of isolated sarcomeres. We thus developed a new experimental preparation in which we mechanically isolated sarcomeres and consistently measured their force and length during activation and relaxation. We derived a force–length (FL) relation, and also evaluated the length dependence of A-band displacement (and thus half-sarcomere dynamics) during contractions. Differently from myofibrils, isolated sarcomeres follow the traditional FL curve, with and ascending limb and a descending limb well defined.

This chapter will present recent results from studies performed in our laboratory with the intention to understand the mechanical behaviour of (1) sarcomeres arranged in series in a myofibril, and (2) sarcomeres mechanically isolated from its adjacent units.

2 Methods

2.1 Isolation of Myofibrils

Myofibrils from the rabbit psoas muscle were prepared as in previous studies (Pavlov et al. 2009b; Rassier 2008). Briefly, small sections of the muscles were dissected and stored in a rigor solution (in mM: 50 Tris, 100 NaC1, 2KC1, 2 $MgC1_2$,

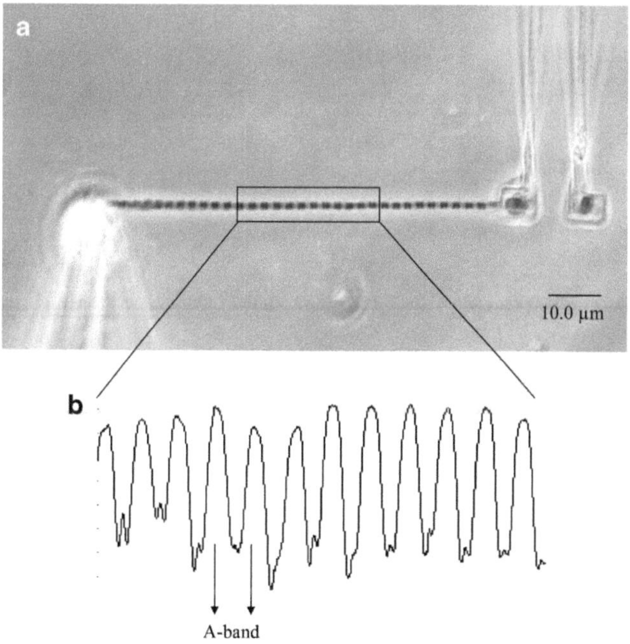

Fig. 1 (**a**) Picture of a typical myofibril used during experiments. The myofibril is attached between a pair of cantilevers at one end and a glass needle at the other end. The cantilevers are used for force measurements. (**b**) One scan obtained from the linear photodiode array, showing signals from the dark (myosin) and light (actin) bands. The distance between the peaks is used for SL measurements

10 EGTA, pH=7.0) for approximately 4 h, after which they were transferred to a rigor:glycerol (50:50) solution with the addition of protease inhibitors (Complete, Roche Diagnostics, Canada) and stored at –20°C for at least 7 days. On the day of experiment, small sections of the muscle sample were cut and homogenized in a rigor solution, leaving a suspension containing individual myofibrils. The sample was transferred into an experimental chamber placed on the stage of an inverted microscope (NIKON Eclipse TE 2000U, Japan). After ~5 min, the rigor solution was exchanged by a relaxing solution (in mM: 10 MOPS, 64.4 K^+ proprionate, 5.23 Mg^{2+} proprionate, 9.45 Na_2SO_4, 10 EGTA, 7 ATP, 10 CrP, pH=7.0; pCa^{2+}=9.0). A myofibril was chosen based on its striation appearance, and was captured by a rigid glass needle at one end, and a pair of micro-fabricated cantilevers at the other end (Rassier 2008) (Fig. 1). The cantilevers had stiffness between 22 and 348 nm μm^{-1}.

2.2 *Isolation of Sarcomeres*

The procedures for isolation of single sarcomeres are explained in details elsewhere (Pavlov et al. 2009a). Briefly, myofibrils were isolated and transferred into the experimental chamber using the same procedures as described above.

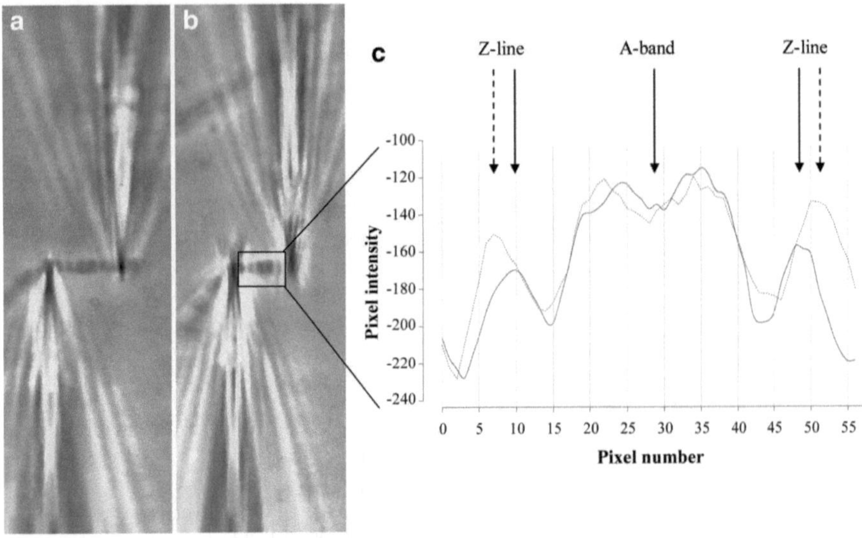

Fig. 2 Photographs of a sarcomere caught between two micro-needles, before (**a**) and after (**b**) activation. When the sarcomere contracts, it produces force and displaces the micro-needles. The initial SL in this experiment was 2.81 μm, and it shortened to 2.35 μm during contraction, producing a maximum force of 112.00 nN μm^{-2}. (**c**) Representative signals from the linear photodiode array collected from an activated sarcomere during the experiment. Superimposed scans collected during relaxation (*traced lines*, corresponding to (**a**)) and activation (*solid lines*, corresponding to (**b**)). Activation brings the two peaks corresponding to the Z-lines closer together

A myofibril was chosen based on its striation appearance. An individual sarcomere from the mid-section of the myofibril was selected and isolated using pre-calibrated glass micro-needles, which were also used for force measurements (Fig. 2). The micro-needles were calibrated by a cross-bending method (Ayittey et al. 2009) using the cantilevers from the myofibril experiments; the final stiffness of the micro-needles was between 200 and 377 nN μm^{-1}. The micro-needles were pierced ~0.2–0.3 μm externally adjacent to each Z-line and lifted off the glass cover slip by ~2–3 μm.

2.3 *Visualization of Myofibrils and Sarcomeres*

Under high magnification (Nikon plan-fluor, 100X, NA 1.30) further enhanced by 1.5X, the images of myofibrils and sarcomeres were projected onto a linear photodiode array (Schafter+Kirchhoff SK10680DJR, Germany) with 10,680 pixels (pixel size: 4×4μm) and were scanned to produce signals of light intensity. The contrast between the micro-needles and/or cantilevers, Z-lines, A-band and I-bands produces a dark-light peak pattern (Figs. 1 and 2) that allows for the evaluation of their position over time during the experiments. The centroids of the peaks were detected by a minimum average risk (Sokolov et al. 2003) for evaluation of SL and force.

2.4 *Activation and Force Measurements*

Once a myofibril or a sarcomere was set for testing, the relaxing solution was rapidly replaced by an activating solution (in mM: 10 MOPS, 45.1 K^+ proprionate, 5.21 Mg^{2+} proprionate, 9.27 Na_2SO_4, 10 EGTA, 7.18 ATP, 10 CrP, pH = 7.0; $pCa^{2+} = 4.75$). Rapid changes in solution were performed by a double-barreled pipette attached to a multichannel perfusion system (VC–6M, Harvard Apparatus, USA) (Rassier 2008; Piroddi et al. 2006). A peristaltic pump (Instech P720, Harvard Apparatus, USA) was used to drag the solutions from the chamber trough a back-channel. All myofibrils and sarcomeres were activated at 15°C, with the exception of some experiments with sarcomeres that were also performed at 20°C. During activation, contraction causes displacement of the cantilevers (myofibrils) or micro-needles (sarcomeres); the force (F) can be determined by calculating their deflection: $F = \Delta d\left(\frac{K_1K_2}{K_1 + K_2}\right)$, where Δd = absolute difference between the initial and the final distance between the cantilevers/micro-needles, K = stiffness, 1 and 2 = cantilevers/micro-needles 1 and 2, respectively.

3 Results

3.1 *Force and SL Changes Produced by Myofibrils*

Figure 3 shows individual SL traces during two contractions produced by a myofibril, at short and long lengths. In both cases all sarcomeres shortened during activation, except for one (Fig. 3a) or two (Fig. 3b) sarcomeres that were stretched (longer sarcomeres in the graphs). In most sarcomeres, the initial shortening was not followed by significant changes in length, showing a highly stable behavior.

Figure 4a shows an example of a myofibril in which two sarcomeres showed a slow, systematic length change during force production, suggesting an unstable behavior. These sarcomeres were next to each other and adjacent to the attachment of the myofibril to the glass needle. Their combined SL change during activation was stable – any instability created in this region does not

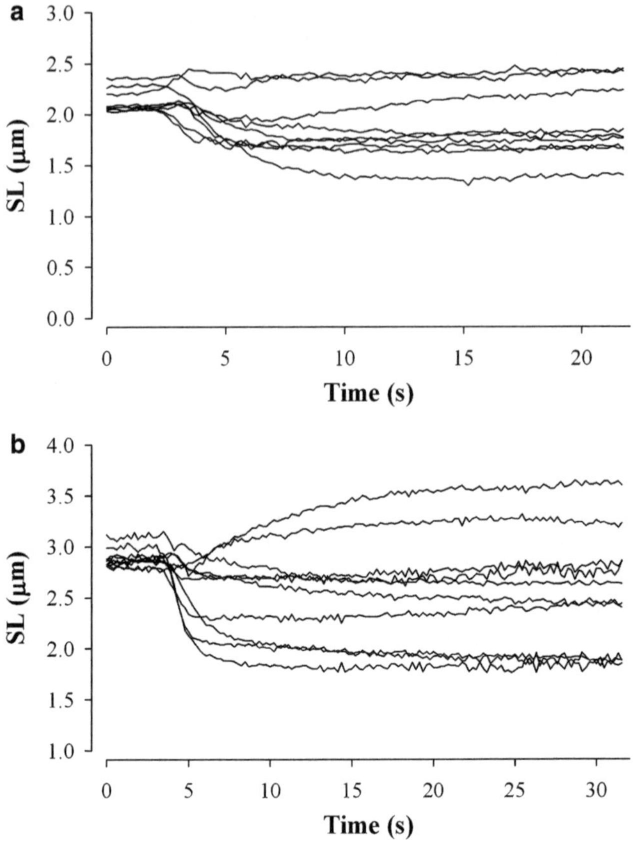

Fig. 3 Individual sarcomere dynamics during and after activation of the myofibrils activated at short (**a**) and long (**b**) average SLs

seem to spread throughout the myofibril (Fig. 4b). From 130 sarcomeres analyzed in all myofibrils used in these experiments, 115 sarcomeres were stable and did not change their length consistently during the isometric period of contractions.

We classified all myofibrils used in this study in two groups, according to the average SL in which they produced maximal force: short (≤2.5 µm) and long (≥2.5), corresponding to the ascending and descending limbs of the FL relation, respectively. Figure 5 shows the relation between the average SL and force produced at rest (F_r) during early force development during the contraction (~3 s after activation, F_e) and at the end of the activation period (~10 s before relaxation, F_l). The difference in force between contractions at short and long lengths were within the range observed in studies describing the force–SL relation for mammalian skeletal muscles, in which maximum forces

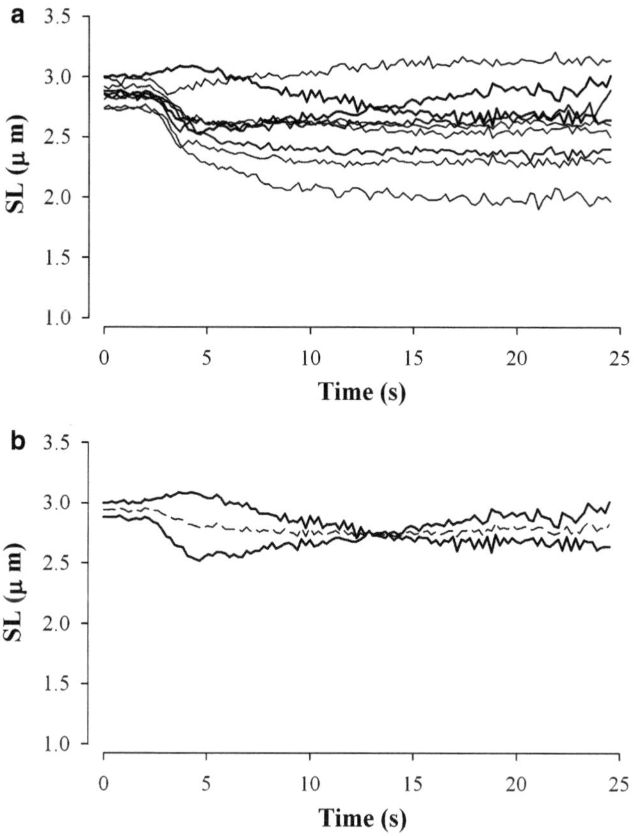

Fig. 4 (**a**) Individual sarcomere dynamics during and after activation of a myofibril with two unstable sarcomeres (*thick lines*). The two sarcomeres were adjacent to each other, and adjacent to the attachment site in the myofibril. (**b**) The average length of the two unstable sarcomeres (*thin line*) is constant throughout the contraction

vary between 2.1 and 2.8 μm (Elmubarak and Ranatunga 1984; Edman 2005; ter Keurs et al. 1984).

Figure 6 shows the degree of SL_{dis} evaluated at the same times as force was measured. Although an overall increase in SL dispersion was observed with initial activation, no further changes were observed throughout the activation period. The SL dispersion in the two groups of myofibrils was not statistically different during the contractions (Fig. 5b).

In summary, individual sarcomeres arranged in series in myofibrils activated at short and long lengths remain relatively stable throughout activation; large instabilities are not observed even if they share the same force. The result suggests a complex, intrinsic coordination among sarcomeres that cannot be explained solely by distinct degrees of filament overlap.

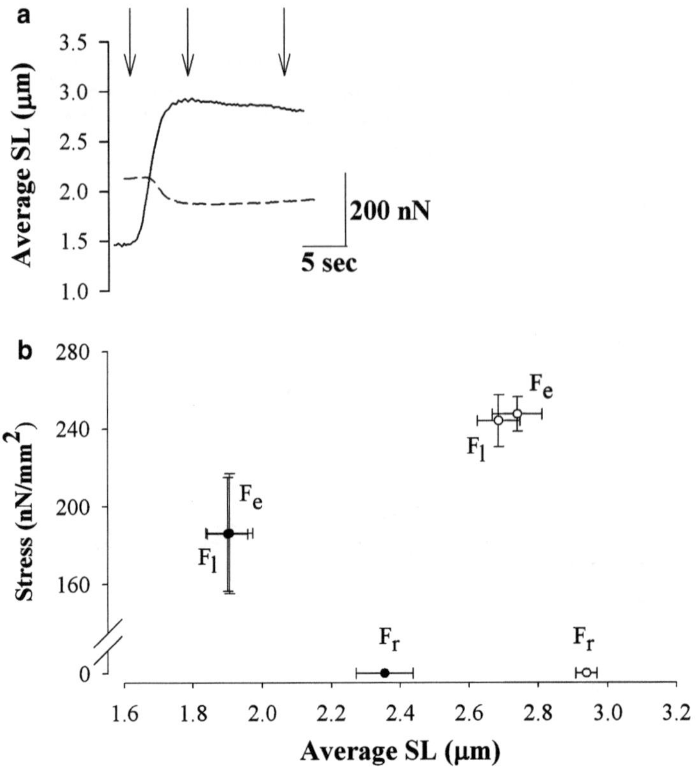

Fig. 5 (**a**) Force (*solid line*) and average SL (*dashed line*) produced during a myofibril contraction. The *arrows* show the times in each force was measured, at rest (F_r), early activation (F_e) and late activation (F_l). (**b**) The relation between the average SL and the force produced by myofibrils activated at short (≤2.5 μm; *solid circles*) and long (≥2.5 μm; *empty circles*) lengths. F_e and F_l are not significantly different in both groups

3.2 *Force and Length Changes Produced by Isolated Sarcomeres*

Figure 7a shows results from an experiment conducted with a mechanically isolated sarcomere that was activated three times, along the plateau of the FL relation or just below the plateau (Gordon et al. 1966; Sosa et al. 1994). The data reflects the high reproducibility of the experiments, as force measurements during the consecutive contractions did not decrease significantly. The force averaged over the last 5 s of contraction was 128.53, 116.94 and 116.81 nN μm^{-2} during full activation of the sarcomeres at lengths of 1.86, 2.28 and 1.99 μm, respectively. Figure 7b shows another sarcomere activated at a long length. The first and third contractions were produced at relatively similar lengths, and the forces were similar. Forces averaged over the last 5 s of contraction of 118.22, 72.26 and 103.78 nN μm^{-2} were produced with full activation at 2.35, 3.13 and 2.28 μm,

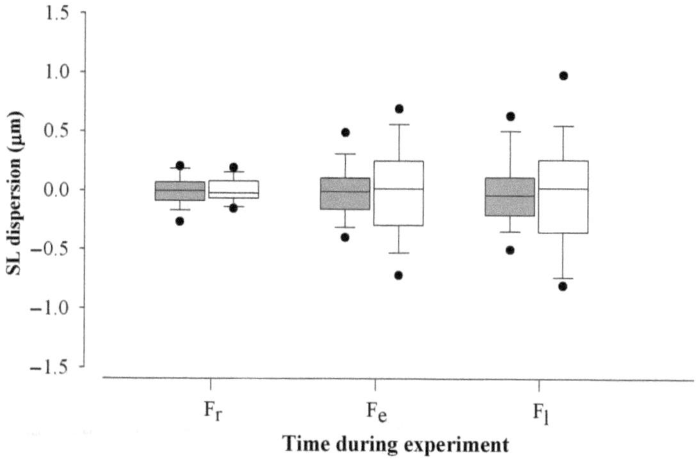

Fig. 6 Boxplots showing SL dispersion during experiments with myofibrils contracting at short (*filled boxes*) and long (*empty boxes*) lengths. Data was taken at the same time as the forces shown in Fig. 5. The *line* inside each *boxplot* represents the median, boundaries represent the 25th and 75th percentiles, error bars represent the 10th and 90th percentiles, and outliners (*dots*) represent the 95% confidence interval. SL dispersion increases from rest to activation and then it remains unchanged. SL dispersion is not different between myofibrils activated at short and long lengths

respectively. Note that when the contraction was produced at 3.13 μm, the decrease in force from the first contraction was of 39%, similar to what would be expected (~42%) based on the loss of filament overlap when the sarcomere is stretched from 2.35 to 3.13 μm (Sosa et al. 1994).

3.3 Force–SL Relation

Figure 8 shows the FL relation obtained during experiments performed with isolated sarcomeres. Only the contractions produced during the first activation-relaxation cycle were plotted in this graph. This approach generated some variability in the forces produced at similar portions of the FL relation curve, which was nonetheless small (~8–10% of force). A mean FL curve was determined from the data by fitting a fourth degree polynomial curve to the experimental values using the least-squares method (continuous thick line in Fig. 8a). The general shape of the FL relation shows a maximum force produced at 2.18 μm, where the ascending and descending limbs are joined. The upper portion of the FL curve (between ~1.8 and ~2.6 μm) is symmetric on either side of the peak. A linear regression for force values produced in lengths between 2.4 and 3.5 μm extrapolates to the abscissa at 3.87 μm (traced line in Fig. 8a); this portion corresponds to the descending limb of the FL relation. We compared our results with the original study of Gordon et al. (1966) by drawing a theoretical FL curve using filament lengths from the rabbit psoas muscle

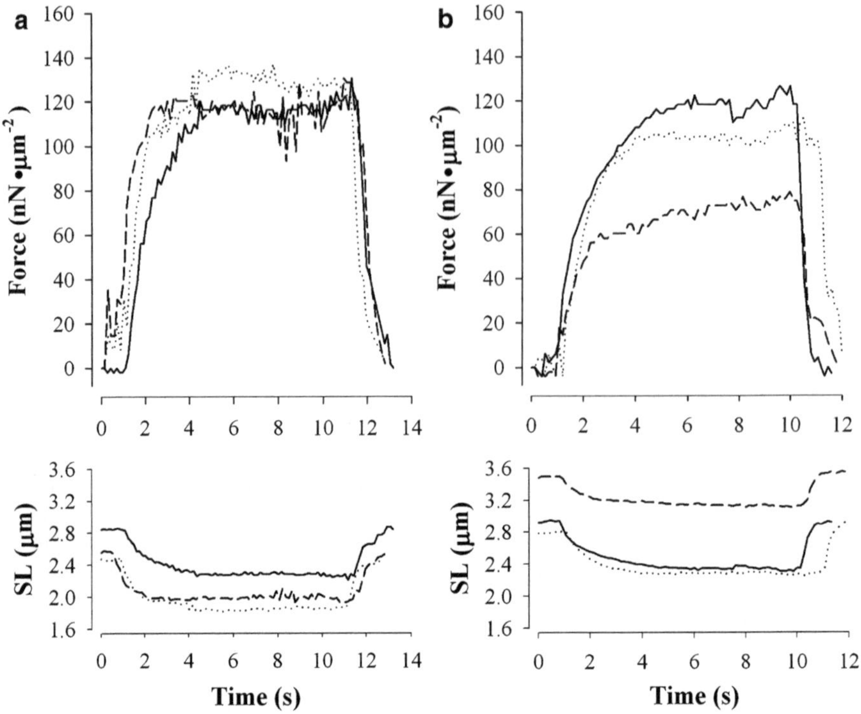

Fig. 7 Contractions produced by two different sarcomeres during experiments produced at 15°C; the *top panels* show the force and the *bottom panels* show the length changes. (**a**) The sarcomere was activated at initial lengths of 2.47, 2.84 and 2.55 µm. The sarcomere produced forces that were relatively similar in these contractions. (**b**) The sarcomere was activated at initial lengths of 2.79, 3.48 and 2.91 µm. The force decreased significantly when the contraction was produced at the longest length, on agreement with the loss in filament overlap observed along the descending limb of the FL relation (Gordon et al. 1966)

(Sosa et al. 1994) (Fig. 8b). The plateau of the FL relation in this case should span from 2.24 to 2.39 µm, the ascending limb should include lengths below 2.24 µm and the descending limb should span lengths from 2.39 to 3.87 µm. The two curves are similar, except by the plateau region, in which the flat portion described by Gordon et al. (1966) contrasts with the rounded shaped plateau derived in our experiments.

3.4 A-Band Movements and Half-Sarcomere Dynamics

We tracked the centroids of the peaks corresponding to the A-bands during the contractions, and measured their distances from the Z-lines. As a result, we obtained a direct measurement of the A-band displacement relative to the center of the sarcomere, and therefore the half-sarcomere dynamics during contractions.

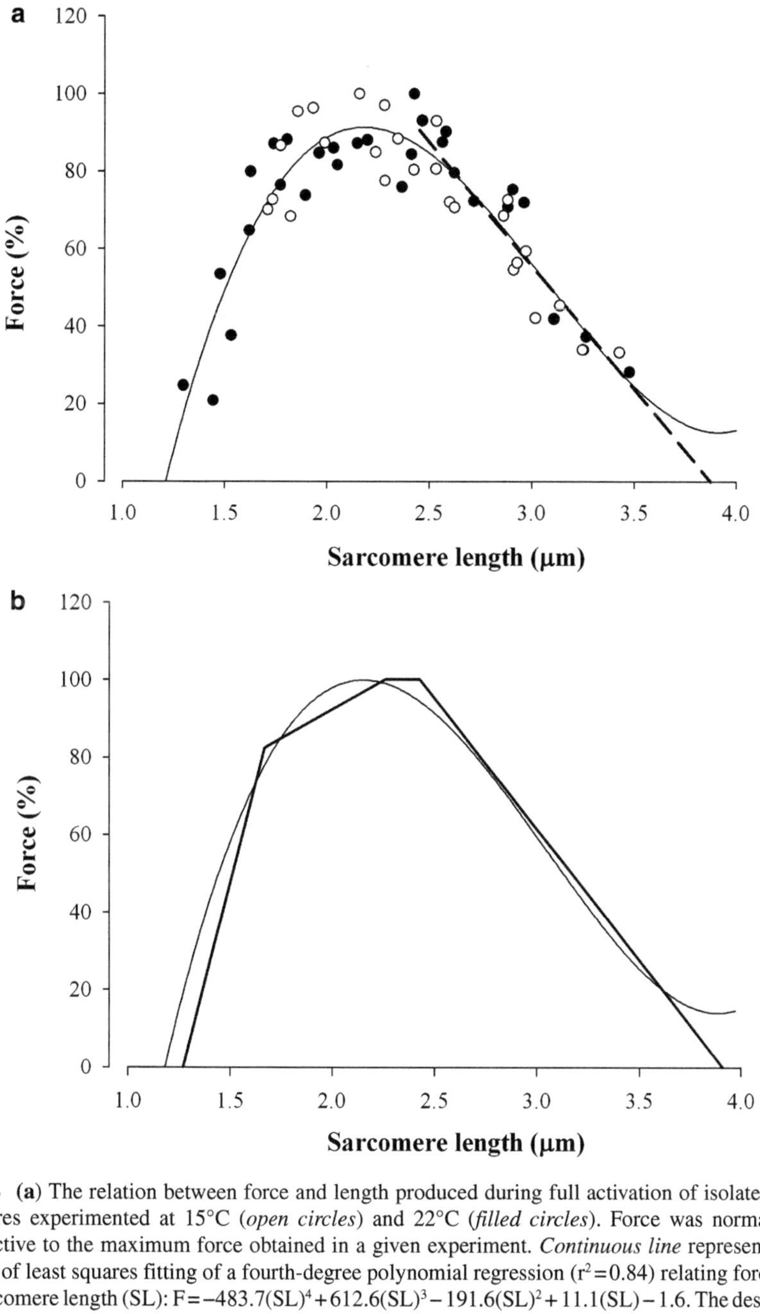

Fig. 8 (**a**) The relation between force and length produced during full activation of isolated sarcomeres experimented at 15°C (*open circles*) and 22°C (*filled circles*). Force was normalized respective to the maximum force obtained in a given experiment. *Continuous line* represents the result of least squares fitting of a fourth-degree polynomial regression ($r^2=0.84$) relating force (F) to sarcomere length (SL): $F=-483.7(SL)^4+612.6(SL)^3-191.6(SL)^2+11.1(SL)-1.6$. The descending limb was also fitted with linear regression for data between 2.4 and 3.5 μm (*traced line*), with an abscissa extrapolating to 3.87 μm. (**b**) The FL relation derived in our experiments and the FL curve derived from theoretical predictions (*thick line*) based on filament lengths of the rabbit psoas muscle (Sosa et al. 1994)

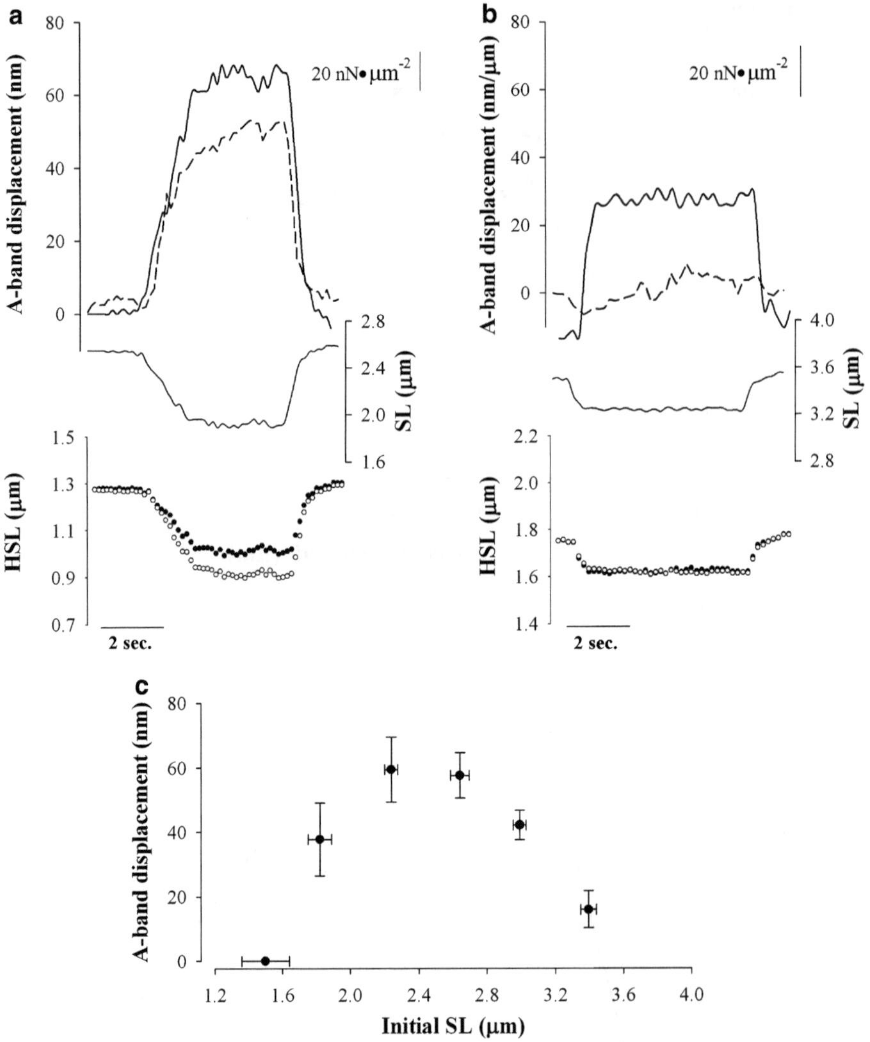

Fig. 9 Measurements of A-band displacements and half-sarcomere dynamics during a typical experiment with a sarcomere activated at 15°C. The *top panels* in (**a**) and (**b**) show the force and A-band positions (*traced line*), the *middle panels* show the length, and the *bottom panels* show the corresponding right (*filled circles*) and left (*open circles*) half-sarcomeres. Changes in A-band position at the short SL follow the same pattern as the force traces, while substantial changes in A-band position are not observed at the longer length. Such behaviour leads to different half-sarcomere dynamics when the contraction is elicited at 2.53 µm but not when the contraction is elicited at 3.51 µm. (**c**) The relation between the initial SL and the maximum amount of A-band displacement during activation. Values are means ± SEM

In Fig. 9a, a sarcomere was activated initially at a length of 2.55 µm, producing a force of 128.31 nN µm^{-2} at a final length of 1.93 µm. The A-band was displaced 52 nm from the center of the sarcomere during activation, which caused different half-sarcomere dynamics during the contraction. The A-band started to move in synchrony with the beginning of force development, and moved back to the center of the sarcomere during relaxation. In Fig. 9b, the sarcomere was activated at an initial length of 3.5 μm, producing a force of 54.08 nN μm^{-2} at a final length of 3.25 μm. In this case, the A-band deviation from the center was minimal, and was constantly adjusted during the contraction, such that it was mostly centralized. As a result, the two half-sarcomeres contracted in a similar way.

The directions and speed of A-band displacements during the contractions were random. In most sarcomeres the A-band movements were unidirectional towards one Z-line throughout the contraction. We measured the maximum A-band displacement, which invariably happened toward the end of the contraction. Figure 9c shows the relation between initial SL and amount of A-band displacement. When the initial SL was below 1.6 μm, movements of A-band were not detected. The magnitude of A-band displacement increased to obtain a maximum at SL of 2.0–2.4 μm. A-band displacements then decreased linearly between SL of 2.4 and 3.6 μm.

4 Discussion

Myofibrils and isolated sarcomeres represent techniques that allow the evaluation of contraction at basic levels or muscle organization. The forces produced by myofibrils and sarcomeres are consistent with experiments performed with larger preparations, but they also show differences that need to be addressed. While sarcomeres arranged in series in the myofibrils are able to contract isometrically and share similar forces at entirely different lengths, isolated sarcomeres depend strongly on the length, and thus filament overlap.

4.1 Myofibril Experiments

The study with myofibrils was aimed at evaluating the mechanics of individual sarcomeres arranged in series during isometric contractions. Although the FL relation has been investigated previously in myofibril experiments, they were obtained after an imposed stretch, which may change the properties of cross-bridge kinetics (Rassier and Herzog 2004) and passive elements (Joumaa et al. 2008b) differently than isometric contractions. Our results show that myofibril activation leads to an increase in SL non-uniformity, as shown in single fibers (ter Keurs et al. 1978; Julian and Morgan 1979a; Edman and Reggiani 1984b), but it does not necessarily lead to instability and continuous SL changes during contractions. Instead, the degree of SL non-uniformity was constant throughout activation, suggesting that sarcomeres are mechanically stable. The result is not trivial, suggesting that sarcomeres with different degrees of filament overlap can share similar forces.

Comparing our results with studies performed with single fibers is difficult, but there have been a few studies that measured the average SL and also suggest that it does not change significantly during isometric contractions (ter Keurs et al. 1978; Tameyasu et al. 1982).

What are the mechanisms underlying sarcomere stability at short and long lengths? Presently our results do not allow conclusive explanations, but there are a few possibilities. First, differences in passive forces could stabilize sarcomeres at different lengths. However, this mechanism is unlikely, as titin – the protein responsible for passive forces – expresses mostly the N2A isoform in the rabbit psoas muscle (Prado et al. 2005; Freiburg et al. 2000). Two subsets of the N2A isoforms (3.3 and 3.4 MD) have been reported, but a major proportion of the shortest isoform is always observed (Prado et al. 2005). Furthermore, passive forces in rabbit psoas myofibrils are ~5–10% of the maximal active forces at SL of ~3.2 μm or longer (Bartoo et al. 1993, 1997), not covering most of the SL investigated in the current study.

Second, half-sarcomeres may shorten differently during activation – non-uniformity in half-SLs induced at the beginning of activation could be increased as the thick filaments (A-band) would be pulled towards one end of the sarcomeres (Page and Huxley 1963; Eisenberg and Eisenberg 1982; Horowits and Podolsky 1987; Telley et al. 2006a). However, our study with isolated sarcomeres (discussion below) shows that A-band displacements at long lengths are small (potentially limited by titin), and thus our results observed at long lengths would not be explained by such mechanism. The mechanism that allows for an intrinsic sarcomere stabilization need further investigation, as new studies suggest an intra-sarcomere coordination which may underlie basic mechanisms of contraction (Shimamoto et al. 2009).

4.2 Sarcomere Experiments

Sarcomeres represent the smallest functional unit of striated muscles. As far as we know, mechanical evaluation of isolated sarcomeres has not been performed before. Therefore, the sarcomere technique presented in this chapter may open many possibilities for investigators in the field of muscle physiology and biophysics. The forces produced by the sarcomeres are in the same range are those produced by fibers and myofibrils when activated in lengths between ~2.2 and 2.4 μm (Gordon et al. 1966; Shimamoto et al. 2008; Rassier 2008). Sarcomeres activated at 15°C could undergo 3–4 consecutive activation–relaxation cycles without a significant decrease in force (>5%), similar to what is commonly observed in myofibrils and permeabilized muscle fibers. Thus, the preparation is highly reliable and consistent.

The FL relation for the sarcomeres is comparable with that observed in studies with muscle fibers (Edman and Reggiani 1984a, 1987; Gordon et al. 1966), and could be described by filament overlap predictions (Gordon et al. 1966). Since there are variations in force, it is hard to know if the data would have presented a flat plateau, similar to that observed in previous studies. However, the upper portion of the FL curve varies little between ~1.8 and ~2.6 μm, similar to previous studies with mammalian muscles showing plateaus varying between 2.1 and 2.8 μm (Elmubarak and Ranatunga 1984; Edman 2005; Stephenson and Williams 1982; ter Keurs et al. 1984).

4.3 Half-Sarcomere Dynamics

Significant displacements of the A-bands during activation, and consequently differences in half-sarcomere dynamics during contractions, have been observed in studies using micrographs of frog muscles (Page and Huxley 1963; Eisenberg and Eisenberg 1982; Bergman 1983; Horowits and Podolsky 1987). This phenomenon is also observed in myofibrils from the rabbit psoas muscle (Telley et al. 2006a). The length dependence of individual A-band displacements, not evaluated previously during contractions, may therefore represent an important characteristic of the FL relation in skeletal muscles.

When sarcomeres were activated at lengths between ~1.8 and ~2.6 μm, significant A-band displacements were observed. When we activated the sarcomeres at long lengths, on the other hand, shifts in the A-bands were not observed. The mechanisms responsible for stabilizing the A-band at the center of the sarcomere at long lengths are elusive, but our results suggest that it may be regulated by titin. It has been proposed that, when titin becomes stiffer at increasing SL, it centralizes the thick filaments (Horowits and Podolsky 1987, 1988; Horowits et al. 1986). Such mechanism is compatible with the inverse relationship between the initial length and A-band displacement along the descending limb of the FL relation. Furthermore, it is known that titin increase its stiffness with Ca^{2+}, and that such an increase becomes more pronounced at increasing lengths (Labeit et al. 2003), which could also stabilize the A-bands.

The A-band displacement started simultaneously with the beginning of force development, different from what was observed by others (Horowits and Podolsky 1987) but similar to results obtained with myofibrils (Telley et al. 2006a). The contrasting results do not necessarily imply a difference in the intrinsic mechanisms of A-band dynamics. The lag between A-band displacements and force development is supposed to be caused by the development of SL non-uniformity upon muscle activation (Horowits and Podolsky 1987, 1988), a phenomenon eliminated with the single sarcomere preparation.

4.4 Summary

Myofibrils and isolated sarcomeres allow the evaluation of the basic mechanisms of force production at the most basic level of organization – the three dimensional lattice containing all contractile and passive proteins involved in contraction. While sarcomeres in series in a myofibrils show a complex behavior during activation, they stabilize at full force development. Isolated sarcomeres, on the other hand, behave similar to larger preparations and show a FL relation that resembles that observed in studies with single muscle fibers. Measurements of individual sarcomeres can be used to solve long-standing issues in the field of muscle physiology: the origin of sarcomere non-uniformity during muscle activation and relaxation (Edman and Reggiani 1984b; Julian and Morgan 1979a, b, Telley et al. 2006a), the force creep commonly observed in contractions performed at long SL (Edman and Reggiani 1984a, b; Gordon et al. 1966; ter Keurs et al. 1978), and the effects of length changes in activated muscles (Edman and Tsuchiya 1996, Edman et al. 1982, 1993; Granzier and Pollack 1989, Rassier et al. 2003). Future studies using isolated sarcomeres are promising.

Acknowledgements This research was supported by the "Canadian Institutes of Health Research" and the "Natural Sciences and Engineering Research Council" of Canada. Dilson Rassier is supported by Fonds de la Recherche en Santé du Quebec, Canada.

References

Ayittey PN, Walker JS, Rice JJ, de Tombe PP (2009) Glass microneedles for force measurements: a finite-element analysis model. Pflugers Arch 457:1415–1422

Bartoo ML, Popov VI, Fearn LA, Pollack GH (1993) Active tension generation in isolated skeletal myofibrils. J Muscle Res Cell Motil 14:498–510

Bartoo ML, Linke WA, Pollack GH (1997) Basis of passive tension and stiffness in isolated rabbit myofibrils. Am J Physiol Cell Physiol 273:C266–C276

Bergman RA (1983) Ultrastructural configuration of sarcomeres in passive and contracted frog sartorius muscle. Am J Anat 166:209–222

de Tombe PP, Belus A, Piroddi N, Scellini B, Walker JS, Martin AF, Tesi C, Poggesi C (2007) Myofilament calcium sensitivity does not affect cross-bridge activation-relaxation kinetics. Am J Physiol Regul Integr Comp Physiol 292:R1129–R1136

Edman KA (2005) Contractile properties of mouse single muscle fibers, a comparison with amphibian muscle fibers. J Exp Biol 208:1905–1913

Edman KA, Flitney FW (1982) Laser diffraction studies of sarcomere dynamics during 'isometric' relaxation in isolated muscle fibres of the frog. J Physiol 329:1–20

Edman KA, Reggiani C (1984a) Absence of plateau of the sarcomere length-tension relation in frog muscle fibres. Acta Physiol Scand 122:213–216

Edman KA, Reggiani C (1984b) Redistribution of sarcomere length during isometric contraction of frog muscle fibres and its relation to tension creep. J Physiol 351:169–198

Edman KA, Reggiani C (1987) The sarcomere length-tension relation determined in short segments of intact muscle fibres of the frog. J Physiol 385:709–732

Edman KA, Tsuchiya T (1996) Strain of passive elements during force enhancement by stretch in frog muscle fibres. J Physiol 490:191–205

Edman KA, Elzinga G, Noble MI (1982) Residual force enhancement after stretch of contracting frog single muscle fibers. J Gen Physiol 80:769–784

Edman KA, Caputo C, Lou F (1993) Depression of tetanic force induced by loaded shortening of frog muscle fibres. J Physiol 466:535–552

Eisenberg BR, Eisenberg RS (1982) The T-SR junction in contracting single skeletal muscle fibers. J Gen Physiol 79:1–19

Elmubarak MH, Ranatunga KW (1984) Temperature sensitivity of tension development in a fast-twitch muscle of the rat. Muscle Nerve 7:298–303
Freiburg A, Trombitas K, Hell W, Cazorla O, Fougerousse F, Centner T, Kolmerer B, Witt C, Beckmann JS, Gregorio CC, Granzier H, Labeit S (2000) Series of exon-skipping events in the elastic spring region of titin as the structural basis for myofibrillar elastic diversity. Circ Res 86:1114–1121
Fukuda N, Sasaki D, Ishiwata S, Kurihara S (2001) Length dependence of tension generation in rat skinned cardiac muscle: role of titin in the Frank-Starling mechanism of the heart. Circulation 104:1639–1645
Gordon AM, Huxley AF, Julian FJ (1966) The variation in isometric tension with sarcomere length in vertebrate muscle fibres. J Physiol 184:170–192
Granzier HL, Pollack GH (1989) Effect of active pre-shortening on isometric and isotonic performance of single frog muscle fibres. J Physiol 415:299–327
Horowits R, Podolsky RJ (1987) The positional stability of thick filaments in activated skeletal muscle depends on sarcomere length: evidence for the role of titin filaments. J Cell Biol 105: 2217–2223
Horowits R, Podolsky RJ (1988) Thick filament movement and isometric tension in activated skeletal muscle. Biophys J 54:165–171
Horowits R, Kempner ES, Bisher ME, Podolsky RJ (1986) A physiological role for titin and nebulin in skeletal muscle. Nature 323:160–164
Joumaa V, Leonard TR, Herzog W (2008a) Residual force enhancement in myofibrils and sarcomeres. Proc Biol Sci 275:1411–1419
Joumaa V, Rassier DE, Leonard TR, Herzog W (2008b) The origin of passive force enhancement in skeletal muscle. Am J Physiol Cell Physiol 294:C74–C78
Julian FJ, Morgan DL (1979a) Intersarcomere dynamics during fixed-end tetanic contractions of frog muscle fibres. J Physiol 293:365–378
Julian FJ, Morgan DL (1979b) The effect on tension of non-uniform distribution of length changes applied to frog muscle fibres. J Physiol 293:379–392
Labeit D, Watanabe K, Witt C, Fujita H, Wu Y, Lahmers S, Funck T, Labeit S, Granzier H (2003) Calcium-dependent molecular spring elements in the giant protein titin. Proc Natl Acad Sci U S A 100:13716–13721
Martyn DA, Gordon AM (2001) Influence of length on force and activation-dependent changes in troponin c structure in skinned cardiac and fast skeletal muscle. Biophys J 80: 2798–2808
Page SG, Huxley HE (1963) Filament lengths in striated muscle. J Cell Biol 19:369–390
Pavlov I, Novinger R, Rassier DE (2009a) The mechanical behavior of individual sarcomeres of myofibrils isolated from rabbit psoas muscle. Am J Physiol Cell Physiol 297:C1211–1219
Pavlov I, Novinger, R, Rassier DE (2009b) Sarcomere dynamics in skeletal muscle myofibrils during isometric contractions. J Biomech 42:2808–2812
Piroddi N, Belus A, Eiras S, Tesi C, van der Velden J, Poggesi C, Stienen GJ (2006) No direct effect of creatine phosphate on the cross-bridge cycle in cardiac myofibrils. Pflugers Arch 452:3–6
Prado LG, Makarenko I, Andresen C, Kruger M, Opitz CA, Linke WA (2005) Isoform diversity of giant proteins in relation to passive and active contractile properties of rabbit skeletal muscles. J Gen Physiol 126:461–480
Rassier DE (2008) Pre-power stroke cross bridges contribute to force during stretch of skeletal muscle myofibrils. Proc Biol Sci 275:2577–2586
Rassier DE, Herzog W (2004) Active force inhibition and stretch-induced force enhancement in frog muscle treated with BDM. J Appl Physiol 97:1395–1400
Rassier DE, Herzog W, Pollack GH (2003) Dynamics of individual sarcomeres during and after stretch in activated single myofibrils. Proc Biol Sci 270:1735–1740
Reconditi M, Linari M, Lucii L, Stewart A, Sun YB, Boesecke P, Narayanan T, Fischetti RF, Irving T, Piazzesi G, Irving M, Lombardi V (2004) The myosin motor in muscle generates a smaller and slower working stroke at higher load. Nature 428:578–581

Shimamoto Y, Suzuki M, Ishiwata S (2008) Length-dependent activation and auto-oscillation in skeletal myofibrils at partial activation by Ca^{2+}. Biochem Biophys Res Commun 366:233–238

Shimamoto Y, Suzuki M, Mikhailenko SV, Yasuda K, Ishiwata S (2009) Inter-sarcomere coordination in muscle revealed through individual sarcomere response to quick stretch. Proc Natl Acad Sci U S A 106:11954–11959

Sokolov SY, Grinko AA, Tourovskaia AV, Reitz FB, Yakovenko O, Pollack GH, Blyakhman FA (2003) 'Minimum average risk' as a new peak-detection algorithm applied to myofibrillar dynamics. Comput Methods Programs Biomed 72:21–26

Sosa H, Popp D, Ouyang G, Huxley HE (1994) Ultrastructure of skeletal muscle fibers studied by a plunge quick freezing method: myofilament lengths. Biophys J 67:283–292

Stehle R, Kruger M, Pfitzer G (2002) Force kinetics and individual sarcomere dynamics in cardiac myofibrils after rapid ca(2+) changes. Biophys J 83:2152–2161

Stephenson DG, Williams DA (1982) Effects of sarcomere length on the force–pCa relation in fast- and slow-twitch skinned muscle fibres from the rat. J Physiol 333:637–653

Tameyasu T, Ishide N, Pollack GH (1982) Discrete sarcomere length distribution in skeletal muscle. Biophys J 37:489–492

Telley IA, Denoth J, Stussi E, Pfitzer G, Stehle R (2006a) Half-sarcomere dynamics in myofibrils during activation and relaxation studied by tracking fluorescent markers. Biophys J 90:514–530

Telley IA, Stehle R, Ranatunga KW, Pfitzer G, Stussi E, Denoth J (2006b) Dynamic behaviour of half-sarcomeres during and after stretch in activated rabbit psoas myofibrils: sarcomere asymmetry but no 'sarcomere popping'. J Physiol 573: 173–185

ter Keurs HE, Iwazumi T, Pollack GH (1978) The sarcomere length-tension relation in skeletal muscle. J Gen Physiol 72:565–592

ter Keurs HE, Luff AR, Luff SE (1984) Force–sarcomere-length relation and filament length in rat extensor digitorum muscle. Adv Exp Med Biol 170, 511–525

Wang Y, Fuchs F (2001) Interfilament spacing, Ca^{2+} sensitivity, and Ca^{2+} binding in skinned bovine cardiac muscle. J Muscle Res Cell Motil 22:251–257

The Force–Length Relationship of Mechanically Isolated Sarcomeres

W. Herzog, V. Joumaa, and T.R. Leonard

Abstract The sarcomere force–length relationship is arguably the most basic property of skeletal muscle force production. It has been accepted as textbook knowledge and is in direct support of the sliding filament and cross-bridge theories of contraction. However, the sarcomere force–length relationship has never been measured directly. Here, we show results of two experiments elucidating the force–length properties of mechanically isolated sarcomeres. We demonstrate that sarcomere forces are greatly dependent on sarcomere lengths for purely isometric conditions, but can take on essentially any steady-state value depending on an individual sarcomere's contractile history. Therefore, we conclude that steady-state isometric forces in isolated sarcomeres do not only depend on sarcomere lengths (or equivalently actin-myosin overlap) but depend crucially on a sarcomere's contractile history. These results have direct implications for our understanding of the molecular mechanisms of muscle contraction.

Keywords Mechanisms of contraction • History dependence • Force–length relationship • Single sarcomeres

1 Introduction

Sarcomeres are the smallest functional contractile units of skeletal and cardiac muscles. Their properties have been associated with corresponding properties on the cell- (fiber) and whole muscle-level. Specifically, the sarcomere force–length relationship, determined first by Gordon et al. (1966), is arguably the most basic

W. Herzog (✉)
Faculty of Kinesiology, University of Calgary, Calgary, AB, Canada T2N1N4
e-mail: walter@kin.ucalgary.ca

D.E. Rassier (ed.), *Muscle Biophysics: From Molecules to Cells*,
Advances in Experimental Medicine and Biology 682,
DOI 10.1007/978-1-4419-6366-6_8,

property associated with sarcomere and muscle function and has been taken as textbook knowledge ever since it was first described. Importantly, it was also in accordance with predictions made by the sliding filament (Huxley and Niedergerke 1954; Huxley and Hanson 1954) and cross-bridge theories of muscle contraction (Huxley 1957).

The sliding filament theory asserted that muscle contraction occurred through the relative sliding of two sets of intricately aligned protein filaments: actin, or the thin filament, and myosin, or the thick filament (Huxley and Niedergerke 1954; Huxley and Hanson 1954). The cross-bridge theory, first proposed in mathematical terms by A. Huxley (1957) described how this relative sliding between actin and myosin occurs. The idea was that extensions from the myosin filaments (cross-bridges) attach cyclically to binding sites on actin, exert force, and pull actin past myosin filaments. Basic assumptions of the cross-bridge theory included that (1) actin and myosin filaments are essentially rigid, that (2) cross-bridges and actin binding sites are arranged uniformly on the thick and thin filament, respectively, and that (3) cross-bridges act independently of each other and exert the same average force.

These assumptions allowed for precise predictions of the steady-state isometric force as a function of sarcomere lengths under maximal activation conditions; the sarcomere force–length relationship. Specifically, maximal overlap between actin and myosin filaments allows for a maximal number of cross-bridges to attach to actin, and thus force is maximal. When overlap between actin and myosin decreases, the number of potential cross-bridge attachment decreases linearly with the loss of overlap, and the isometric steady-state force is predicted to decrease linearly until actin-myosin overlap is completely lost and active force becomes zero. Knowing the actin and myosin filament lengths, the sarcomere force–length relationship can be predicted theoretically (e.g., Herzog et al. 1992; Huxley and Peachey 1961; Walker and Schrodt 1973). The sarcomere force–length relationship determined experimentally by Gordon et al. (1966) matched the theoretically predicted force–length relationship perfectly (Fig. 1), thereby establishing the sliding filament and cross-bridge theories on firm ground.

An interesting aspect of the classic cross-bridge theory (Huxley 1957, 1969), and later derivations; Huxley and Simmons 1971; Rayment et al. 1993), is that cross-bridge attachment/detachment from actin is governed by rate constants that exclusively depend on Huxley's so-called "x"-distance, which is defined as the distance from the cross-bridge equilibrium position to the nearest actin attachment site (Fig. 2). This definition of the rate functions immediately implies that steady-state isometric forces for a sarcomere, fiber or whole muscle are exclusively determined by the amount of actin-myosin filament overlap. Thus, for constant activation, isometric steady-state forces are uniquely associated with sarcomere lengths. In particular, the contractile history (any sarcomere shortening or stretch) preceding the steady-state isometric conditions, does not affect force (Huxley 1957).

However, the sarcomere force–length relationship determined by Gordon et al. (1966) had several uncertainties and limitations. Most importantly, sarcomere force and sarcomere length, the two basic variables of the force–length relationship, were not measured directly but were inferred using non-trivial assumptions.

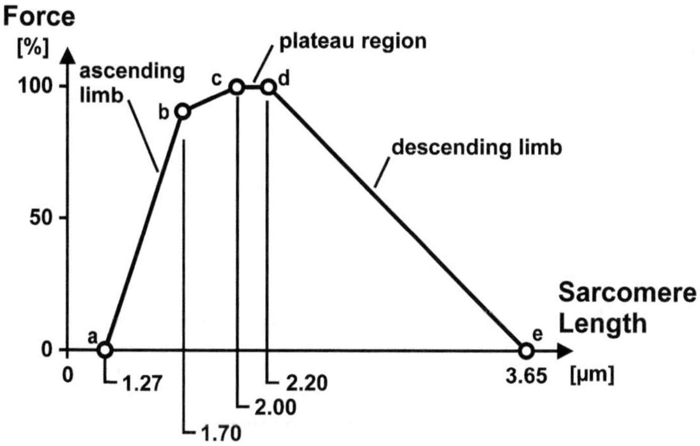

Fig. 1 Classic sarcomere force–length relationship for frog skeletal muscle as first described by Gordon et al. (1966). Note, the linear decrease of the descending limb of the force–length relationship where loss of force with increasing sarcomere lengths is associated with a decrease in actin-myosin filament overlap and the associated loss of cross-bridge attachment sites

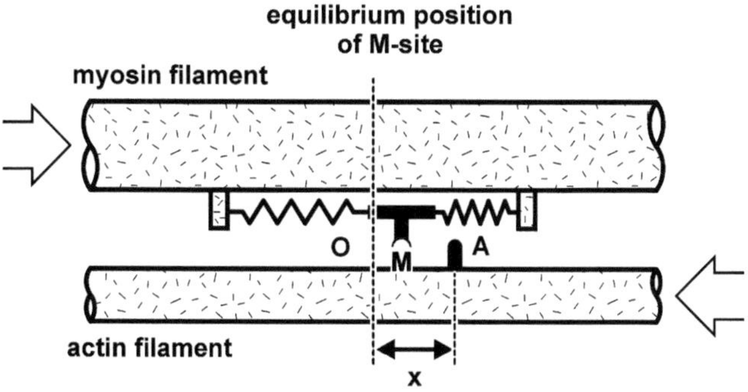

Fig. 2 Classic cross-bridge model with definition of Huxley's "x"-distance. The "x"-distance is defined as the distance from the cross-bridge (M) equilibrium position to the nearest attachment site (A) on the actin filament

Force, in the experiments by Gordon et al. (1966), was measured at the end of a single isolated frog fiber. Such fibers contain millions of sarcomeres intricately arranged and connected to each other in parallel and in series, thus the contribution of a given sarcomere to the force measured at the end of the fiber cannot be determined as such a system is statically indeterminate. In addition, forces in their preparations often did not reach a steady-state (constant) value, especially at fiber lengths corresponding to the descending limb of the force–length relationship. This so-called force creep (slow and continuous increase in force – Fig. 3) was associated with the redistribution of sarcomere lengths on the descending limb of the force–length relationship, and was presumed to be caused by sarcomere lengths instabilities, as first suggested by Hill (1953). Force, in these situations, was approximated by the

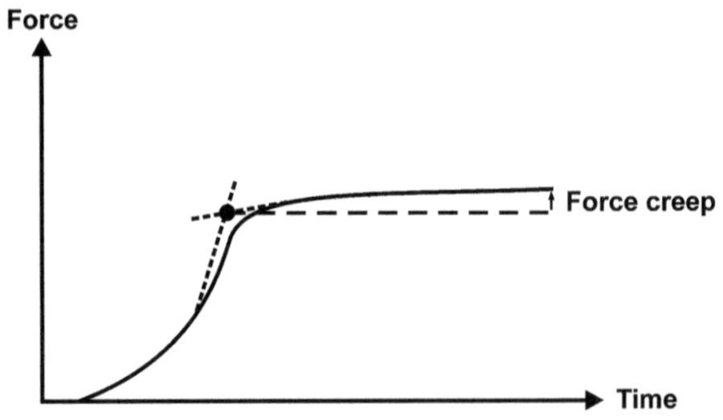

Fig. 3 Force–time curve obtained by Gordon et al. (1966) and the associated "creep" phase (slow increase in force following full activation). The "*dot*" indicates the force used by Gordon et al. (1966) in their attempt to capture the steady-state force value

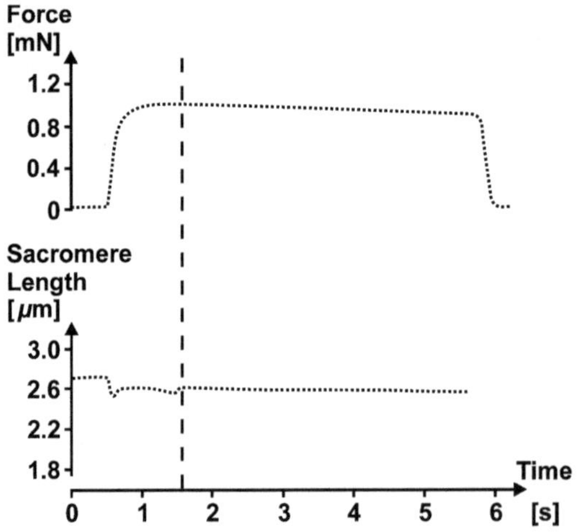

Fig. 4 Force–time and sarcomere length–time histories of an isometric contraction of a single frog fiber fixed at its ends. Note the lack of any force "creep" behavior of the isometric contraction on the descending limb of the force–length relationship (sarcomere length about 2.6 μm). Note also the steady sarcomere length traces following the transient phase of activation indicating that there is no need to use a segment clamp approach for steady-state conditions on the descending limb of the force–length relationship

intersection of two straight lines fitted to the activation part and the creep part of the unsteady force–time traces (Fig. 3). Later experiments demonstrated that single fiber forces are perfectly stable on the descending limb of the force length relationship, and sarcomere lengths remain perfectly constant if the fiber is held fixed at its ends (Lee and Herzog 2008, Fig. 4). In the study by Gordon et al. (1966), one end of the fiber was not fixed, but its position was continuously changed to hold an apparently changing mid-section fiber length constant. It is assumed that the continuous change in fiber length that was part of this experiment caused the force "creep" rather than any sarcomere length instability. But whatever the detailed reason for the force "creep", steady-state force was not achieved in their preparations (Gordon et al. 1966), and thus the force results must be viewed with this limitation in mind.

Similarly, sarcomere length was not measured directly, but was approximated for a mid-segment, by dividing the length of the mid-segment by the estimated number of longitudinally arranged sarcomeres within that segment. Sarcomere lengths in other parts of the fiber were allowed to change and were, to a certain degree, dependent on the fiber length changes that were continuously imposed on the preparation to keep the mid-section of the fiber at a constant length. Thus, sarcomere lengths within this preparation were not at steady-state, and direct measurement of sarcomere lengths within the mid-segment were not made, thus some could have shortened at the expense of others. In fact, Gordon et al. (1966) attributed some of the force creep to possible rearrangements of sarcomere lengths within the clamped mid-section of the fiber.

In summary, Gordon et al. (1966) did not measure sarcomere forces, but rather they measured a constantly changing force at the end of a fiber. Furthermore, they did not measure sarcomere lengths (or their distribution) within the target segment, and they did not control length changes outside the target segment. Therefore, their results, which were taken as representing the sarcomere force–length properties, must be interpreted with caution and should be checked with independent and direct measurements of mechanically isolated sarcomeres.

Furthermore, an overwhelming amount of experimental evidence suggests that the force–length properties of muscles and fibers are not unique but depend on the history of contraction (e.g., Abbott and Aubert 1952; Edman et al. 1982; Herzog and Leonard 2002), a result that cannot be accommodated within the frame work of the traditional cross-bridge theory (Huxley 1957; Walcott and Herzog 2008; Huxley and Simmons 1971). The word "traditional" refers to all cross-bridge models with rate constants of attachment and detachment that are dependent on cross-bridge strain, or Huxley's "x" distance, independent of the number of cross-bridge states. If the force–length property for sarcomeres was also not unique, but would depend, for example, on the contractile history, then one of the most basic assumptions underlying the cross-bridge theory, the idea that steady-state isometric force is uniquely determined by sarcomere length, would be violated.

Therefore, the purpose of this series of studies was to determine steady-state isometric forces of sarcomeres at different lengths and for different contractile histories preceding the steady-state conditions. All studies were performed using isolated myofibrils from the rabbit psoas. Myofibrils are sub-cellular organelles with all sarcomeres arranged mechanically in series (Fig. 5). Therefore, forces measured at the ends of a myofibril represent the forces of all and each sarcomere, and individual sarcomere lengths (or half-sarcomere lengths; (Telley et al. 2006a, b) can be measured readily with our setup. Therefore, the force in a specific sarcomere can be related to the corresponding sarcomere length, and if steady-state forces and sarcomere lengths are achieved, the corresponding single sarcomere force–length relationship can be measured directly.

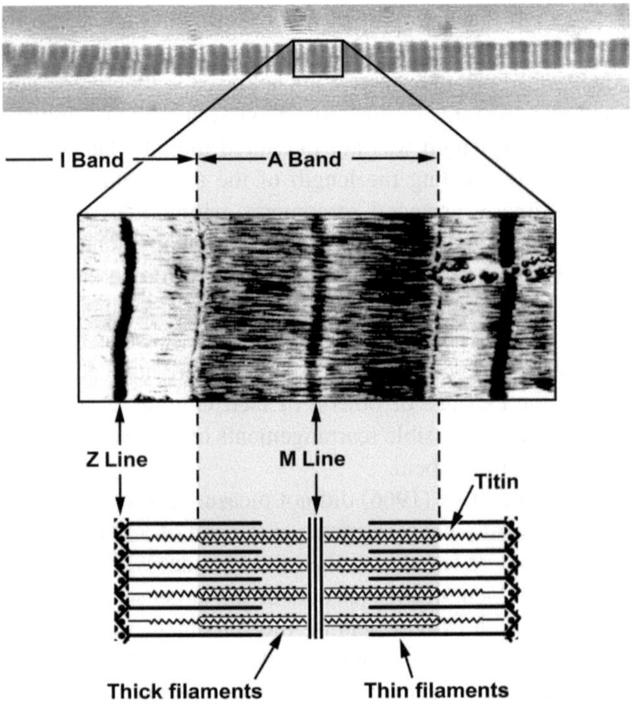

Fig. 5 Micrograph of a myofibril (*top panel*) showing the in series arrangement of sarcomeres in this preparation. A single sarcomere is expanded in the *middle panel* showing the z-lines and the I- and A-bands. The *bottom panel* illustrates the arrangement of the thick (myosin) and thin (actin) contractile filaments and the structural protein titin within the sarcomere

The specific question we wanted to answer was: is sarcomere length (or equivalently, the amount of actin and myosin overlap) on the descending limb of the force–length relationship uniquely related to the isometric, steady-state force as predicted by the cross-bridge theory? Based on results obtained on whole muscle (Abbott and Aubert 1952; Lee and Herzog 2002; Rousanoglou et al. 2007; Maréchal and Plaghki 1979) and single fiber preparations (Sugi and Tsuchiya 1988; Edman et al. 1978, 1982), we hypothesized that sarcomere forces were not uniquely determined by sarcomere length (for given activation conditions) but depended crucially on the contractile history: steady-state isometric sarcomere forces following stretch were hypothesized to be greater and those following shortening were hypothesized to be smaller than the corresponding forces for purely isometric reference contractions.

2 Methods

Extraction of single myofibrils and experimental set-up: Strips of rabbit psoas muscle were dissected and tied to small wooden sticks. These samples were stored in a rigor/glycerol (50/50) solution at –20°C. On the day of the experiments, the muscle strips were cut into pieces of about 2 mm length using a razor blade, and subsequently blended using previously described protocols (Rassier et al. 2003). The blended muscle was then put into a chamber whose bottom was a glass cover slip placed on top of an inverted microscope (Zeiss, Axiovert 200M, Germany). After sufficient time for stabilization (5–10 min), the rigor solution was replaced with a relaxing solution, and myofibrils in suspension were washed away leaving those attached to the bottom of the experimental chamber. A myofibril with a good striation pattern was then lifted from the glass coverslip and attached to a glass needle and motor at one end, and to one nano-lever of a pair (Bartoo et al. 1993) which allowed for myofibril force measurements at the other end. Forces were determined by the deflection of the nano-lever to which the myofibril was attached relative to its non-attached reference pair, and the known stiffness of the levers.

The image of the attached myofibril was projected onto a high-density linear photodiode array (Schafter and Kirschoff Model SK10680DJR, Hamburg, Germany, resolution of 6 nm) to give tracings of the myofibrillar striation pattern for identification of the A- and I-bands and the Z-lines. Sarcomere and half sarcomere lengths were calculated from Z-line to Z-line or from Z-line to the centroid of the A-band, respectively, or when the Z-lines could not be identified reliably, sarcomere lengths were determined from the centroids of adjacent A-bands.

Protocol: Once a myofibril was ready for mechanical testing, and a clear striation pattern could be observed, a ten-minute rest was given and then the relaxing solution was replaced by the activating solution causing contraction of the myofibril. Six myofibrils were tested isometrically at an average sarcomere length of 2.4 and 3.4 μm and the active and passive forces were determined. A further twelve myofibrils were activated isometrically at a short length, stretched while activated, and then held isometrically for another 30s until force transients had disappeared. Six of these myofibrils were stretched from a nominal average sarcomere length of 2.4–3.4 μm, while the remaining six myofibrils underwent a series of stretches, starting at different average sarcomere lengths and being subjected to variable stretch magnitudes (between 12 and 38% of the initial sarcomere length). For all myofibrils, the mid-diameter was measured using a calibrated eyepiece, and myofibril cross-sectional areas were calculated assuming a cylindrical myofibril shape.

Like the stretch tests, shortening tests were performed at room temperature (~ 22°C). Myofibrils (n = 11) were activated at an average nominal sarcomere length of 2.8 μm and then shortened at a speed of 0.1 μm/s/sarcomere to an average nominal sarcomere length of about 2.4 μm. Myofibrils were then held isometrically for 1 min, and deactivated. Stiffness was calculated during a quick stretch-release cycle (0.1 μm/sarcomere at a speed of 0.5 μm/s/sarcomere) performed just prior to deactivation. After a rest period of 5 min, myofibrils were reactivated at the final sarcomere length (2.4 μm) and stiffness was measured as above.

Forces were normalized by myofibril cross-sectional area and expressed as stress (kN/m^2). Stiffness was evaluated by determining the elastic modulus of the myofibrils, by dividing the myofibril stress by its strain (kN/m^2).

Force depression for a myofibril was defined as the difference in the steady-state isometric force following shortening, and the purely isometric reference contraction at 2.4 μm sarcomere length. Passive forces were not measured separately, as they have been shown to be negligible in this range of sarcomere lengths (<2.8 μm) (Bartoo et al. 1993; Joumaa et al. 2008a, b). Since all comparisons of reference and shortening test contractions were made at the same sarcomere length, passive forces, if present, would have been the same for the experimental and reference contractions.

Since myofibrils are formed of sarcomeres arranged in series, measuring the force at the end of a myofibril gives the instantaneous force in each sarcomere. Therefore, by measuring individual sarcomere lengths and individual sarcomere shortening, we could also determine force depression and stiffness for each individual sarcomere contained in the myofibrils. Also, half-sarcomeres have been shown to form independent units behaving non-uniformly within the same sarcomere (Telley et al. 2006a, b), thus we further investigated force depression and non-uniformity at the half-sarcomere level in myofibrils where all the Z-lines and A-bands were perfectly visible during the entire experiment.

Solutions: The rigor, relaxing and activating solutions were identical to those described previously in our studies (e.g., Rassier et al. 2003).

Analysis: Isometric, steady-state myofibril forces, and the associated individual sarcomere forces, were determined and plotted against individual sarcomere lengths for the purely isometric contractions, the isometric contractions preceding and the isometric contractions following myofibril stretch and shortening. For comparison across myofibrils and sarcomeres, forces were normalized relative to the cross-sectional area of the myofibrils and were expressed as a percentage of the purely isometric reference forces obtained at optimal sarcomere length (2.26–2.43 μm) for rabbit psoas myofibrils. Theoretically predicted sarcomere forces were calculated based on myofilament overlap (Gordon et al. 1966) and the known myofilament lengths of rabbit psoas (1.08 μm for actin and 1.65 μm for myosin – (Huxley and Peachey 1961)).

3 Results

3.1 *Force Enhancement*

We confirmed in this study that sarcomeres on the descending limb of the force–length relationship are stable, not only for purely isometric contractions, but also following myofibril stretch onto the descending limb of the force–length relationship (Fig. 6). We further confirmed most recent findings that the isometric steady-state forces following myofibril stretch are greater than the forces obtained without prior stretching at that same length (Fig. 7). New to the literature is the finding that individual sarcomere forces following stretch vastly exceed the purely isometric reference forces and the forces predicted theoretically based on the cross-bridge theory and

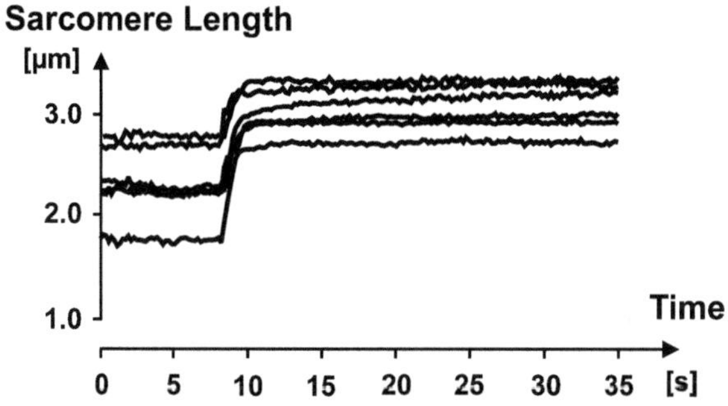

Fig. 6 Sarcomere lengths as a function of time for six sarcomeres of one myofibril. Shown are an initial isometric state (flat sarcomere traces), followed by myofibril stretching of about 0.8 μm/sarcomere (increasing sarcomere lengths), followed by a second isometric state (flat sarcomere traces). Note that following stretch all sarcomeres are on the descending limb of the force–length relationship. Note further that all sarcomere lengths are stable (constant) following stretch, and do not show the instabilities "predicted" in the literature

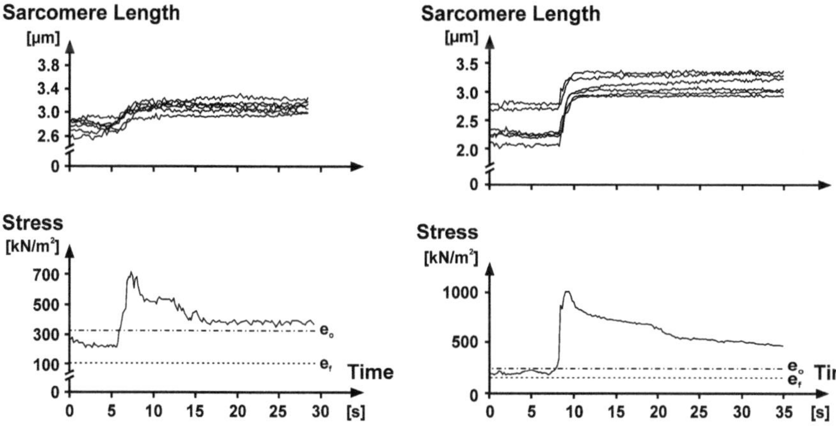

Fig. 7 Sarcomere lengths and forces normalized to cross-sectional area (stress) for two representative myofibrils that were stretched onto the descending limb of the force–length relationship. Note that the isometric force prior to stretch is smaller than the isometric force obtained at optimal sarcomere length (e_0 – plateau of the force–length relationship). Note further that force following stretch is greater than the isometric force obtained at the final (average) sarcomere length (e_f) and the optimal sarcomere length (e_0)

reported by Gordon et al. (1966). Steady-state sarcomere forces following stretch also exceeded the purely isometric forces on the plateau of the force–length relationship consistently (Fig. 8). Across all tests, sarcomere forces following stretch did not follow a trend of decreasing force with decreasing myofilament overlap, rather sarcomere forces of a magnitude equivalent to the purely isometric reference forces at optimal length could be observed for the entire range of the descending limb of the force–length relationship (Fig. 8). Similarly, the isometric steady state sarcomere forces at a given sarcomere length could take on a large range of values if they were

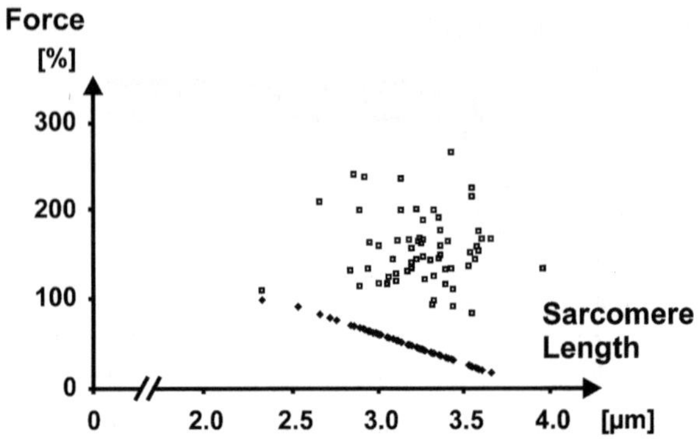

Fig. 8 Predicted sarcomere force–length relationship based on the sliding filament theory (*filled diamonds* – (Gordon et al. 1966)) and experimentally obtained isometric steady-state forces of individual sarcomeres following stretches of various magnitudes (*open squares*). Note that the isometric steady-state forces following stretch are always greater than the corresponding purely isometric reference forces. Note further, that the isometric forces following sarcomere stretching often exceed the purely isometric reference forces at optimal sarcomere length (100% force)

stretched prior to the isometric contraction. Finally, the amount of force enhancement, as well as the enhancement of forces above the plateau of the force–length relationship, often was in the hundreds of percent, values that were much larger than ever reported before in single fiber (Edman et al. 1978, 1982) or whole muscle preparations (Abbott and Aubert 1952; Lee and Herzog 2002).

3.2 *Force Depression*

The isometric steady-state forces following shortening were significantly smaller than the corresponding isometric reference forces (mean 31 ± 4%) for all myofibrils (Fig. 9), thereby showing the characteristic force depression previously observed in muscle (Maréchal and Plaghki 1979; Herzog and Leonard 1997) and single fiber preparations (Sugi and Tsuchiya 1988). The average steady-state isometric forces following shortening were also 9.5% smaller than the isometric forces preceding shortening, despite the fact that following shortening, actin-myosin filament overlap was vastly increased (Gordon et al. 1966; Huxley and Peachey 1961).

Following activation, sarcomere striation patterns remained perfect in five myofibrils containing 62 sarcomeres; these sarcomeres were used for force depression analysis of individual sarcomeres. Sixty of these 62 sarcomeres showed force depression (Fig. 10). Because of differences in the amount of shortening between sarcomeres, sarcomeric force depression varied within the same myofibril and between myofibrils, ranging from 0 to over 50% of the corresponding isometric reference forces. Furthermore, force depression in individual sarcomeres was positively correlated to the sarcomere shortening magnitude; increasing with increasing shortening magnitude (Fig. 11).

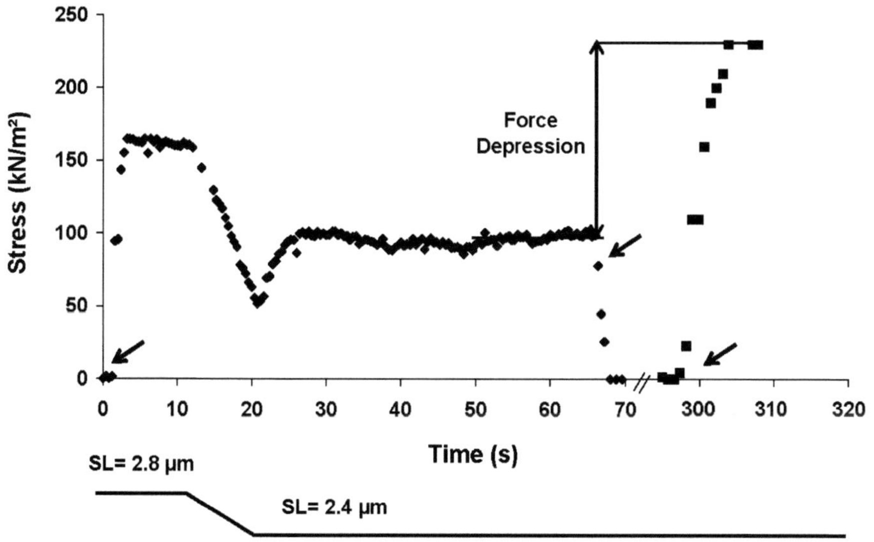

Fig. 9 Typical myofibril response when activated at an average sarcomere length of about 2.8 μm, then shortened to an average sarcomere length of about 2.4 μm, deactivated and then activated again. Note that the force after active shortening is smaller than the purely isometric force at the same (final) length and the (initial) length preceding shortening. *Arrows* from left to right along the time axis indicate the time of activation (initial length = 2.8 μm), deactivation (final length = 2.4 μm) and re-activation at the final length (2.4 μm)

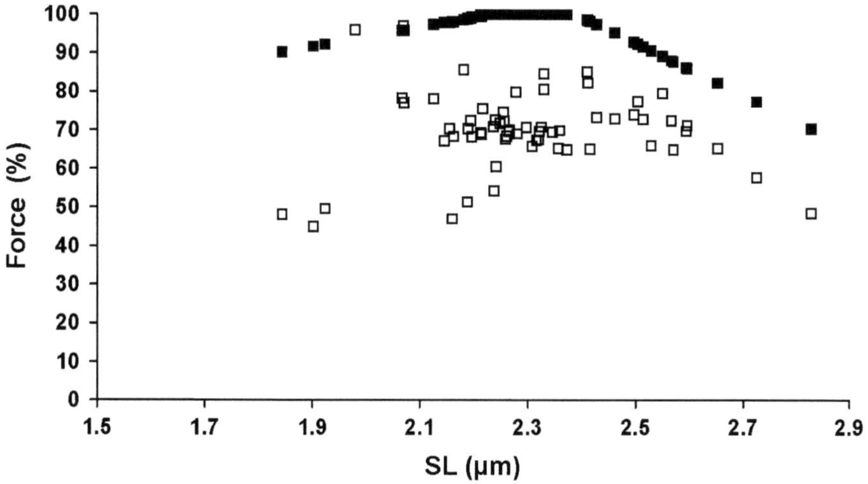

Fig. 10 Expected isometric steady-state forces for single sarcomeres (n = 62) based on the sliding filament and cross-bridge theory and the actin and myosin filament lengths for rabbit skeletal muscle (*filled squares*), and the corresponding sarcomere forces following myofibril shortening (*open squares*). Steady-state isometric forces are smaller in all but two sarcomeres following shortening, compared to the expected and the measured isometric reference forces

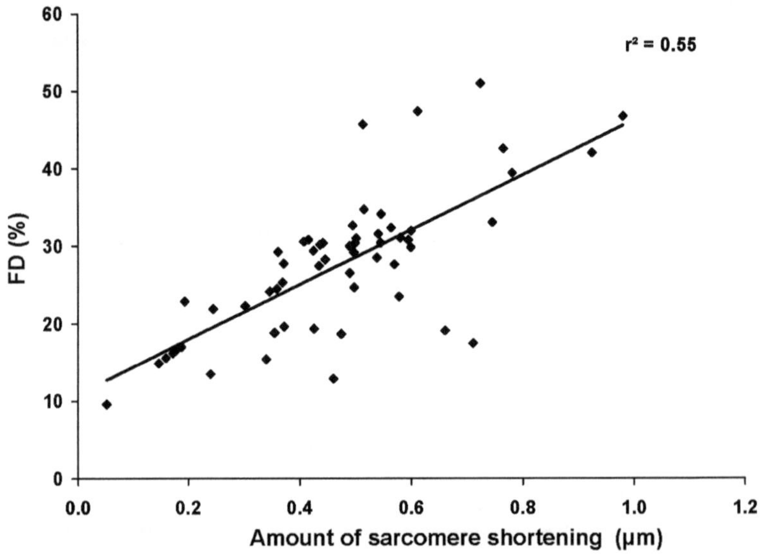

Fig. 11 Force depression as a function of shortening magnitude for individual sarcomeres (n = 60; the two sarcomeres that did not show force depression are excluded). There is a significant correlation ($r^2 = 0.55$, $p < 0.05$) between force depression and shortening magnitude

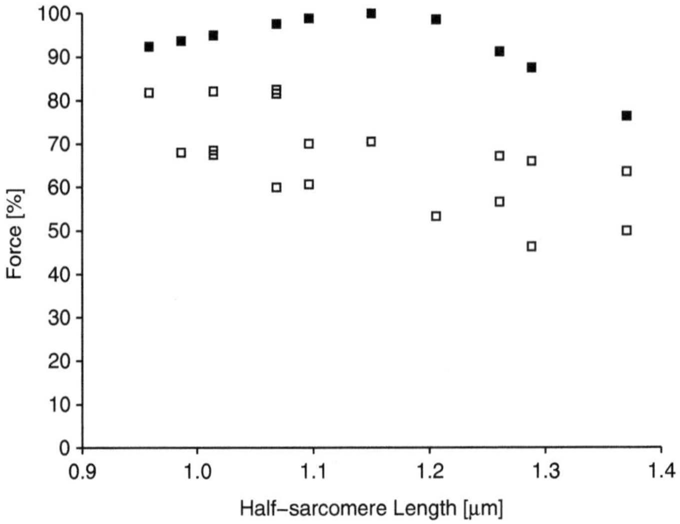

Fig. 12 Expected isometric steady-state forces for single half-sarcomeres (n = 18) based on the sliding filament and cross-bridge theory (*filled squares*), and the corresponding half-sarcomere forces following myofibril shortening (*open squares*). Steady-state forces following shortening are smaller in all half-sarcomeres compared to the corresponding purely isometric reference forces at the same lengths

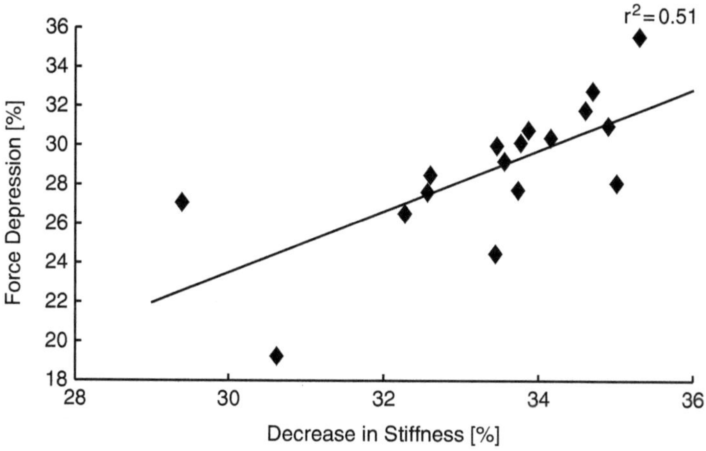

Fig. 13 Force depression as a function of decreases in stiffness for individual sarcomeres (n = 18). The striation pattern could not be tracked in all individual sarcomeres after activation and during quick stretches and therefore the stiffness measurements could only be done in 18 sarcomeres from two myofibrils. There is a correlation between the decrease in stiffness and force depression ($r^2 = 0.51$, $p < 0.05$)

During shortening, all sarcomeres shortened, but not by the same amount. Sarcomere lengths before and after shortening were non-uniform with the same mean standard deviation (0.11 ± 0.06 μm) before and after shortening.

Half-sarcomeres (n = 18), obtained from one myofibril, also showed force depression (Fig. 12). As observed for the full sarcomeres, the half sarcomere force depression was positively related to the half sarcomere shortening magnitude ($r^2 = 0.55$). Non-uniformities of half sarcomere lengths were the same before (SD = 0.13 μm) and after (SD = 0.14 μm) shortening, and half sarcomere shortening tended to be greater for initially long half sarcomeres, consistent with the observations on the full sarcomere level.

Stiffness in the force depressed state was significantly smaller (329 ± 36 kN/m^2) than that measured for the purely isometric reference contractions (470 ± 83 kN/m^2). Stiffness in individual sarcomeres was also decreased and was correlated with the amount of force depression (Fig. 13).

4 Discussion

The relationship between the steady-state isometric force and length on the sarcomere level was determined in the classic study by Gordon et al. (1966). The results of that study agreed perfectly with the theoretical predictions of the sliding filament (Huxley and Niedergerke 1954; Huxley and Hanson 1954) and cross-bridge theories (Huxley 1957; Huxley and Simmons 1971; Huxley 1969), and have been accepted as textbook knowledge. Specifically, the plateau and linear descending limb of the force–length relationship agreed with the idea that isometric steady-state force was uniquely determined by sarcomere length. However, neither sarcomere force nor sarcomere lengths were directly measured in that study, and neither force nor sarcomere lengths along the fiber reached steady-state. Therefore, the linear descending limb observed by Gordon et al. (1966) does not necessarily represent the relationship between an individual sarcomere's length and its associated steady-state isometric force.

Here, we measured for the first time the steady-state isometric forces of mechanically isolated sarcomeres and the associated sarcomere lengths in isolated myofibril preparations. Measurements were obtained when sarcomere lengths and the associated myofibril forces had reached a constant value after its transient changes either following activation or following myofibril length changes. Therefore, true steady-state isometric conditions had been achieved at the time of measurement for all sarcomeres in the myofibrils. Since sarcomeres in myofibrils are strictly arranged mechanically in series, this setup is equivalent to a mechanical isolation of individual sarcomeres, and the forces measured at the end of the myofibril are sustained by each and all sarcomeres in the preparation Therefore, forces of mechanically isolated sarcomeres, and the associated sarcomere and half sarcomere lengths, could be directly related to each other, thus providing direct measures of force in single sarcomeres and the corresponding (half-) sarcomere lengths.

The results of our studies indicate that there is a vast region of possible steady-state isometric forces at a given sarcomere length, and that sarcomeres of vastly different lengths can easily produce the same amount of force, and therefore can co-exist in a steady-state on the descending limb of the force–length relationship (Figs. 6 and 7). The idea that different sarcomere lengths on the descending limb of the force–length relationship result in an unstable contraction (Hill 1953; Allinger et al. 1996; Zahalak 1997; Denoth et al. 2002; Morgan 1994; Morgan et al. 2000) was not supported, and dynamic sarcomere length changes producing "temporary" stability by virtue of the force–velocity properties (Hill 1938) were also not observed, and were not required. Rather, sarcomeres of vastly different lengths, and thus vastly different myofilament overlap regions, could perfectly co-exist at steady-state on the descending limb of the force–length relationship. This result implies that there must be a mechanism that determines steady-state isometric sarcomere forces that is independent of the myofilament overlap.

There are several trivial reasons of how sarcomeres of different lengths could co-exist at steady-state on the descending limb of the force–length relationship. First, adjacent sarcomeres could have different amounts of contractile proteins. For example, let's assume that one sarcomere has 20% more contractile proteins than its neighbouring sarcomere. Thus, the bigger sarcomere would be expected to have approximately the same number of attached cross-bridges as the smaller sarcomere if its myofilament overlap was reduced by about 20% compared to the smaller sarcomere, thus its sarcomere length would be about 20% greater. If a contractile protein imbalance between sarcomeres was the reason for the observed sarcomere length non-uniformity, then it would follow that the relative lengths of these two sarcomeres on the descending limb would remain approximately constant.

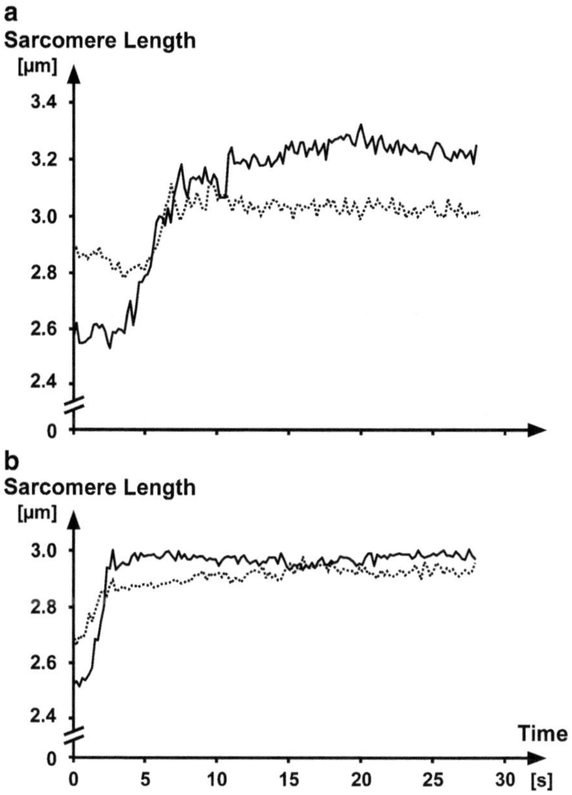

Fig. 14 Sarcomere length as a function of time for two sarcomeres each from the same myofibril. The sarcomeres are stretched on the descending limb of the force–length relationship, and while stretched, the initially short sarcomere (solid line) becomes long relative to the other (initially long, dotted line) sarcomere. This sarcomere length reversal suggests that passive forces, or a difference in the amount of contractile filaments, cannot explain why sarcomeres of different length on the descending limb of the force–length relationship can support the same force

However, that is clearly not the case as we observed sarcomere length switching during myofibril stretching on the descending limb of the force–length relationship: that is, an initially short sarcomere became longer during stretch than a neighbouring initially long sarcomere (Fig. 14). Clearly, such observations of switching sarcomere lengths, only two of which are illustrated as an example in Fig. 14, eliminates the theory of sarcomere lengths non-uniformities caused by an imbalance of contractile proteins.

A further argument against the idea that a contractile protein imbalance can explain equal force production for sarcomeres of vastly different lengths is the observation of half-sarcomere length non-uniformities. For example, in Fig. 12, we show half-sarcomere forces that are similar across half-sarcomere lengths ranging from 1.0 to 1.4 μm. Half-sarcomere length non-uniformities for half-sarcomeres belonging to the same sarcomere have also been observed previously (Telley et al. 2006a, b).

However, half-sarcomeres of the same sarcomere share the same myosin filaments, and the ratio of myosin to actin filaments is constant (Squire 1992), thus corresponding half-sarcomeres must contain the same number of contractile proteins, but corresponding half-sarcomeres of vastly different lengths can support the same stead-state isometric force for given activation conditions. An imbalance of contractile proteins obviously cannot be the cause for this observation.

A second trivial reason for the observed sarcomere length non-uniformities at steady state on the descending limb of the force–length relationship could be a systematic difference in the passive forces of adjacent sarcomeres. Obviously, a sarcomere with a high passive force would need less active force than a sarcomere with a small passive force. Therefore, the sarcomere with the high passive force could have less myofilament overlap (or equivalently greater sarcomere length) and still be able to produce the same force as the sarcomere with the small passive force and greater myofilament overlap. However, as argued above, if differences in passive forces were responsible for the observed sarcomere length differences on the descending limb of the force–length relationship, one would not expect to see sarcomere length reversals from one steady-state situation to the next, as illustrated in Fig. 14. Furthermore, passive forces in rabbit psoas myofibrils are small and reach about 5–10% of the maximal isometric forces at sarcomere lengths of approximately 3.2 μm (Bartoo et al. 1997; Joumaa et al. 2007, 2008a). Since we observed similar forces for individual sarcomeres at optimal length, 2.4 μm, and up to lengths where myofilament overlap was lost (about 3.8–4.0 μm; Fig. 8), passive forces cannot account for the expected differences in the active forces based on myofilament overlap considerations.

Although this study was not aimed at elucidating the mechanisms for the multitude of possible steady-state forces at a given sarcomere length, it is fair to speculate that the contractile history plays an important role in explaining the results. When myofibrils were held isometrically, a force–length relationship approximating that predicted by Gordon et al. (1966) was obtained, while stretch prior to isometric contractions resulted in increases and shortening resulted in decreases of force compared to isometric. Such observations of steady-state residual force enhancement following stretching of muscle fibers or whole muscles, and force depression following shortening, have been observed for a long time (e.g., Abbott and Aubert 1952; Edman et al. 1978, 1982, Herzog and Leonard 2002; Herzog and Leonard 1997). However the detailed mechanisms underlying this history dependence of steady-state isometric force production remain unclear.

One potential mechanism of force enhancement following muscle stretching has been associated with the engagement of an "additional" passive force. In support of this argument, it has been demonstrated that the structural protein titin, which acts as a molecular spring in muscle tissue and produces much of the passive force in rabbit myofibrils, has a calcium (and therefore activation) dependent regulation of stiffness and so contributes to the increased sarcomere force following stretch (Joumaa et al. 2007, 2008b, Labeit et al. 2003). However, it has been ascertained that this "passive" contribution to the increased sarcomere forces is small and may account for a few percent of the total force enhancement but cannot account for the hundreds of percent of force increase observed here following sarcomere stretch. Thus, it is reasonable to assume that active mechanisms must play an important role (Joumaa et al. 2008a; Herzog et al. 2006; Herzog and Leonard 2006). Since force enhancement following stretch is not associated with a systematic increase in muscle, fiber or myofibril stiffness (e.g., Sugi and Tsuchiya 1988), it may be assumed that the increase in force is associated with an increase in the force/attached cross-bridge. This would imply that cross-bridges are able to enter an attached state through stretching that cannot be reached systematically for isometric or shortening contractions. Such cross-bridge behavior, of course, is perfectly permissible for the transient states of classic cross-bridge models (Huxley 1957; Huxley and Simmons 1971), but has not been considered to be part of regular cross-bridge behavior at steady-state (Walcott and Herzog 2008).

Another possibility for the vast force enhancement observed here for mechanically isolated sarcomeres and myofibrils is that there is an engagement of a passive force that is not only calcium (and thus activation) dependent, as argued above, but also depends either on active force or cross-bridge attachment to actin binding sites and the associated configurational changes of the regulatory proteins troponin and tropomyosin. We have strong evidence that much of the sarcomeric force enhancement is indeed associated with such a passive, force- or attachment-dependent increase in steady-state sarcomere force (Leonard et al. 2009), and have determined that such a force is critically dependent on the presence of titin. Therefore, a preliminary hypothesis suggests that titin is attaching to actin in a force- or cross-bridge-attachment-dependent manner (Fig. 15). Preliminary data confirm this hypothesis and they suggest that titin attachment decreases the spring length, and thus increases titin's stiffness by a factor of 3 or 4, thereby providing most of the force enhancement observed in single sarcomeres.

In contrast to the results on force enhancement, the decrease in sarcomere force following myofibril shortening, and force depression in single fiber and muscle preparations, is associated with a concomitant decrease in stiffness (Sugi and Tsuchiya 1988), thereby suggesting that the loss of force following active muscle shortening is caused by a decrease in the proportion of attached cross-bridges, rather than a decrease in the force/attached cross-bridge. In agreement with these observations in fiber and muscle preparations, myofibril and sarcomere stiffness decreased by 30% in proportion with force depression (31%) in our studies, suggesting that the same mechanism also holds true on the sarcomere level, and the proportionality in decrease of force and stiffness further suggests that all of the force depression is caused by a decreased proportion of attached cross-bridges.

Maréchal and Plaghki (1979) argued 30 years ago that force depression was caused by a stress-dependent inhibition of cross-bridge attachment in the myofilament overlap zone that was newly formed during muscle shortening. If so, forces in the depressed state should not be smaller than the forces observed at the initial length, that is the length prior to shortening. However, we observed an almost 10% decrease in the depressed force compared to the steady-state, isometric force at the initial length, thereby suggesting that Maréchal and Plaghki's (1979) suggestion cannot be entirely correct and that an additional force depressing factor must be at play,

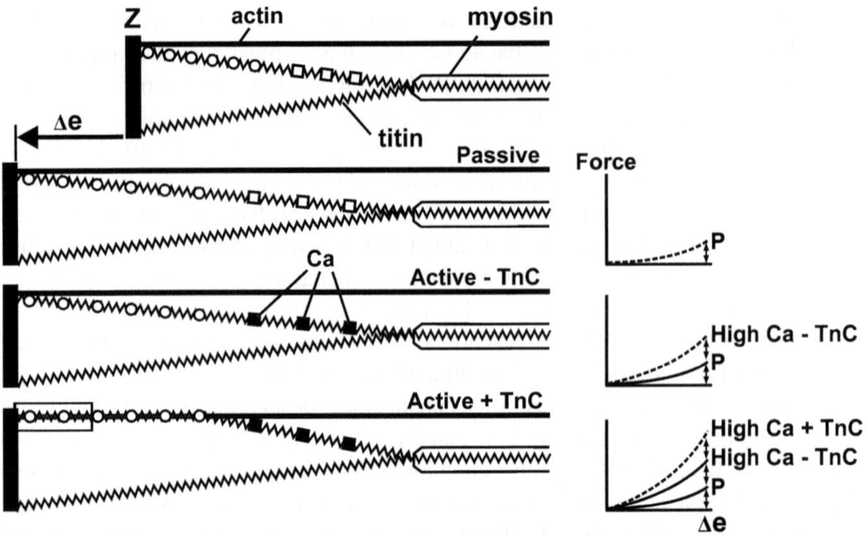

Fig. 15 Preliminary hypothesis about titin's role in force enhancement. The *four panels* show schematic representations of half sarcomeres with the contractile proteins, actin and myosin, and the structural protein, titin. The *top panel* shows the half sarcomere at a short length. The *second panel* illustrates a passive stretch of the half sarcomere by an amount of Δe. There is some passive force. The *third panel* illustrates an active stretch (high calcium concentration) but no active force production (because of troponin C depletion from actin). The force measured for this condition is higher than for the purely passive stretch possibly because of calcium binding to titin and an associated increase in titin stiffness. Finally, the *bottom panel* illustrates an active stretch (high calcium concentration) with active force production and the passive (only) force is increased again compared to the previous condition. We speculate that this further increase in passive force is associated with a force-dependent attachment of titin to actin, as illustrated schematically in the *bottom panel*, thereby reducing titin's spring length, and thus making titin stiffer upon stretch

at least in isolated myofibrils and individual sarcomeres, as tested here. Stiffness of myofibrils and sarcomeres in the force depressed and initial steady-state was about the same. In combination with the force results, these observations suggest that force depression is caused by an inhibition of cross-bridge attachment and a decrease of the average force per cross-bridge, a result that is also consistent with recent observations on intact frog fibers treated with 2,3-butanedione monoxime (Lee and Herzog 2009).

In summary, we measured first ever steady-state isometric sarcomere forces on the descending limb of the force–length relationship where forces and sarcomere length were measured directly in mechanically isolated sarcomeres and contractile history was changed in a systematic way. We found that, in contrast to the predictions of the sliding filament and cross-bridge theory, sarcomere forces are not uniquely determined at a given sarcomere length and actin-myosin overlap, but depend strongly on the contractile history. Although the detailed mechanisms for this sarcomere force enhancement and force depression remain unclear, the idea of a linear descending limb of the isometric sarcomere force–length property, as described in the classic work by Gordon et al. (1966), does not hold for single sarcomeres. Rather, there are an infinite number of isometric, steady-state sarcomere force–length relationships that can be obtained by changing the contractile conditions. Increasing forces are obtained with increasing stretch magnitudes and decreasing forces are obtained with increasing shortening magnitudes and/or increasing forces during the shortening phase.

The force enhancements observed in individual sarcomeres were much greater (by a factor of ten on average) than ever observed in other preparations and this result needs careful attention in the future. We hypothesize, and have strong preliminary evidence, that force enhancement is caused to a large extent by a force-dependent attachment of titin to actin, thereby shortening its resting spring length and thus increasing its stiffness. However, it is not clear why this mechanism should produce so much greater force increases in single sarcomeres than single fibers or muscles.

Force depression in mechanically isolated sarcomeres and single myofibrils is associated with a proportional decrease in sarcomere and myofibril stiffness, suggesting that much of the force depression following active sarcomere shortening can be explained by a decrease in the proportion of attached cross-bridges. In accordance with Maréchal and Plaghki (1979), we hypothesize that active shortening is associated with a stress-induced inhibition of cross-bridge attachments to actin, possibly because of actin deformations (Daniel et al. 1998), but more likely because of stress-related changes in the regulatory proteins, troponin and tropomyosin.

Acknowledgements The Canadian Institutes of Health Research, The Canada Research Chair Program (for Molecular and Cellular Biomechanics), the Natural Sciences and Engineering Research Council of Canada, and the Canada Foundation for Innovation.

References

Abbott BC, Aubert XM (1952) The force exerted by active striated muscle during and after change of length. J Physiol 117:77–86

Allinger TL, Herzog W, Epstein M (1996) Force–length properties in stable skeletal muscle fibers – theoretical considerations. J Biomech 29:1235–1240

Bartoo ML, Popov VI, Fearn LA, Pollack GH (1993) Active tension generation in isolated skeletal myofibrils. J Muscle Res Cell Motil 14:498–510

Bartoo ML, Linke WA, Pollack GH (1997) Basis of passive tension and stiffness in isolated rabbit myofibrils. Am J Physiol Cell Physiol 273:C266–C276

Daniel TL, Trimble AC, Chase PB (1998) Compliant realignment of binding sites in muscle: transient behaviour and mechanical tuning. Biophys J 74:1611–1621

Denoth J, Stussi E, Csucs G, Danuser G (2002) Single muscle fibre contraction is dictated by inter-sarcomere dynamics. J Theor Biol 216:101–122

Edman KAP, Elzinga G, Noble MIM (1978) Enhancement of mechanical performance by stretch during tetanic contractions of vertebrate skeletal muscle fibres. J Physiol 281:139–155

Edman KAP, Elzinga G, Noble MIM (1982) Residual force enhancement after stretch of contracting frog single muscle fibers. J Gen Physiol 80:769–784

Gordon AM, Huxley AF, Julian FJ (1966) The variation in isometric tension with sarcomere length in vertebrate muscle fibres. J Physiol 184:170–192

Herzog W, Leonard TR (1997) Depression of cat soleus forces following isokintic shortening. J Biomech 30(9):865–872
Herzog W, Leonard TR (2002) Force enhancement following stretching of skeletal muscle: a new mechanism. J Exp Biol 205:1275–1283
Herzog W, Leonard TR (2006) Response to the (Morgan and Proske) letter to the editor by Walter Herzog (on behalf of the authors) and Tim Leonard. J Physiol 578.2:617–620
Herzog W, Kamal S, Clarke HD (1992) Myofilament lengths of cat skeletal muscle: theoretical considerations and functional implications. J Biomech 25:945–948
Herzog W, Lee EJ, Rassier DE (2006) Residual force enhancement in skeletal muscle. J Physiol 574:635–642
Hill AV (1938) The heat of shortening and the dynamic constants of muscle. Proc R Soc Lond 126:136–195
Hill AV (1953) The mechanics of active muscle. Proc R Soc Lond 141:104–117
Huxley AF (1957) Muscle structure and theories of contraction. Prog Biophys Biophys Chem 7:255–318
Huxley HE (1969) The mechanism of muscular contraction. Science 164:1356–1366
Huxley HE, Hanson J (1954) Changes in cross-striations of muscle during contraction and stretch and their structural implications. Nature 173:973–976
Huxley AF, Niedergerke R (1954) Structural changes in muscle during contraction. Interference microscopy of living muscle fibres. Nature 173:971–973
Huxley AF, Peachey LD (1961) The maximum length for contraction in vertebrate striated muscle. J Physiol 156:150–165
Huxley AF, Simmons RM (1971) Proposed mechanism of force generation in striated muscle. Nature 233:533–538
Joumaa V, Rassier DE, Leonard TR, Herzog W (2007) Passive force enhancement in single myofibrils. Pflügers Arch – Eur J Physiol 455:367–371
Joumaa V, Leonard TR, Herzog W (2008a) Residual force enhancement in myofibrils and sarcomeres. Proc R Soc B 275:1411–1419
Joumaa V, Rassier DE, Leonard TR, Herzog W (2008b) The origin of passive force enhancement in skeletal muscle. Am J Physiol Cell Physiol 294:C74–C78
Labeit D, Watanabe K, Witt C et al. (2003) Calcium-dependent molecular spring elements in the giant protein titin. Proc Natl Acad Sci U S A 100:13716–13721
Lee HD, Herzog W (2002) Force enhancement following muscle stretch of electrically and voluntarily activated human adductor pollicis. J Physiol 545:321–330
Lee EJ, Herzog W (2008) Residual force enhancement exceeds the isometric force at optimal sarcomere length for optimized stretch conditions. J Appl Physiol 105:457–462
Lee EJ, Herzog W (2009) Shortening-induced force depression is primarily caused by cross-bridges in strongly bound states. J Biomech 42(14):2336–2340
Leonard TR, Herzog W (2010) Regulation of muscle force in the absence of actin-myosin based cross-bridge interaction. Am J Physiol Cell Physiol [Epub ahead of print, doi:10.1152/ajpcell.00049.2010]
Maréchal G, Plaghki L (1979) The deficit of the isometric tetanic tension redeveloped after a release of frog muscle at a constant velocity. J Gen Physiol 73:453–467
Morgan DL (1994) An explanation for residual increased tension in striated muscle after stretch during contraction. Exp Physiol 79:831–838
Morgan DL, Whitehead NP, Wise AK, Gregory JE, Proske U (2000) Tension changes in the cat soleus muscle following slow stretch or shortening of the contracting muscle. J Physiol 522.3:503–513
Rassier DE, Herzog W, Pollack GH (2003) Dynamics of individual sarcomeres during and after stretch in activated single myofibrils. Proc R Soc Lond B 270:1735–1740
Rayment I, Holden HM, Whittaker M et al. (1993) Structure of the actin-myosin complex and its implications for muscle contraction. Science 261:58–65
Rousanoglou EN, Oskouei AE, Herzog W (2007) Force depression following muscle shortening in sub-maximal voluntary contractions of human adductor pollicis. J Biomech 40:1–8

Squire JM (1992) Muscle filament lattices and stretch-activation: the match-mismatch model reassessed. J Muscle Res Cell Motil 13:183–189

Sugi H, Tsuchiya T (1988) Stiffness changes during enhancement and deficit of isometric force by slow length changes in frog skeletal muscle fibres. J Physiol 407:215–229

Telley IA, Denoth J, Stussi E, Pfitzer G, Stehle R (2006a) Half-sarcomere dynamics in myofibrils during activation and relaxation studied by tracking fluorescent markers. Biophys J 90:514–530

Telley IA, Stehle R, Ranatunga KW, Pfitzer G, Stussi E, Denoth J (2006b) Dynamic behaviour of half-sarcomeres during and after stretch in activated rabbit psoas myofibrils: sarcomere asymmetry but no 'sarcomere popping'. J Physiol 573:173–185

TR Leonard and W Herzog (2010) Regulation of muscle force in the absence of actin-myosin based cross-bridge interaction. Am.J Physiol Cell Physiol [Epub ahead of print, doi:10.1152/ajpcell.00049.2010]

Walcott S, Herzog W (2008) Modeling residual force enhancement with generic cross-bridge models. Math Biosci 216:172–186

Walker SM, Schrodt GR (1973) I segment lengths and thin filament periods in skeletal muscle fibers of the Rhesus monkey and the human. Anatomical Record 178:63–81

Zahalak GI (1997) Can muscle fibers be stable on the descending limbs of their sarcomere length-tension relations? J Biomech 30:1179–1182

Extraction and Replacement of the Tropomyosin–Troponin Complex in Isolated Myofibrils

Beatrice Scellini, Nicoletta Piroddi, Corrado Poggesi, and Chiara Tesi

Abstract Tropomyosin (Tm) is an essential component in the regulation of striated muscle contraction. Questions about Tm functional role have been difficult to study because sarcomere Tm content is not as easily manipulated as Troponin (Tn). Here we describe the method we recently developed to replace Tm-Tn of skeletal and cardiac myofibrils from animals and humans to generate an experimental model of homogeneous Tm composition and giving the possibility to measure a wide range of mechanical parameters of contraction (e.g. maximal force and kinetics of force generation). The success of the exchange was determined by SDS–PAGE and by mechanical measurements of calcium dependent force activation on the reconstituted myofibrils. In skeletal and cardiac myofibrils, the percentage of Tm replacement was higher than 90%. Maximal isometric tension was 30–35% lower in the reconstituted myofibrils than in control myofibrils but the rate of force activation (k_{ACT}) and that of force redevelopment (k_{TR}) were not significantly changed. Preliminary results show the effectiveness of Tm replacement in human cardiac myofibrils. This approach can be used to test the functional impact of Tm mutations responsible for human cardiomyopathies.

Keywords Regulatory proteins • Tropomyosin • Troponin • Myofibrils • Cardiac • Skeletal

1 Introduction

Contraction and relaxation of striated muscles are regulated by the thin filament proteins tropomyosin (Tm) and troponin (Tn). Tm is a dimer of two α-helical chains (each 284 residues long and about 33 kDa molecular mass) that wrap

C. Tesi (✉)
Dipartimento di Scienze Fisiologiche, Università di Firenze, Viale Morgagni 63, Firenze I-50134, Italy
e-mail: chiara.tesi@unifi.it

D.E. Rassier (ed.), *Muscle Biophysics: From Molecules to Cells*, Advances in Experimental Medicine and Biology 682,
DOI 10.1007/978-1-4419-6366-6_9, © Springer Science+Business Media, LLC 2010

around one-another to form a continuous unbroken coiled-coil cable that winds around the actin helix. Tn is an etherotrimeric complex made of the calcium-binding subunit Troponin C (TnC), the inhibitory subunit Troponin I (TnI), and the TnC-TnI-Tm binding subunit Troponin T (TnT). Tn binds to Tm 1:1. The Tm filaments which are formed by head to tail assembly with a short overlap play a pivotal role in regulating the actin-myosin interaction in striated muscle and stabilizing actin structure (Gordon et al. 2000). In the last years, several methods proved to efficiently replace whole Tn complex or its subunits in both skeletal and cardiac muscle preparations (Brandt et al. 1984; Moss et al. 1986; Brenner et al. 1999; Piroddi et al. 2003; Kruger et al. 2005; Narolska et al. 2006; Yang et al. 2009). A significant advancement in understanding the thin filament regulatory system should be achieved then by extending to native Tm a replacement protocol for Tm isoforms and mutants, associated with myopathies (or engineered forms). As Tm is a filamentous protein strongly attached to actin, extracting and replacing with exogenous proteins is a challenge and the few attempts made met with little success (Chang et al. 2005). Recently, a successful method for the Tm-Tn replacement has been developed for skeletal myofibrils that preserve the viability of the preparation for mechanical studies (Siththanandan et al. 2009).

Replacement of regulatory proteins in single myofibrils offers a number of advantages, as compared to more conventional muscle preparations (She et al. 2000). The smaller diffusion distances allow a more complete and homogeneous exchange of proteins in a much shorter time (Piroddi et al. 2003). Furthermore, fast solution switching methods in single myofibrils (Tesi et al. 1999) can be used (1) to abruptly change the concentration of Ca^{2+} and investigate activation and relaxation kinetics (2) to abruptly change the concentrations of substrate and products of the acto-myosin ATPase and dissect some specific steps of the cross-bridge (CB) mechanical cycle (Tesi et al. 1999, 2000, 2002a, b).

A major goal of our recent work has been to learn how to completely replace myofibril endogenous Tm with isoforms and mutants (associated with myopathies or engineered) to get mechanistic insight on Tm's role in regulation. To this aim, we developed an effective Tm-Tn extraction-reconstitution technique for both skeletal and cardiac myofibrils, producing sarcomeres homogeneous for Tm composition; these allow measurement of a wide range of mechanical parameters of contraction (maximal force, kinetics of force generation and relaxation at maximal and sub-maximal Ca^{2+}-activation, myofilament Ca^{2+}-sensitivity and cooperativity). This will improve the resolution of previous studies performed in vitro (Clemmens et al. 2005) or in skinned fibers from transgenic mouse models (Pieples et al. 2002; Jagatheesan et al. 2009) and gelsolin treated/reconstituted systems (Fujita et al. 2002).

With this experimental approach we will measure the functional effects of: (1) selected Tm isoforms in the form of homogeneous dimers (alpha-alpha or beta-beta), (2) cardiac and skeletal myopathy-linked Tm mutants, (3) engineered Tm mutants with modified structural properties.

2 Methods and Results

Single myofibrils are the smallest units of the contractile apparatus of striated muscle that retain the organized myofilament lattice and ensemble of associated proteins. Single myofibrils or bundles of two to three myofibrils were prepared from fast skeletal or cardiac muscle by homogenization in rigor solution of glycerinated (rabbit psoas) or triton treated (mouse left ventricle) fiber bundles or tissue strips (Tesi et al. 1999; Piroddi et al. 2006). Recently, we also used myofibrils isolated from human atrial and ventricular biopsies prepared as previously described (Piroddi et al. 2007). All solutions to which the samples and myofibrils were exposed contained a cocktail of protease inhibitors including leupeptin (10 μM), pepstatin (5 μM), phenylmethylsulphonylfluoride (200 μM), E64 (10 μM), NaN_3 (500 μM), and dithiothreitol (500 μm).

Myofibril suspensions were centrifuged (1,600×*g*, 8 min, 4°C) and resuspended in a low ionic strength solution (2 mM Tris–HCl, pH 8) to remove native Tm and Tn. To increase the amount of extraction, the cycle was repeated five to seven times (or more in cardiac samples which were found more resistant to Tm removal). Extracted myofibrils were then washed in a 200 mM ionic strength rigor solution (100 mM KCl, 2 mM $MgCl_2$, 1 mM EGTA, 50 mM Tris–HCl, pH 7) and reconstituted with exogenous Tm and Tn following a 2-steps protocol.

First, the extracted myofibrils were collected and incubated in rigor solution containing 5 μM Tm (0°C, 1 h at least). Tm was either rabbit skeletal (sTm), bovine cardiac (cTm) and human cardiac (hcTm), depending on the experimental model used.

Secondly, Tm-reintroduced myofibrils were incubated in rigor solution containing 2 μM Tn (0°C, 1 h at least). Tn was either rabbit skeletal (sTn), bovine cardiac (cTn) or human cardiac (hcTn), depending on the experimental model used. Skeletal and cardiac Tn and Tm, both extracted and purified from rabbit fast skeletal and bovine cardiac muscles were kindly provided by Dr. E. Homsher (UCLA University, Los Angeles, USA). Human cardiac Tm was purified from human samples; recombinant human cardiac Tn was kindly provided by Dr. F. Matsumoto (Quantum Beam Science Directorate, Ibaraki, Japan). Finally, reconstituted myofibrils were washed and stored in 200 mM ionic strength rigor solution at 4°C, and used within 4 days. At each stage of the protocol, samples were retained from both supernatant and pellet fractions and then used to determine the extent of the Tm-Tn extraction and replacement procedure by 12% SDS–PAGE analysis (de Tombe et al. 2007).

Measurements of myofibrillar maximal isometric force and kinetics of force development (k_{ACT}) and redevelopment (k_{TR}) were also used to evaluate the mechanical viability of myofibrils following the reconstitution, as well as to establish the effectiveness of the technique.

All activating (pCa 4.5) and relaxing solutions (pCa 9.0), which were calculated as described previously (Tesi et al. 2000), were at pH 7.0. The solutions contained: 10 mM total EGTA (CaEGTA/EGTA ratio set to obtain the different values of pCa),

5 mM MgATP, 1 mM free Mg^{2+}, 10 mM MOPS, propionate and sulphate to adjust the final solution to an ionic strength of 200 mM and a monovalent cation concentration of 155 mM. Although continuous solution flow minimizes alterations in the concentration of MgATP and its hydrolysis products in the myofibrillar space, the measurements were made in the presence of creatine phosphate (10 mM) and creatine kinase (200 units ml^{-1}) to prevent any ADP gradients. Contaminant [Pi] (around 170 mM in standard solutions) was reduced in some experiments to less than 5 mM (Pi-free solutions) by a Pi-scavenging enzyme system (purine-nucleoside-phosphorylase with substrate 7-methyl-guanosine; Tesi et al. 2000).

After the Tm-Tn depletion step, both skeletal and cardiac myofibrils remained stable in rigor solution but underwent an irreversible Ca^{2+}-independent contraction when perfused with Mg-ATP containing solutions ("relaxing solution", pCa 9.0). This behavior is expected from a successful depletion of Tn. After the two steps protocol for Tm-Tn replacement, skeletal and cardiac myofibrils were stable in relaxing solution and could be activated only by the concomitant presence of micromolar Ca^{2+} ("activating solution", pCa 4.5). These observations are confirmed by the SDS–PAGE analysis.

Figure 1 shows representative gels of skeletal (a) and cardiac (b) myofibril samples retained at each step of the extraction-reconstitution protocol, run on a 12% Tris–HCl SDS–PAGE gel, after Comassie blue staining. For each lane, the amount of protein loaded (4–10 μg) is different due to protein loss during the preparation. The degree of Tm replacement in each lane was assessed from the intensity profile of scanned gels using UN-SCAN-IT gel 6.0 software (Silk Scientific, Inc., UT, USA) and determining the ratio of α-Tm to actin band intensities. As in Siththanandan et al. (2009), only α-Tm is used in these measurements because β-Tm (which is significantly present in skeletal myofibrils) migrates more slowly, within the TnT band (see lane 1 and 4; Cummins and Perry 1973; Salviati et al. 1982). For extracted rabbit skeletal myofibrils (Fig. 1a), the α-Tm/actin ratio (lane 2, Ext) is 0.04 compared to 0.35 in the controls (lane 1, Ctrl) and 0.33 in the reconstituted myofibrils (lane 3, Rec), representing about 89% removal and replacement. For mouse extracted cardiac myofibrils (Fig. 1b), the α-Tm/actin ratio is 0.03 (lane 2, Ext) compared to 0.21 in the controls (lane 1, Ctrl) and 0.19 in the reconstituted myofibrils (lane 3, Rec), representing about 86% removal and replacement.

As shown in Fig. 2, Tm (and Tn) replacement works equally well in skeletal (a) and cardiac (b) myofibrils. Mean values of Tm extraction-replacement were: sTm extraction 90% ± 2 and sTm reintroduction 96% ± 3 for rabbit skeletal myofibrils (mean ± SEM; n = 5); cTm extraction 90% ± 5 and cTm reintroduction 91% ± 5 in mouse cardiac myofibrils (mean ± SEM; n = 5). In both cases, Tm was always expressed relative to actin.

The critical test to assess the success of the Tm and Tn removal and reconstitution protocol using exogenous proteins is the preservation of the mechanical properties of myofibril contraction and its overall Ca^{2+} switching mechanism. To this aim, small bundles of rabbit skeletal and mouse cardiac myofibrils were mounted in a force recording apparatus in an experimental bath filled with relaxing solution of pCa 9.0. The system used to record force from single myofibrils and for rapid solution changes

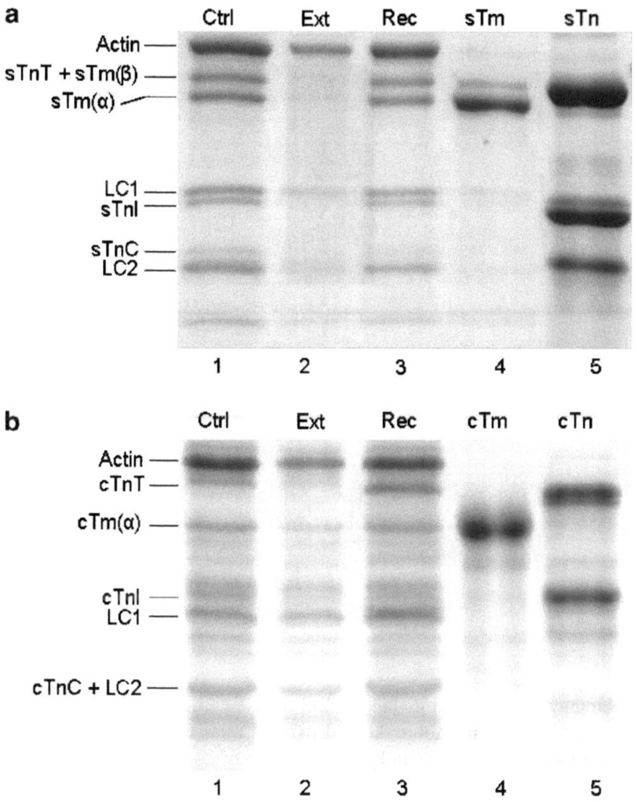

Fig. 1 12 percent Tris–HCl SDS–PAGE gel analysis of Tm-Tn extraction and reconstitution in rabbit skeletal and mouse left ventricle myofibrils. (**a**) Comassie-stained SDS–PAGE analysis of myofilament proteins from rabbit psoas myofibrils before (*lane 1*; Ctrl) and after (*lane 2*; Ext) removal of endogenous sTm-sTn complex and following reconstitution (*lane 3*; Rec) with exogenous sTm-sTn. sTm (*lane 4*); sTn complex (*lane 5*). (**b**) Comassie-stained SDS–PAGE analysis of myofilament proteins from mouse ventricle myofibrils before (*lane 1*; Ctrl) and after (*lane 2*; Ext) removal of endogenous cTm-cTn complex and following reconstitution (*lane 3*; Rec) with exogenous cTm-cTn. cTm (*lane 4*); cTn complex (*lane 5*). cTnC and LC2 bands are here fused. Only parts of gels containing Tm and Tn subunits are shown

has been described earlier (Colomo et al. 1997; Tesi et al. 1999). Briefly, myofibrils selected for use were mounted horizontally between two glass micro-tools. One tool is a calibrated cantilevered force probe while the other is connected to a length control motor. The myofibrils strongly adhered to the glass tools, which were positioned using micromanipulators to maximize the attachment area. The initial length of reconstituted myofibrils between attachments averaged 63.8 ± 2.7 μm, at a sarcomere length of 2.38 ± 0.03 μm (mean ± SEM, n = 24) for rabbit psoas myofibrils and 65.1 ± 5.3 μm at a sarcomere length of 2.17 ± 0.02 μm (mean ± SEM, n = 18) for mouse ventricle myofibrils. Isometric force was measured from the deflection of the force probe shadow projected on a split photodiode. Myofibrils were activated and relaxed by rapid

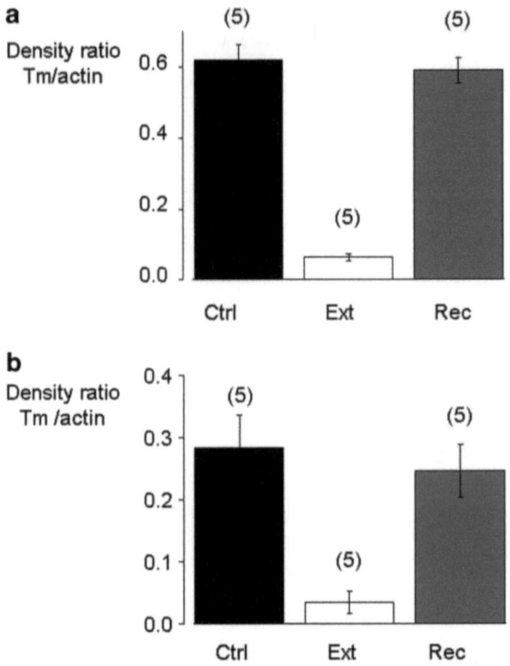

Fig. 2 Replacement of Tm in rabbit skeletal and mouse cardiac myofibrils. Mean values of Tm extraction and reconstitution in rabbit skeletal (**a**) and mouse ventricle (**b**) myofibrils from the evaluation of the ratio of α-Tm to actin band intensities of correspondent gel lanes (*Ctrl* control; *Ext* extracted; *Rec* reconstituted). Values are given as mean ± SEM of five gels. Average levels of sTm extraction and replacement in rabbit psoas samples were 90% ± 2 and 96% ± 3, respectively. Average levels of cTm extraction and replacement in mouse ventricle samples were 90% ± 5 and 91% ± 5, respectively

translation between two continuous streams of relaxing (pCa 9.0) and activating (pCa 4.5) solutions flowing by gravity from a double-barreled glass pipette placed at a right angle to, and within 1 mm of, the preparation. Solution changes after the start of the paired-pipette movement (driven by a stepper-motor controlled system) occurred with a time constant of 2–4 ms and were complete within 10 ms (Tesi et al. 1999, 2000). Experiments were performed at 15°C. Release-restretch protocols were used to measure k_{TR} (Brenner 1988). Values of the rate of rise of tension following a step decrease in pCa by fast solution switching (k_{ACT}) were estimated from the time required to reach 50% of the maximal isometric force.

Typical force records of Fig. 3 show that both rabbit skeletal (a, b) and mouse cardiac (c, d) Tm-Tn replaced myofibrils preserve the basic property of being completely relaxed at high pCa, as expected from the proper reintroduction of the Tn calcium switch in the sarcomere suggested by SDS–PAGE analysis.

Mean data of force development in control and Tm-Tn reconstituted rabbit skeletal and mouse cardiac myofibrils are reported in Table 1. Following calcium activation, maximal isometric tension was significantly lower (about 30–35%; $p < 0.05$ – Student t test) in both the Tm-Tn reconstituted skeletal and cardiac myofibrils than in

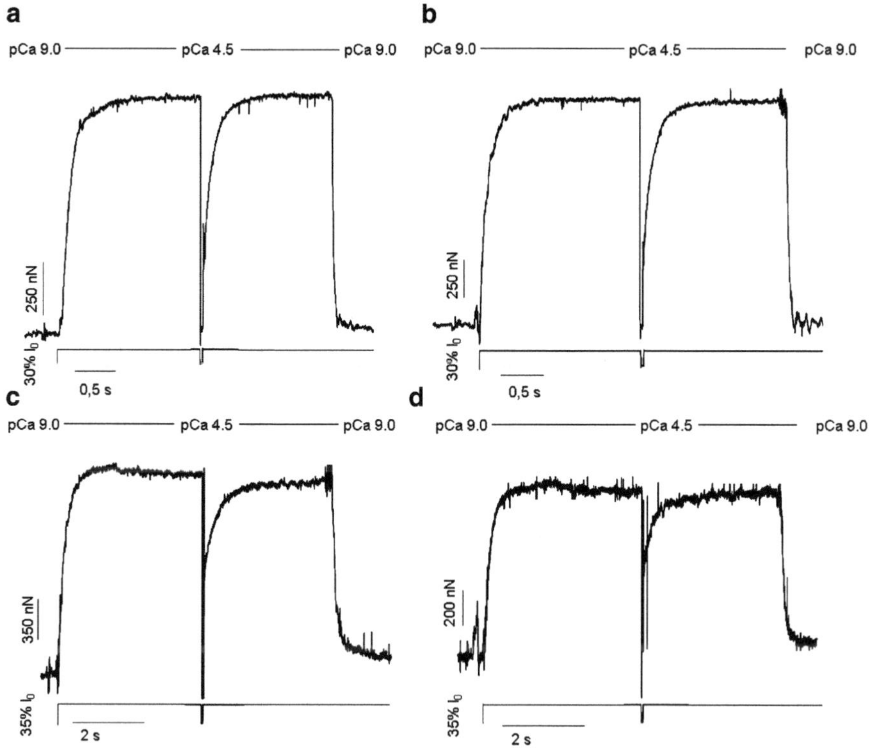

Fig. 3 Maximal calcium activated tension in rabbit skeletal and mouse cardiac myofibrils before and after Tm-Tn replacement at 15°C. Representative force records from myofibrils maximally Ca^{2+} activated by rapid solution switching (pCa changes marked by *lines over traces*). The rate of tension generation (k_{ACT}) was measured form the kinetics of force development following fast Ca^{2+} activation. Fast length changes (*bottom traces*, release-restretch protocol) are applied to myofibrils under conditions of steady tension generation in order to measure the rate constant of tension redevelopment (k_{TR}). *Upper panels*: Representative force responses from rabbit psoas myofibrils. (**a**) Control myofibril (sl 2.5 μm, P_0 420 mN mm^{-2}, k_{ACT} 7.2 s^{-1}, k_{TR} 7.3 s^{-1}). (**b**) sTm/sTn-reconstituted myofibril (sl 2.4 μm, P_0 406 mN mm^{-2}, k_{ACT} 7.7 s^{-1}, k_{TR} 7.9 s^{-1}). *Lower panels*: Representative force responses from mouse ventricle myofibrils. (**c**) Control myofibril (sl 2.1 μm, $P_0$122 mN mm^{-2}, k_{ACT} 5.6 s^{-1}, k_{TR} 4.4 s^{-1}).; (**d**) cTm/cTn-reconstituted myofibril (sl 2.1 μm, P_0 79 mN mm^{-2}, k_{ACT} 5.2 s^{-1}, k_{TR} 4.1 s^{-1}). Time scale, force and length calibrations as indicated by bars

Table 1 Effect of Tm-Tn extraction reconstitution on tension generation in rabbit psoas and mouse cardiac myofibrils at 15°C

Myofibril batch	P_0 (mN mm^{-2})	k_{ACT} (s^{-1})	k_{TR} (s^{-1})
Rabbit skeletal			
Control	355 ± 20 (13)	9.00 ± 0.50 (13)	9.06 ± 0.46 (13)
Reconstituted	263 ± 14* (24)	8.68 ± 0.45 (24)	8.23 ± 0.46 (24)
Mouse cardiac			
Control	82 ± 8 (10)	5.63 ± 0.60 (10)	4.65 ± 0.66 (10)
Reconstituted	61 ± 4* (18)	4.28 ± 0.40 (18)	3.70 ± 0.27 (18)

All values are given as mean ± SEM; values in parentheses are the myofibril numbers. P_0 maximum isometric tension; k_{ACT} rate constant for force development; k_{TR} rate constant for force redevelopment following release restretch of maximally activated myofibrils

*$P<0.05$ (Student's *t* test) versus the same parameter measured in control myofibrils

correspondent controls. In both experimental models, following Tm-Tn reconstitution, no significant difference was found for the rate of force activation (k_{ACT}) and that of force redevelopment (k_{TR}), showing that skeletal and cardiac myofibril remains mechanically viable and capable of producing substantial forces with normal kinetics. Sarcomere length, maximal calcium activated force, k_{ACT} and k_{TR} of reconstituted myofibrils were in the range of control values previously observed in skeletal (Piroddi et al. 2003) and cardiac (Piroddi et al. 2006) myofibrils.

Recently, the method worked also in human cardiac myofibrils from left ventricular and atrial biopsies. Protein exchange assessment in human cardiac myofibrils, treated as described above and replaced with hcTm-hcTn, is shown by SDS–PAGE gel electrophoresis of Fig. 4a. Preliminary results indicate that the level of hcTm

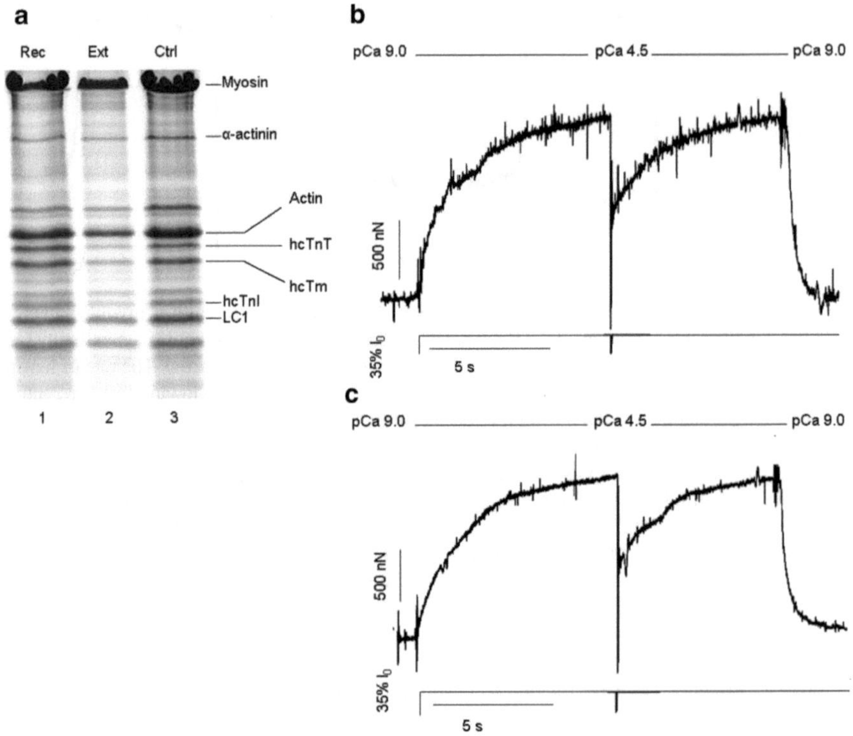

Fig. 4 Tm-Tn extraction and reconstitution in human ventricle myofibrils. (**a**) Comassie-stained 12% Tris–HCl SDS–PAGE gel of human ventricle myofilament proteins before (*lane 3*; Ctrl) and after (*lane 2*; Ext) removal of endogenous hcTm-hcTn complex and following reconstitution (*lane 1*; Rec) with exogenous hcTm-hcTn. *Right panels*: Representative recordings of tension development following maximal Ca^{2+}-activation by rapid solution switching (*upper traces*; pCa changes marked by *lines over traces*) from a control myofibril (**b**) and from a myofibril whose endogenous Tm and Tn had been extracted and replaced with exogenous hcTm and recombinant hcTn (**c**). *Bottom traces*: length changes. Control myofibril: sl 2.2 μm, P_0 104 mN mm^{-2}, k_{ACT} 0.98 s^{-1}, k_{TR} 0.70 s^{-1}. hcTm/hcTn-reconstituted myofibril: sl 2.1 μm, P_0 83 mN mm^{-2}, k_{ACT} 0.64 s^{-1}, k_{TR} 0.50 s^{-1}. Time scale, force and length calibrations as indicated by bars

extraction ranged 65% and that of hcTm reconstitution 96%. Preliminary data indicate that after reconstitution, human cardiac myofibrils (both atrial and ventricular) can develop substantial amount of calcium activated tension, as shown in Fig 4b and c with kinetics unchanged by the Tm-Tn extraction and reconstitution procedure. These results in human cardiac myofibrils are promising for the full development of a method for the study of functional impact of Tm mutations associated with cardiomyopathies in humans.

3 Discussion

The present work demonstrates that we can substantially replace Tm-Tn in skeletal and cardiac myofibrils, from both animal and human source, with low impact on myofibril mechanics. This is important as little is known about the functional role of Tm in the cooperative mechanisms of force activation and relaxation in cardiac and skeletal muscle. It is also important because several mutations in the protein have been associated with cardiomyopathies and skeletal muscle disease (Gunning et al. 2008) such as nemaline myopathy or distal arthrogriposis (Ochala et al. 2008). Production of an experimental model with homogeneous Tm composition gives us a system in which to compare Tm forms over a wide range of mechanical parameters of contraction (maximal force, kinetics of force generation and relaxation at maximal and sub-maximal Ca^{2+}-activation, myofilament Ca^{2+}-sensitivity and cooperativity) in both skeletal and cardiac muscle.

This will improve the resolution of previous studies performed in vitro (Clemmens et al. 2005) or in skinned fibers from transgenic models (Pieples et al. 2002) and gelsolin treated/reconstituted systems (Fujita et al. 2002). Until recently, replacement of Tm in a skinned cardiac muscle preparation was only possible using a technique that selectively removed the entire thin filaments by gelsolin (Fujita et al. 1996, 2002) followed by sequential reconstitution of the actin filaments with exogenous G actin and then with Tm and Tn. Though the technique is powerful, control of the length of the reconstituted thin filaments is quite empirical and the final structure of the sarcomere may be rather different than the original. Genetic techniques, making use of transgenic animals for Tm isoforms (Wolska et al. 1999) or mutants (Muthuchamy et al. 1999; see also Gunning et al. 2008) also show important limitations regarding the control of the expression level of exogenous transgenes that can dramatically influence the morphological and/or physiological phenotype of the tissue.

Several decades ago, a method to deplete native Tm from myofibrils was introduced by Perry and Corsi (1958) using low ionic strength alkaline solutions. In an attempt to extract all proteins from myofibrils, Perry and Corsi (1958) found that the total amount of protein extracted increased with the alkalinity of the buffers used but not for actin. Kominz and Yoshioka (1969) later used sodium carbonate/bicarbonate buffers to remove Tm from actomyosin preparations.

In the last years, the successful development in our lab of a method for substantial and non-invasive Tn replacement in myofibrils, suggested an approach to the challenging aim of replacing the whole Tm-Tn complex in skeletal and cardiac myofibrils. We based our initial attempts on a protocol published in 2005 (Chang et al. 2005) that claimed to selectively remove and replace all thin filament regulatory proteins in porcine cardiac myofibrils. Though myofibril suspensions treated by this Tm-Tn extraction and reconstitution method exhibited normal biochemical activity (Chang et al. 2005), in our hands they were no longer able to develop substantial levels of Ca^{2+}-activated tension (unpublished observations). Only very recently, starting from Chang et al. (2005) approach but modifying the pH buffer (Tris instead of bicarbonate) and reducing alkalinity (pH 8 instead of 9 for the Tm removal step), Homsher and collaborators (Siththanandan et al. 2009) succeeded in designing a Tm-Tn depletion/reconstitution protocol producing skeletal rabbit psoas myofibrils viable for mechanical studies. Here we report improvements of this method for the replacement of native Tm and Tn with exogenous regulatory proteins not only in skeletal but also in mouse and human cardiac myofibrils with a significant recovery of the tension generation capability.

This Tm-Tn protocol has a certain degree of variability with an overall reduction of maximal force to about 65–70%. This could be a consequence of the reconstitution process itself and of a possible loss in actomyosin during the extraction stage (Siththanandan et al. 2009). When only Tn (or its subunits) are replaced in myofibrils, low or no decrease in maximal force relative to controls is observed (de Tombe et al. 2007; Kreutziger et al. 2008), especially when a "replacement by exchange" approach instead of an "extraction-reconstitution" protocol is used. In the first case, instead of extracting Tn, wash myofibril suspension, and then replace/reconstitute the thin filament with an exogenous protein, native Tn is replaced by mass action by incubating myofibrils in a rigor solution containing 3 mM calcium in presence of an excess (0.5 mg/ml) of the exogenous Tn isoform (pH 6.9, 4°C, overnight; de Tombe et al. 2007). This treatment is much milder and reduces myofibril damage because at no time are the myofibrils without their complement of all the Tn subunits. We are now working on a further development of the Tm-Tn protocol towards finding conditions of pH, pH buffering and ionic strength promoting a whole Tm-Tn complex replacement by exchange. A milder treatment to replace endogenous Tm and Tn, besides better preserving myofibrils and their force generating capacity, could be particularly relevant in the case of human cardiac myofibrils because of the limited availability of suitable control tissue samples.

In conclusion, the development of this Tm-Tn replacement protocol in striated sarcomere with exogenous Tm or Tn of different structure and properties in association with the advantages offered by the use of single myofibrils for mechanical studies will offer unique opportunities (1) to better understand the structural mechanism through which the regulation signal of Ca^{2+} binding passes between Tn subunits to the Tm-actin filament, also in its cooperative aspects (2) to investigate whether and how regulatory proteins modulate CB function.

Acknowledgement This work was supported by 7th Framework Programs of the European Union (STREP Project "BIG-HEART", grant agreement 241577) and by Telethon-Italy (GGP07133). Authors gratefully acknowledge Dr. Earl Homsher for helpful discussion and technical advice.

References

Brandt PW, Diamond MS, Schachat FH (1984) The thin filament of vertebrate skeletal muscle co-operatively activates as a unit. J Mol Biol 180:379–84

Brenner B (1988) Effect of Ca^{2+} on cross-bridge turnover kinetics in skinned single rabbit psoas fibers: implications for regulation of muscle contraction. Proc Natl Acad Sci U S A 85:3265–3269

Brenner B, Kraft T, Yu L, Chalovich JM (1999) Thin filament activation probed by fluorescence of N-((2-(Iodoacetoxy)ethyl)-N-methyl)amino-7-nitrobenz-2-oxa-1, 3-diazole-labeled troponin I incorporated into skinned fibers of rabbit psoas muscle. Biophys J 77:2677–2691

Chang AN, Harada K, Ackerman MJ, Potter JD (2005) Functional consequences of hypertrophic and dilated cardiomyopathy-causing mutations in alpha-tropomyosin. J Biol Chem 280:34343–3439

Clemmens EW, Entezari M, Martyn DA, Regnier M (2005) Different effects of cardiac versus skeletal muscle regulatory proteins on in vitro measures of actin filament speed and force. J Physiol 566:737–46

Colomo F, Piroddi N, Poggesi C, te Kronnie G, Tesi C (1997) Active and passive forces of isolated myofibrils from cardiac and fast skeletal muscle of the frog. J Physiol 500:535–548

Cummins P, Perry SV (1973) The subunits and biological activity of polymorphic forms of tropomyosin. Biochem J 133:765–777

de Tombe PP, Belus A, Piroddi N, Scellini B, Walker JS, Martin AF, Tesi C, Poggesi C. (2007) Myofilament calcium sensitivity does not affect cross-bridge activation-relaxation kinetics. Am J Physiol Regul Integr Comp Physiol 292:R1129–1136

Fujita H, Yasuda K, Niitsu S, Funatsu T, Ishiwata S (1996) Structural and functional reconstitution of thin filaments in the contractile apparatus of cardiac muscle. Biophys J 71:2307–2318

Fujita H, Sasaki D, Ishiwata S, Kawai M (2002) Elementary steps of the cross-bridge cycle in bovine myocardium with and without regulatory proteins. Biophys J 82:915–28

Gordon AM, Homsher E, Regnier M (2000) Regulation of contraction in striated muscle. Physiol Rev 80:853–924

Gunning P, O'Neill G, Hardeman E (2008) Tropomyosin-based regulation of the actin cytoskeleton in time and space. Physiol Rev 88: 1–835

Jagatheesan G, Rajan S, Schulz EM, Ahmed RP, Petrashevskaya N, Schwartz A, Boivin GP, Arteaga GM, Wang T, Wang YG, Ashraf M, Liggett SB, Lorenz J, Solaro RJ, Wieczorek DF (2009) An internal domain of {beta}-tropomyosin increases myofilament Ca^{2+} sensitivity. Am J Physiol Heart Circ Physiol 297:H181–190

Kominz DR, Yoshioka K (1969) The influence of native tropomyosin on the ATP threshold for turbidity development of actomyosin and myofibril suspensions. Arch Biochem Biophys 129:609–614

Kreutziger KL, Piroddi N, Scellini B, Tesi C, Poggesi C, Regnier M (2008) Thin filament Ca^{2+} binding properties and regulatory unit interactions alter kinetics of tension development and relaxation in rabbit skeletal muscle. J Physiol 586:3683–700

Kruger M, Zittrich S, Redwood C, Blaudeck N, James J, Robbins J, Pfitzer G, Stehle R (2005) Functional consequences of hypertrophic and dilated cardiomyopathy-causing mutations in alpha-tropomyosin. J Physiol 564:347–357

Moss RL, Allen JD, Greaser ML (1986) Effects of partial extraction of troponin complex upon the tension-pCa relation in rabbit skeletal muscle. Further evidence that tension development involves cooperative effects within the thin filament. J Gen Physiol 87:761–74

Muthuchamy M, Pieples K, Rethinasamy P, Hoit B, Grupp IL, Boivin GP, Wolska B, Evans C, Solaro RJ, Wieczorek DF (1999) Mouse model of a familial hypertrophic cardiomyopathy mutation in alpha-tropomyosin manifests cardiac dysfunction. Circ Res 85:47–56

Narolska NA, Piroddi N, Belus A, Boontje NM, Scellini B, Deppermann S, Zaremba R, Musters RJ, dos Remedios C, Jaquet K, Foster DB, Murphy AM, van Eyk JE, Tesi C, Poggesi C, van der Velden J, Stienen GJ (2006) Effects of the mutation R145G in human cardiac troponin I on the kinetics of the contraction-relaxation cycle in isolated cardiac myofibrils. Circ Res 99:1012–1020

Ochala J, Li M, Ohlsson M, Oldfors A, Larsson L (2008) Defective regulation of contractile function in muscle fibres carrying an E41K beta-tropomyosin mutation. J Physiol 586:2993–3004

Perry SV, Corsi A (1958) Extraction of proteins other than myosin from the isolated rabbit myofibril. Biochem J. 68:5–12

Pieples K, Arteaga G, Solaro RJ, Grupp I, Lorenz JN, Boivin GP, Jagatheesan G, Labitzke E, DeTombe PP, Konhilas JP, Irving TC, Wieczorek DF (2002) Tropomyosin 3 expression leads to hypercontractility and attenuates myofilament length-dependent Ca(2+) activation. Am J Physiol Heart Circ Physiol 283:H1344–1353

Piroddi N, Tesi C, Pellegrino MA, Tobacman LS, Homsher E, Poggesi C (2003) Contractile effects of the exchange of cardiac troponin for fast skeletal troponin in rabbit psoas single myofibrils. J Physiol 552:917–931

Piroddi N, Belus A, Eiras S, Tesi C, van der Velden J, Poggesi C, Stienen GJ (2006) No direct effect of creatine phosphate on the cross-bridge cycle in cardiac myofibrils. Pflugers Arch 452:3–6

Piroddi N, Belus A, Scellini B, Tesi C, Giunti G, Cerbai E, Mugelli A, Poggesi C (2007) Tension generation and relaxation in single myofibrils from human atrial and ventricular myocardium. Pflügers Arch 454:63–73

Salviati G, Betto R, Danieli Betto D (1982) Polymorphism of myofibrillar proteins of rabbit skeletal-muscle fibres. An electrophoretic study of single fibres. Biochem J. 207:261–272

She M, Trimble D, Yu L, Chalovich JM (2000) Factors contributing to troponin exchange in myofibrils and in solution. J Mus Res Cell Mot 21:737–745

Siththanandan VB, Tobacman LS, Van Gorder N, Homsher E (2009) Mechanical and kinetic effects of shortened tropomyosin reconstituted into myofibrils. Pflugers Arch 458:761–776

Tesi C, Colomo F, Nencini S, Piroddi N, Poggesi C (1999) Modulation by substrate concentration of maximal shortening velocity and isometric force in single myofibrils from frog and rabbit fast skeletal muscle. J Physiol 516:847–853

Tesi C, Colomo F, Nencini S, Piroddi N, Poggesi C (2000) The effect of inorganic phosphate on force generation in single myofibrils from rabbit skeletal muscle. Biophys J 78:3081–3092

Tesi C, Colomo F, Piroddi N, Poggesi C (2002a) Characterization of the cross-bridge force-generating step using inorganic phosphate and BDM in myofibrils from rabbit skeletal muscles. J Physiol. 541:187–99

Tesi C, Piroddi N, Colomo F, Poggesi C (2002b) Relaxation kinetics following sudden Ca(2+) reduction in single myofibrils from skeletal muscle. Biophys J 83:2142–2151

Wolska BM, Keller RS, Evans CC, Palmiter KA, Phillips RM, Muthuchamy M, Oehlenschlager J, Wieczorek DF, de Tombe PP, Solaro RJ (1999) Correlation between myofilament response to Ca^{2+} and altered dynamics of contraction and relaxation in transgenic cardiac cells that express beta-tropomyosin. Circ Res 1999 84:745–51

Yang Z, Yamazaki M, Shen QW, Swartz DR (2009) Differences between cardiac and skeletal troponin interaction with the thin filament probed by troponin exchange in skeletal myofibrils. Biophys J 97:183–94

Stretch and Shortening of Skeletal Muscles Activated Along the Ascending Limb of the Force–Length Relation

Dilson E. Rassier and Clara Pun

Abstract There is a history dependence of skeletal muscle contraction. When muscles are activated and subsequently stretched, they produce a long lasting force enhancement. When muscles are activated and subsequently shortened, they produce a long-lasting force depression. The purposes of the studies shown in this chapter were (1) to evaluate if force enhancement and force depression are present along the ascending limb of the force–length (FL) relation, (2) to evaluate if the history-dependent properties of force production are associated with sarcomere length (SL) non-uniformity, and (3) to determine the effects of cross-bridge (de) activation on force depression. Isolated myofibrils were activated by either Ca^{2+} or MgADP and were subjected to consecutive stretches or shortenings along the ascending limb of the FL relation, separated by periods (~5 s) of isometric contraction. Force after stretch was higher than force after shortening when the contractions were produced at similar SLs. The difference in force could not be explained by SL non-uniformity. After shortening, MgADP activation produced forces that were higher than Ca^{2+} activation. Since MgADP induces the formation of strongly bound cross-bridges, the result suggests that force depression following shortening is associated with cross-bridge deactivation.

Keywords Sarcomeres • ADP • Calcium • Myosin • Cross-bridges

1 Introduction

There is a history dependence of force production in skeletal muscles. The force produced following stretching or shortening of activated muscles is higher (force enhancement) or lower (force depression), respectively, than the force produced during isometric contractions

---

D.E. Rassier (✉)
Department of Kinesiology and Physical Education, McGill University,
Pine Avenue West 475, Montreal, QC, Canada H2W1S4
e-mail: dilson.rassier@mcgill.ca

D.E. Rassier (ed.), *Muscle Biophysics: From Molecules to Cells*,
Advances in Experimental Medicine and Biology 682,
DOI 10.1007/978-1-4419-6366-6_10,

at the corresponding length (Edman et al. 1993; Granzier and Pollack 1989; Maréchal and Plaghki 1979; Sugi and Tsuchiya 1988; Rassier et al. 2003b; Julian and Morgan 1979). Although this phenomenon has been observed for more than 50 years (Abbot and Aubert 1952), its mechanism remains unknown and has generated much debate (Morgan and Proske 2006, 2007; Herzog and Leonard 2007). The lack of an accepted mechanism represents an important gap in our understanding of muscle contraction, as history dependence of contraction cannot be explained by the original cross-bridge model of contraction (Huxley 1957), or the sliding filament theory (Huxley and Niedergerke 1954; Huxley and Hanson 1954).

History-dependent properties have been observed mostly along the descending limb of the force–length (FL) relation, which is commonly associated with sarcomere instability (Morgan 1994; Zahalak 1997). Based on this observation, a hypothesis has been raised and largely accepted by scientists: force enhancement and force depression would be caused by SL non-uniformity that develops when muscles are activated (Julian and Morgan 1979; Morgan 1994; Morgan et al. 2000). During stretch or shortening of activated muscle fibers, some sarcomeres would not change their lengths considerably or would remain mostly isometric, while others would stretch or shorten considerably. Such non-uniformity would place sarcomeres at distinct regions of the FL relation (Morgan 1994) – a situations that could equilibrate sarcomeres at force levels considerably different from those obtained during isometric contractions (Morgan 1994).

Although the SL non-uniformity hypothesis received much support, there is evidence suggesting that it cannot explain solely history-dependent properties of force production (Rassier et al. 2003a; Joumaa et al. 2008; Pinniger et al. 2006; Roots et al. 2007; Telley et al. 2006b). An alternative mechanism to explain force depression is associated with cross-bridge properties: force depression would be caused by an inhibition of cross-bridges that enter a new overlap zone during shortening (Maréchal and Plaghki 1979). The inhibition could be caused by actin filament deformation during activation (Daniel et al. 1998). The cross-bridge inhibition hypothesis has not been rejected yet, but comprehensive studies with SL measurements are lacking.

In this chapter we will present a series of experiments performed with isolated myofibrils evaluating the history-dependent properties of muscle contraction. The experiments were aimed at evaluating the hypotheses that force enhancement and force depression are caused by SL non-uniformities. Furthermore, we will present results from experiments suggesting that force depression may be caused by a shortening-induced cross-bridge inhibition. We will focus on experiments conducted along the ascending limb of the FL relation, which presents an advantage: this region is conceptually stable and large SL non-uniformities are not expected.

2 Methods

2.1 Preparation of Myofibrils

The experiments in this series of studies were performed with skeletal muscle myofibrils. Small sections of rabbit psoas muscle were dissected and tied to wood sticks. The samples were stored in rigor solution (pH=7.0) for 4 h, after which they were transferred to a rigor:glycerol (50:50) solution for 15 h. Specimens were subsequently placed in a fresh rigor:glycerol (50:50) solution with the addition of protease inhibitors (Roche Diagnostics) and stored in a freezer (−20°C) for at least 7 days. On the day of the experiment, single myofibrils were obtained by cutting a small section (~2 mm in length) and homogenizing the muscle samples (Troemer/VWR Power Max AHS 250, Oxford, USA) following standard procedures (Rassier 2008).

2.2 *Solutions*

The rigor solution (pH 7.0) was composed of (in mM): 50 Tris, 100 NaCl, 2 KCl, 2 $MgCl_2$, and 10 EGTA. The relaxing solution (pH 7.0) was composed of (in mM): 10 MOPS, 64.4 K^+ propionate, 5.23 Mg^{2+} propionate, 9.45 Na_2SO_4, 10 EGTA, 7 ATP, and 10 creatine phosphate. The Ca^{2+}-activating solution (pH 7.0; $pCa^{2+}=4.5$) was composed of (in mM): 10 MOPS, 45.1 K^+ propionate, 5.21 Mg^{2+} propionate, 9.27 Na_2SO_4, 10 EGTA, 7.18 ATP, and 10 creatine phosphate. For experiments performed with MgADP activation (see below), the activation solution (pH 7.0) was composed of (in mM): 2 Mg^{2+}, 20 MOPS, 4 EGTA, 1 ATP, and 10 MgADP. The concentration of MgADP was chosen after pilot experiments comparing the force levels obtained with Ca^{2+} (n=4) and 5 mM (n=2), 10 mM (n=2) and 20 mM (n=2) of MgADP. The force produced by 10 mM MgADP closely approximated forces produced by the Ca^{2+}-activated myofibrils during isometric contractions at similar SLs.

2.3 *Experimental Setup*

The homogenized muscle sample was transferred into an experimental chamber built with the bottom made with a vacuum-sealed glass cover slip (thickness: ~0.15 mm). The chamber was kept at a constant temperature of 15°C during the experiments. The sample was rinsed several times with the rigor solution, after which the relaxing solution was added to the chamber. Myofibrils were chosen based on striation pattern and number of sarcomeres (between 10 and 20), and were grabbed using two glass micro-needles produced with a pipette puller (KOPF 720, David Kopf Inst, Tujunga, USA) (Fig. 1a). The myofibrils were then lifted off the chamber bottom by ~2–3 μm. One microneedle used during the experiments was rigid, and was connected to a motor arm, allowing for computer-controlled length changes in the myofibrils. The other microneedle was flexible and pre-calibrated to allow force measurements (stiffness between 210 and 295 nN μm^{-1}). Under high magnification (Nikon plan-fluor, 60×, NA 0.70) further increased 1.5× internally in the microscope, the image of the myofibril was projected onto a linear photodiode array (Schafter+Kirchhoff SK4096DJR, Germany) with 4,096 pixels (pixel size: 10×10 μm) and was scanned to produce peak tracings of light intensity. The contrast between the dark bands of myosin (A-bands) and the light bands of actin (I-bands)

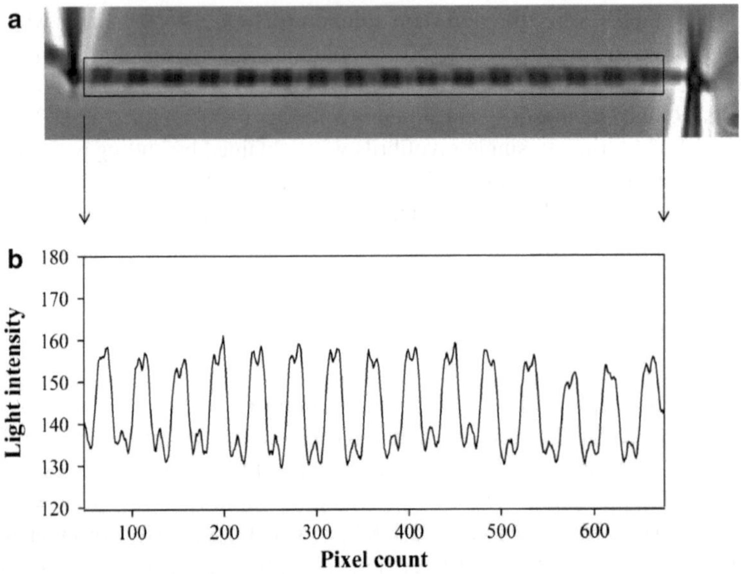

Fig. 1 (**a**) Photograph of a myofibril attached between a pre-calibrated, flexible glass needle and a rigid glass needle. The striation pattern of the myofibril is produced by the arrangement of myosin and actin filaments, and is projected onto a linear photodiode array, represented by the box around the myofibril. (**b**) One representative scan of the same myofibril. The striation pattern results in a light intensity pattern, in which the big peaks correspond to the dark A-bands, the valleys correspond to the I-bands. Note that even the Z-lines and M-lines are visible in some sarcomeres. By tracking the position of the intensity peaks in consecutive scans produced during the experiments, changes in SL can be analyzed as a function of time

provided a dark-light intensity pattern representing the striation pattern produced by the sarcomeres (Fig. 1b). The peaks produced by the micro-needles were also tracked by the linear photodiode array for force measurements. When the myofibril contracted, the flexible micro-needle was displaced and the force (F) was calculated based on the known stiffness and amount of micro-needle displacement: $F = \Delta d\left(\frac{K_1 K_2}{K_1 + K_2}\right)$, where Δd = absolute difference between the initial and the final distance between the two micro-needles, K = stiffness, 1 and 2 = micro-needles 1 and 2, respectively. SL was calculated using a peak-detection algorithm based on the distance between adjacent A-band centroids (Rassier 2008; Sokolov et al. 2003). The image of the myofibril was also collected by a CCD camera (Go-3, QImaging, USA; pixel size: 3.2 × 3.2 μm) used for calculation of diameter, radius (r) and cross-sectional area (CSA), which was estimated assuming circular symmetry. The final resolution of the camera is 0.035 μm, which makes the error on the diameter measurements small (~3.5%).

After the myofibrils were suspended from the bottom of the chamber, the relaxing solution was flown through the myofibrils for ~10 s. The relaxing solution was then replaced by the activating solution. Exchange of solutions was achieved using a computer-controlled, multi-channel perfusion system (VC–6M, Harvard Apparatus,

USA) attached to a double-barrelled pipette (Rassier 2008; de Tombe et al. 2007). The pipette was polished to reach an inner diameter of ~600 μm, each of the two channels with an inner diameter of ~300 μm. The pipette was placed close to the myofibrils (~100 μm) with the two channels having solutions flowing uninterruptedly throughout the experiments. The pipette was moved horizontally to the myofibrils' plane to produce quick solution exchanges. The solutions were dragged out of chamber by a peristaltic pump (Instech P720, Harvard Apparatus, USA); the flow rate achieved with this system was ~5 μl s^{-1}.

2.4 Experimental Protocol

Five sets of experiments were performed in this study: (1) Ca^{2+}-activated myofibrils performed isometric contractions (ISO, n=4), (2) Ca^{2+}-activated myofibrils underwent three stretches at full force development, with periods of isometric contractions between stretches (Ca^{2+}-STR, n=8), (3) Ca^{2+}-activated myofibrils underwent three shortenings at full force development, with periods of isometric contractions between shortenings (Ca^{2+}-SHO, n=7), (4) MgADP-activated myofibrils underwent three stretches at full force development, with periods of isometric contractions between stretches (ADP-STR, n=8), and (5) MgADP-activated myofibrils underwent three shortenings at full force development, with periods of isometric contractions between shortenings (ADP-SHO, n=7). MgADP activation generates strong binding of cross-bridges and actin independently of Ca^{2+} (Bremel and Weber 1972; Greene et al. 1987; Zhang et al. 2000), allowing the evaluation of cross-bridge (de)activation during force production.

For experiment (1), the initial average SL was adjusted to cover distinct portions of the FL relation. After full activation, the myofibrils were held isometric for 10–15 s, after which they were relaxed. For experiments (2–5), the initial average SL was adjusted so that myofibrils would produce full force development along the ascending limb of the FL relation. After activation, the myofibrils were held isometric for at least 5 s, after which three successive length changes (stretching or shortening) were imposed. Each length step was separated by ~5 s to achieve steady-state forces. Length changes were made at nominal amplitudes of 4% of the initial rest SL, at a speed of ~0.07 μm s^{-1} sarcomere^{-1}. After the final isometric period of contraction, the myofibril was relaxed.

2.5 Data Analysis

Force and SL were calculated before activation, after the initial activation, during steady-state force development at ~4.6 s after length changes (stretch or shortening), and after myofibril relaxation. The degree of SL dispersion (SL_{dis}) in each myofibril was calculated by subtracting the individual SL_i from the average SL, and it was used as an indication of SL non-uniformity.

3 Results

3.1 *Myofibril Activation and History Dependence of Force Production*

Figure 2 shows isometric contractions produced by activation induced by Ca^{2+} and MgADP. When 10 mM MgADP is used, the forces were similar to those produced by the Ca^{2+}-activated myofibrils at a corresponding SL. Figure 3 shows two typical experiments performed by myofibrils that were stretched or shortened during full activation produced by Ca^{2+} or MgADP, respectively. In both experiments, all sarcomeres shortened during initial activation, albeit by different amounts, which causes a small increase in the degree of SL_{dis} (Rassier 2008; Telley et al. 2006a). The degree of SL non-uniformity remained relatively constant throughout the consecutive length changes in both contractions. Each stretch step imposed to the myofibrils caused an increase in force, which was maintained elevated after the stretch (Fig. 3a). Each shortening step imposed to the myofibrils caused a decrease in force, which was maintained lower during the isometric contractions (Fig. 3b). The results of Fig. 3 were confirmed when the average value of all experiments are plotted (Fig. 4).

3.2 *The Force–SL Relation and SL Dispersion*

Figure 5 shows the dynamic FL relation – i.e. isometric forces achieved after length perturbations – for all conditions investigated in this study. It also shows the

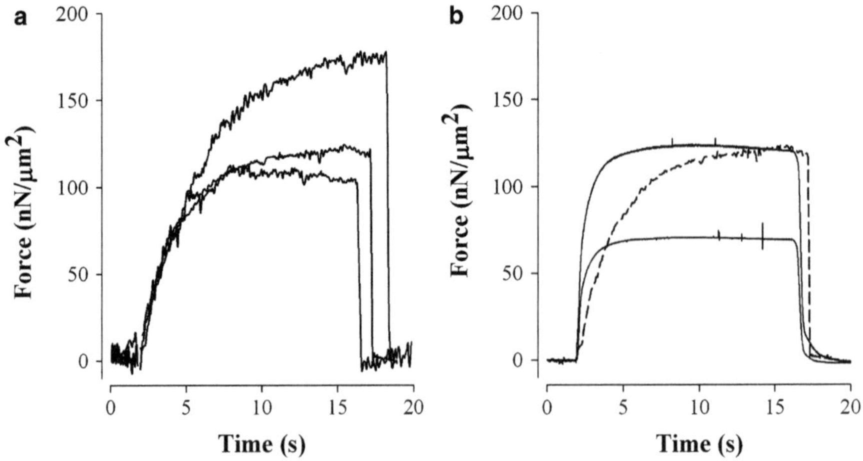

Fig. 2 (**a**) Forces produced by a myofibril activated with 5, 10 and 20 mM of MgADP at an average rest SL of 2.5 μm. (**b**) Forces produced by a myofibril activated with Ca^{2+} at average SLs of 2.5 and 1.9 μm (*high* and *low* forces shown with *solid lines*). The figure also shows the contraction induced with 10 mM of MgADP (*traced line*); after full development, the force is similar to that produced with Ca^{2+} at a similar SL

theoretical FL (Gordon et al. 1966) based on the filament lengths from the rabbit psoas (Sosa et al. 1994). Forces were normalized for the isometric forces produced during initial activation. Forces produced after stretches were always higher than forces produced after shortenings. Furthermore, forces produced after shortening were higher in the experiments in which activation was induced by MgADP than

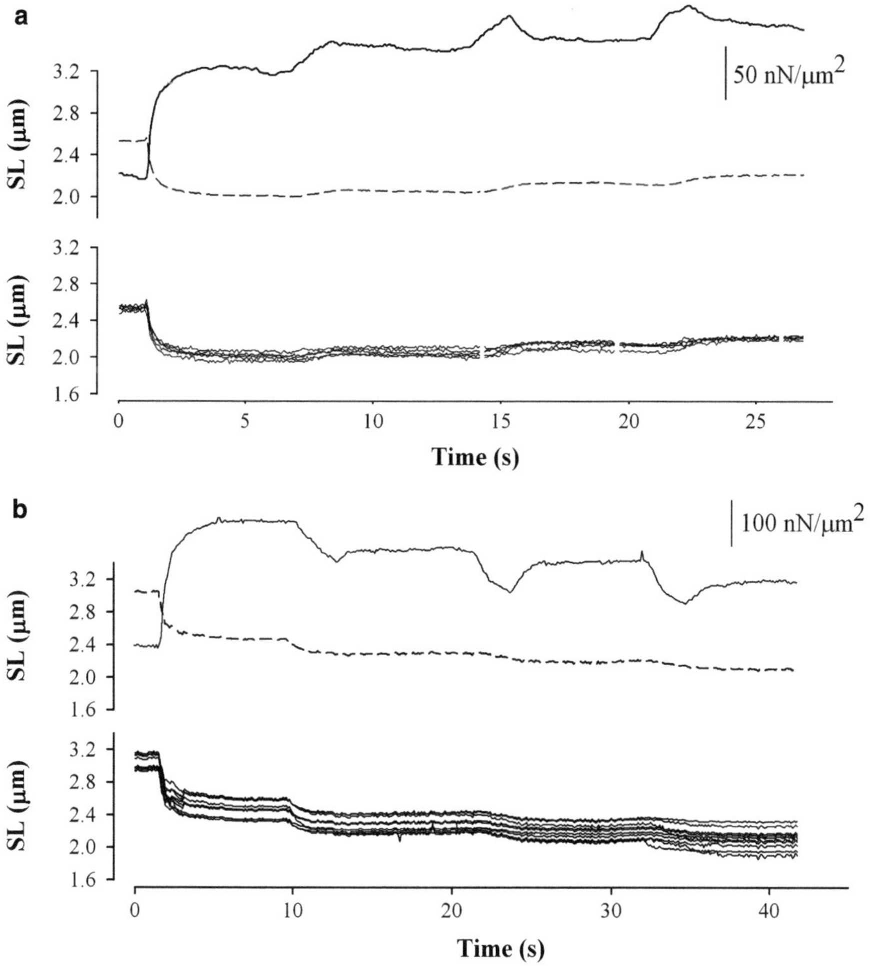

Fig. 3 (**a**) Experiment performed with a myofibril activated with Ca^{2+} and stretched along the ascending limb of the FL relation. *Solid line* represents the force and the dashed line represents the average SL. During stretches, force increases and remains elevated during the isometric period of the contractions. The *bottom panel* shows individual SLs from the same myofibril. The SL_{dis} increases during activation but it does not change significantly after the stretches. (**b**) Experiment performed with a myofibril activated with MgADP and shortened along the ascending limb of the FL relation. Solid line represents the force and the dashed line represents the average SL. During shortening, force decreases and remains low during the isometric period of the contractions. *Bottom panel* shows individual SLs from the same myofibril. The SL_{dis} increases during activation but it remains relatively constant after the shortenings

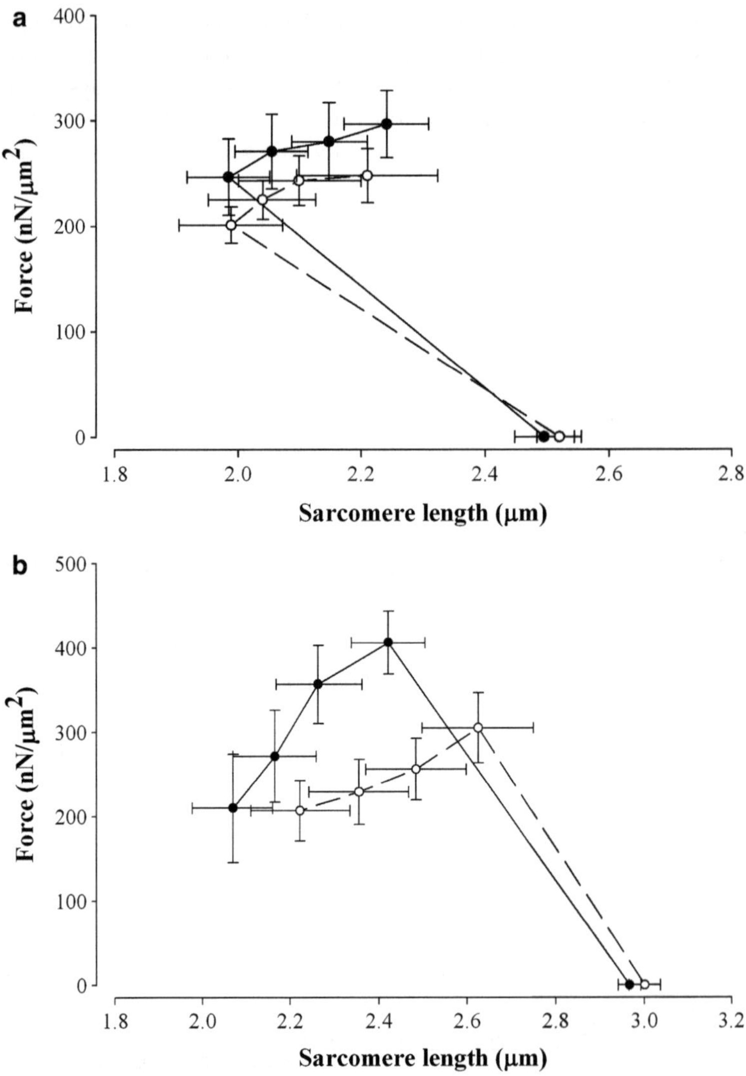

Fig. 4 Average force values for all experiments performed with myofibrils activated with Ca^{2+} (*open symbols*) and MgADP (*closed symbols*) at rest (low forces, close to zero), activation, and after three stretches (**a**) or shortenings (**b**). All results are mean ± SEM

in experiments in which activation was induced by Ca^{2+}. When the data is normalized for the maximum force produced in each experiment, either at the end of the stretch protocol or at the beginning of the shortening protocol (Fig. 6a), the FL relation also differs when constructed through shortening or stretch contractions.

SL_{dis} was calculated by subtracting the deviation of individual SL (SL_i) from the average SL in each myofibril. An overall increase in the SD_{dis} from rest to activation was observed in all experiments, but differences were not observed between shortening and stretching contractions, with either Ca^{2+} of MgADP activation (Fig. 6b). When the

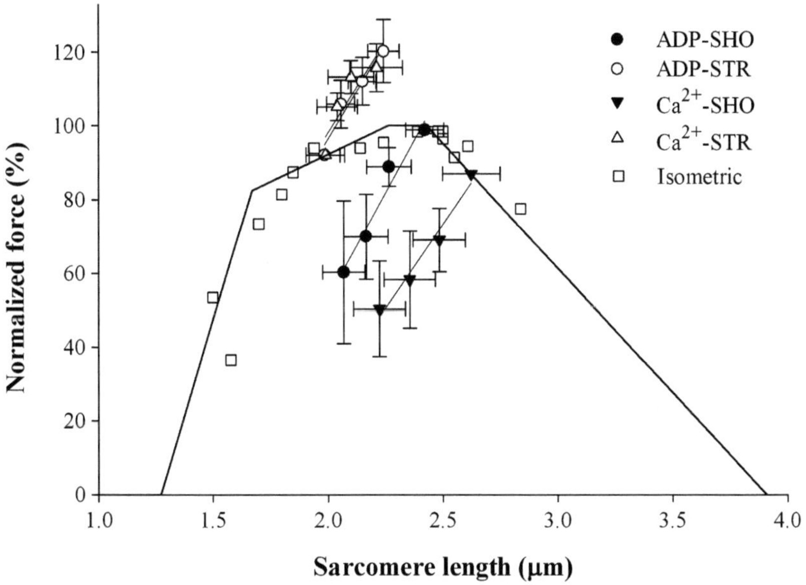

Fig. 5 The dynamic FL relation produced by the five groups investigated in this study. Force was normalized by the isometric forces produced before the first length change in each experiment (~5 s after the initial activation). Solid line shows the theoretical FL relation (1966) using the filament lengths from the rabbit psoas muscle (Sosa et al. 1994). The shapes of the dynamic FL relations are significantly different, except by the two stretch conditions that were similar to each other

SL_{dis} is analyzed for the length steps induced during the experiments, independent from the absolute SL (Fig. 7), it is clear that an overall increase in SL_{dis} observed during initial activation is not accompanied by further changes throughout the contractions.

4 Discussion

The main findings of the studies looking at the history dependence of force production were: (1) stretch and shortening of skeletal muscle myofibrils activated along the ascending limb of the FL relation produced force enhancement and force depression, respectively, (2) force enhancement and force depression were not directly associated with the degree of SL non-uniformity, and (3) force depression was inhibited by MgADP-induced cross-bridge activation. Each of these findings will be discussed in the following paragraphs.

4.1 History-Dependent Properties

The effects of previous stretch and shortening on isometric forces have been broadly investigated on the descending-limb of the FL relation. Little is known

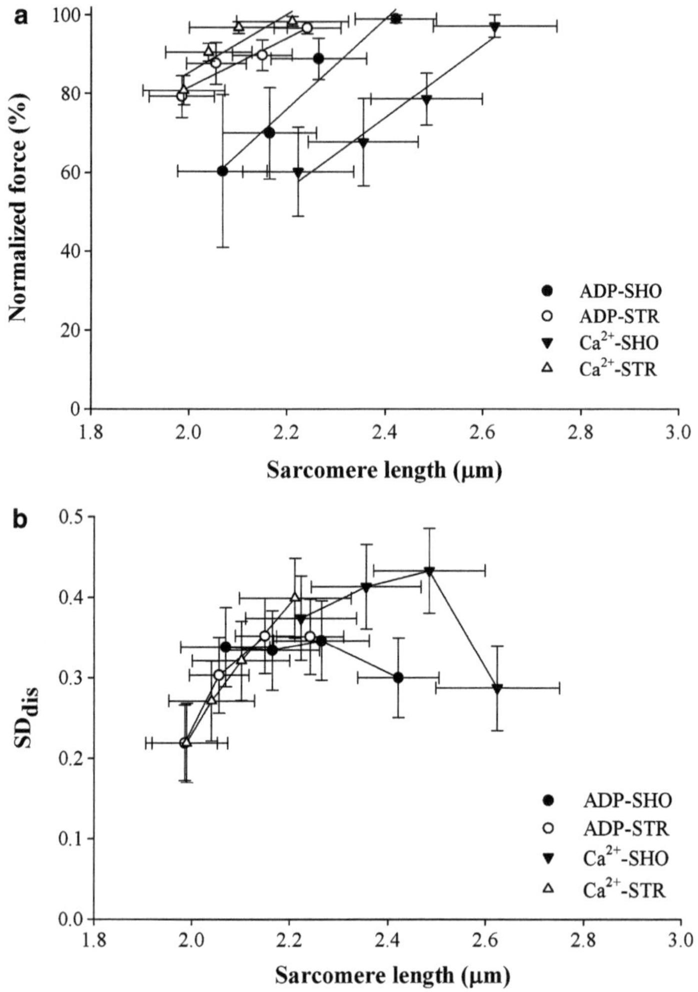

Fig. 6 (**a**) Force normalized by the maximal force produced during the different protocols. The shapes of the FL relation are significantly different, except by the two stretch conditions that were similar to each other. (**b**) The standard deviation of the SL_{dis} (SD_{dis}) calculated during the experiments. SD_{dis} increased from rest to activation, but it did not change throughout contractions and it was not significantly different among conditions. All results are means ± SEM

about history-dependent properties of contractions performed along the ascending limb of the FL relation. A few studies with single muscle fibers show contradictory results; while some observed force enhancement (Peterson et al. 2004) and force depression (Granzier and Pollack 1989), others detect small effects of muscle shortening and/or stretching on force (Edman et al. 1993; Julian and Morgan 1979). The discrepant results may be associated with the experimental preparations used in past studies – muscle fibers allow measurements of the average SL profile, taken from millions of sarcomeres. Since SL shows a high degree of variability and dispersion

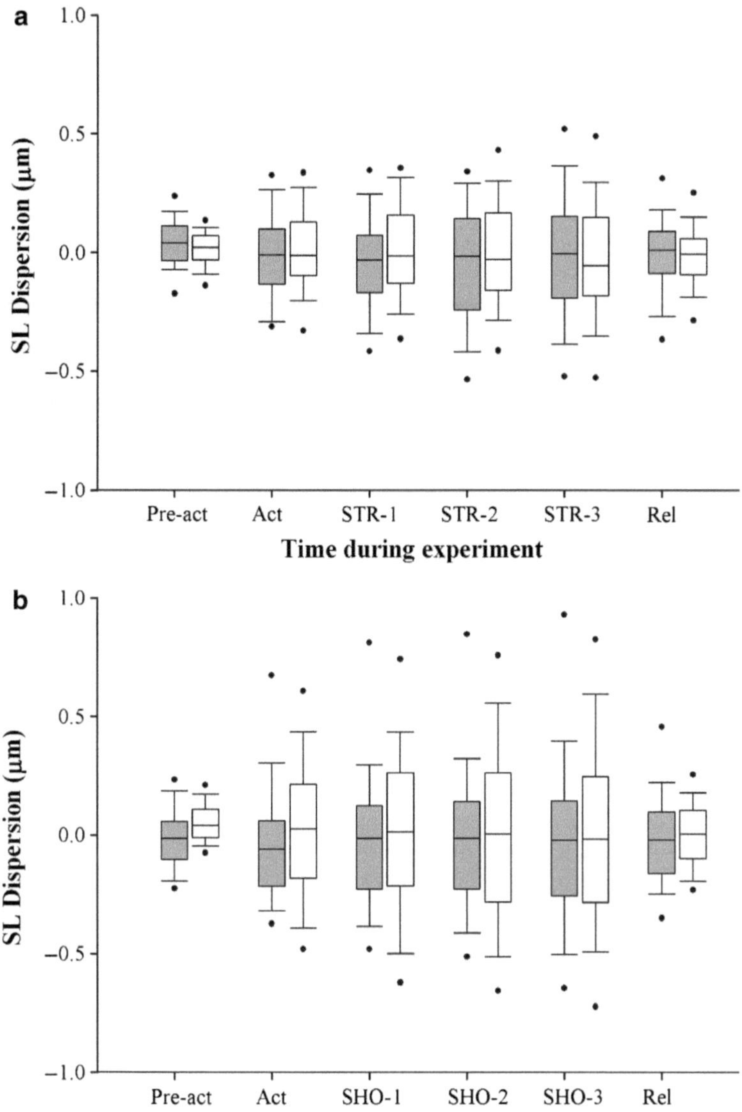

Fig. 7 SL_{dis} plotted against the step change during the experiments with myofibrils activated with Ca^{2+} (*empty boxes*) and MgADP (*filled boxes*), after stretch (**a**) and shortening (**b**). SL_{dis} increases with activation but it remains constant throughout the contractions, with no significant differences between treatments. All sarcomeres from all myofibrils are represented in the boxplots. The line inside each boxplot represents the median, the boundaries represent the 25th and 75th percentiles, and the outliners (*dots*) represent the 95% confidence interval. *Pre* pre-activation, *Act* activation, *Rel* relaxation, *STR* stretch; *SHO* shortening

during muscle activation (Telley et al. 2006a, b; Rassier et al. 2003a; Edman and Reggiani 1984; Julian and Morgan 1979), the results are difficult to interpret. Myofibrils, on the other hand, allow measurements of individual sarcomeres, and thus a direct evaluation of SL non-uniformities.

We observed that forces are in fact dependent on the history of contraction along the ascending limb of the FL relation. The finding is important because force enhancement and depression have been associated with the unstable behaviour assumed to exist uniquely on the descending limb of the FL relation due to its negative slope.

4.2 SL_{dis} and Non-Uniformity

It has been proposed that SL non-uniformity caused by mechanical instability is responsible for the history dependence of force production (Julian and Morgan 1979; Morgan 1994; Morgan et al. 2000). When muscles are stretched or shortened during activation, the degree of SL non-uniformity developed at the beginning of contraction would increase due to an intrinsic instability of the sarcomeres – strong sarcomeres would continuously shorten at the expense of sarcomeres that would stretch. As a result, sarcomeres would contract at lengths that are different from those induced during purely isometric contractions, providing conditions that cause force enhancement and force depression (Morgan 1994; Morgan et al. 2000). This mechanism was not confirmed in our experiments. The degree of SL_{dis} increased with activation, but it did not change significantly after length changes. Furthermore, the degree of SL_{dis} was not different among the different activation methods used in the experiments, which lead to different force levels.

It must be noted that small changes in SLs after the perturbations were observed (e.g. Figs. 3b and 4b), which may change the force, as sarcomeres may never reach a truly steady-state. However, these changes were significantly smaller than those predicted in models that incorporate SL_{dis} and instability to explain force enhancement and force depression (Morgan 1994). Overall, the results of these experiments confirm previous studies in which stretches were applied to myofibrils contracting along the plateau and descending limb of the FL relation (Rassier et al. 2003a, b; Joumaa et al. 2008; Telley et al. 2006b; Rassier 2008). SL non-uniformity and SL_{dis} cannot explain, at least entirely, the changes in force produced by shortening and/or stretch.

4.3 Cross-Bridges Deactivation and Force Depression

Maréchal and Plaghki (1979) suggested that force depression is caused by an inhibition of cross-bridge attachment in the overlap zone newly formed during muscle shortening. When muscle is activated, stress may cause angular distortions in the actin filaments, misaligning myosin binding sites (Daniel et al. 1998). When the muscle is subsequently shortened, if the stress in the thin filaments is maintained, the probability of cross-bridge attachment in the newly formed overlap zone would be reduced. Our results suggest that such mechanism works: when myofibrils were activated with MgADP, which potentially decreases the inhibition of cross-bridge attachment in the newly formed overlap zone after shortening, the degree of force depression was reduced. Consequently, there was a leftward shift in the FL relation when compared to that induced by Ca^{2+} activation; for a given SL the force during MgADP activation was higher than that produced by Ca^{2+} activation in the shortening experiments.

Activation with MgADP through strongly bound cross-bridges is a cooperative process. Studies with isolated filaments show that binding of myosin cross-bridges to actin induces the binding of adjacent cross-bridges (Bremel and Weber 1972; Greene and Eisenberg 1980). Conceptually, MgADP provokes strong binding of cross-bridges to actin, causing conformational changes in the thin filaments that may enhance P_i release and/or other step(s) involved in the isomerization of the AM-ADP-P_i complex. Furthermore, the myofilament lattice is compressed when the number of strongly-bound cross-bridges increases upon muscle activation (Brenner and Yu 1985), approximating thick and thin filaments and increasing the probability of cross-bridge attachment to actin. Finally, the rigidity of the actin filament is likely to be reduced upon binding of myosin cross-bridges (Yanagida et al. 1984), which would release the stress caused by activation. When myofibrils are activated with continuous flow of MgADP, the strong binding of cross-bridges that are responsible for cooperativity will be enhanced while decreasing the lattice spacing and the rigidity of the actin filaments. The newly formed overlap zone may be better aligned for myosin-actin binding in this situation.

4.4 Summary

The force produced by skeletal muscle myofibrils is affected by the history of contraction: consecutive stretches lead to force enhancement and consecutive shortenings lead to force depression. Activation leads to the development of SL non-uniformities, which does not increase significantly after myofibrils are stretched or shortened. Thus, SL non-uniformity does not seem to be correlated with the force produced after stretch or shortening. Instead, our data indicate that force depression is directly linked to changes in the properties of myosin-actin attachment; shortening causes an inhibition of myosin binding probability that decreases the force produced by the muscles.

Acknowledgements This study was supported by the "Canadian Institutes of Health Research". Clara Pun is supported by the "Natural Sciences and Engineering Research Council" of Canada. Dilson Rassier is supported by "Fonds de la Recherche en Santé du Quebec" of Canada.

References

Abbot BC, Aubert X (1952) The force exerted by active striated muscle during and after change of length. J Physiol 117:77–86

Bremel RD, Weber A (1972) Cooperation within actin filament in vertebrate skeletal muscle. Nat New Biol 238:97–101

Brenner B, Yu LC (1985) Equatorial x-ray diffraction from single skinned rabbit psoas fibers at various degrees of activation. Changes in intensities and lattice spacing. Biophys J 48:829–834

Daniel TL, Trimble AC, Chase PB (1998) Compliant realignment of binding sites in muscle: transient behavior and mechanical tuning. Biophys J 74:1611–1621

de Tombe PP, Belus A, Piroddi N, Scellini B, Walker JS, Martin AF, Tesi C, Poggesi C (2007) Myofilament calcium sensitivity does not affect cross-bridge activation-relaxation kinetics. Am J Physiol Regul Integr Comp Physiol 292:R1129–R1136

Edman KA, Reggiani C (1984) Redistribution of sarcomere length during isometric contraction of frog muscle fibres and its relation to tension creep. J Physiol 351:169–198

Edman KA, Caputo C, Lou F (1993) Depression of tetanic force induced by loaded shortening of frog muscle fibres. J Physiol 466:535–552

Gordon AM, Huxley AF, Julian FJ (1966) The variation in isometric tension with sarcomere length in vertebrate muscle fibres. J Physiol 184:170–192

Granzier HL, Pollack GH (1989) Effect of active pre-shortening on isometric and isotonic performance of single frog muscle fibres. J Physiol 415:299–327

Greene LE, Eisenberg E (1980) Cooperative binding of myosin subfragment-1 to the actin-troponin-tropomyosin complex. Proc Natl Acad Sci U S A 77:2616–2620

Greene LE, Williams DL, Jr., Eisenberg E (1987) Regulation of actomyosin ATPase activity by troponin-tropomyosin: effect of the binding of the myosin subfragment 1 (S-1).ATP complex. Proc Natl Acad Sci U S A 84:3102–3106

Herzog W, Leonard TR (2007) Response to the (Morgan and Proske) Letter to the Editor by Walter Herzog (on behalf of the authors) and Tim Leonard. J Physiol 578:617–620

Huxley AF (1957) Muscle structure and theories of contraction. Prog Biophys Biophys Chem 7:255–318

Huxley H, Hanson J (1954) Changes in the cross-striations of muscle during contraction and stretch and their structural interpretation. Nature 173:973–976

Huxley AF, Niedergerke R (1954) Structural changes in muscle during contraction; interference microscopy of living muscle fibres. Nature 173:971–973

Joumaa V, Leonard TR, Herzog W (2008) Residual force enhancement in myofibrils and sarcomeres. Proc Biol Sci 275:1411–1419

Julian FJ, Morgan DL (1979) The effect on tension of non-uniform distribution of length changes applied to frog muscle fibres. J Physiol 293:379–392

Maréchal G, Plaghki L (1979) The deficit of the isometric tetanic tension redeveloped after a release of frog muscle at a constant velocity. J Gen Physiol 73:453–467

Morgan DL (1994) An explanation for residual increased tension in striated muscle after stretch during contraction. Exp Physiol 79:831–838

Morgan DL, Proske U (2006) Sarcomere popping requires stretch over a range where total tension decreases with length. J Physiol 574:627–628

Morgan DL, Proske U (2007) Can all residual force enhancement be explained by sarcomere non-uniformities? J Physiol 578:613–615

Morgan DL, Whitehead NP, Wise AK, Gregory JE, Proske U (2000) Tension changes in the cat soleus muscle following slow stretch or shortening of the contracting muscle. J Physiol 522:503–513

Peterson DR, Rassier DE, Herzog W (2004) Force enhancement in single skeletal muscle fibres on the ascending limb of the force–length relationship. J Exp Biol 207:2787–2791

Pinniger GJ, Ranatunga KW, Offer GW (2006) Crossbridge and non-crossbridge contributions to tension in lengthening rat muscle: force-induced reversal of the power stroke. J Physiol 573:627–643

Rassier DE (2008) Pre-power stroke cross bridges contribute to force during stretch of skeletal muscle myofibrils. Proc Biol Sci 275:2577–2586

Rassier DE, Herzog W, Pollack GH (2003a) Dynamics of individual sarcomeres during and after stretch in activated single myofibrils. Proc Biol Sci 270:1735–1740

Rassier DE, Herzog W, Pollack GH (2003b) Stretch-induced force enhancement and stability of skeletal muscle myofibrils. Adv Exp Med Biol 538:501–515

Roots H, Offer GW, Ranatunga KW (2007) Comparison of the tension responses to ramp shortening and lengthening in intact mammalian muscle fibres: crossbridge and non-crossbridge contributions. J Muscle Res Cell Motil 28:123–139

Sokolov SY, Grinko AA, Tourovskaia AV, Reitz FB, Yakovenko O, Pollack GH, Blyakhman FA (2003) 'Minimum average risk' as a new peak-detection algorithm applied to myofibrillar dynamics. Comput Methods Programs Biomed 72:21–26

Sosa H, Popp D, Ouyang G, Huxley HE (1994) Ultrastructure of skeletal muscle fibers studied by a plunge quick freezing method: myofilament lengths. Biophys J 67:283–292

Sugi H, Tsuchiya T (1988) Stiffness changes during enhancement and deficit of isometric force by slow length changes in frog skeletal muscle fibres. J Physiol 407:215–229

Telley IA, Denoth J, Stussi E, Pfitzer G, Stehle R (2006a) Half-sarcomere dynamics in myofibrils during activation and relaxation studied by tracking fluorescent markers. Biophys J 90:514–530

Telley IA, Stehle R, Ranatunga KW, Pfitzer G, Stussi E, Denoth J (2006b) Dynamic behaviour of half-sarcomeres during and after stretch in activated rabbit psoas myofibrils: sarcomere asymmetry but no 'sarcomere popping'. J Physiol 573:173–185

Yanagida T, Nakase M, Nishiyama K, Oosawa F (1984) Direct observation of motion of single F-actin filaments in the presence of myosin. Nature 307:58–60

Zahalak GI (1997) Can muscle fibers be stable on the descending limbs of their sarcomere length-tension relations? J Biomech 30:1179–1182

Zhang D, Yancey KW, Swartz DR (2000) Influence of ADP on cross-bridge-dependent activation of myofibrillar thin filaments. Biophys J 78:3103–3111

Cross-Bridge Properties in Single Intact Frog Fibers Studied by Fast Stretches

Barbara Colombini, Marta Nocella, Giulia Benelli, Giovanni Cecchi, and M. Angela Bagni

Abstract Cross-bridges properties were measured under different experimental conditions by applying fast stretches to activated skeletal frog muscle fiber to forcibly detach the cross-bridge ensemble. This allowed to measure the tension needed to detach the cross-bridges, P_c, and the sarcomere elongation at the rupture force, L_c. These two parameters are expected to be correlated with cross-bridges number (P_c) and their mean extension (L_c). Conditions investigated were: tetanus rise and plateau under normal Ringer and Ringer containing different BDM concentrations, hyper (1.4T) and hypotonic (0.8T) solutions, 5 and 14°C temperature. P_c was linearly correlated with the tension (P) developed by the fibers under all the conditions examined, however the ratio P_c/P changed depending on conditions being greater at low temperature and higher tonicity. These results indicate that, (a) P_c can be used as a measure of attached cross-bridge number and (b) the force developed by the individual cross-bridge increases at high temperature and low tonicity. L_c was not affected by tension developed, however it changed under different conditions, being greater at low temperature and high tonicity. These findings, suggests, in agreement with P_c data, that cross-bridge extension is smaller at low temperature and high tonicity. By comparing these data with tetanic tension we concluded that potentiation or depression induced on tetanic force by tonicity or temperature changes are entirely accounted for by changes of the force developed by the individual cross-bridge.

Keywords Cross-bridge rupture • Fast stretches • Skeletal muscle • Intact muscle fiber

M.A. Bagni (✉)
Dipartimento di Scienze Fisiologiche and Istituto Interuniversitario di Miologia, Università degli Studi di Firenze, Viale G.B. Morgagni 63, I-50134 Firenze, Italy
e-mail: mangela.bagni@unifi.it

D.E. Rassier (ed.), *Muscle Biophysics: From Molecules to Cells*, Advances in Experimental Medicine and Biology 682,
DOI 10.1007/978-1-4419-6366-6_11,

1 Introduction

This paper is a short review of our recent experiments made to investigate the properties of the actomyosin interaction in skeletal muscle fibers using the "fast stretch technique". This technique, derived from that used by Flitney and Hirst (1978a) to investigate the "give" phenomena (Katz 1939), consists of the application of great and fast stretches to activated single muscle fibers to forcibly detach the cross-bridge ensemble. This allows the measurements of the force needed to induce the detachment, the critical tension, P_c and the sarcomere elongation at critical tension, the critical length, L_c. P_c and L_c are expected to be related to important properties of the contractile apparatus such as the number of attached cross-bridges (proportional to P_c) and their mean extension, inversely proportional to L_c (Flitney and Hirst 1978a; Bagni et al. 2005).

It is well known that when a ramp stretch is applied to an activated skeletal muscle, tension rises almost linearly up to a point at which the force peaks and falls, or rises more slowly or stays constant, depending on the stretching speed (Edman et al. 1978; Flitney and Hirst 1978a, b; Julian and Morgan 1981; Lombardi and Piazzesi 1990; Stienen et al. 1992; Cavagna 1993; Getz et al. 1998). For stretching speed above ~1 fiber length (l_0) s^{-1}, tension reaches a peak (critical tension, P_c) and then falls quickly in spite of the continued stretching, indicating that the muscle has became suddenly very compliant. This increases of muscle compliance, which causes the sarcomere extension at tension peak to rise steeply (sarcomere "give"; Flitney and Hirst 1978a), is caused by the fast cross-bridge detachment induced by the stretch. Since P_c was directly proportional to the overlap between actin and myosin filaments, as tension and stiffness, Flitney and Hirst (1978a) suggested that critical tension was directly proportional to the number of attached cross-bridges. Therefore, by measuring the rupture force of the cross-bridge ensemble we can measure the cross-bridge number in activated muscle fibers. In contrast to stiffness, another putative cross-bridge number measurement, P_c is not affected by either the quick recovery mechanism or filament compliance (Bagni et al. 2005).

Critical tension, is reached when the sarcomeres are elongated by a given amount, L_c, related to the mean cross-bridge extension at the time of the stretch. If a cross-bridge is less strained, a greater elongation would be needed to bring its force up to the rupture point. In contrast to P_c, critical length depends on the compliance in series with cross-bridges including filament compliance.

In the experiments reported here the number and the mean extension of attached cross-bridges, were determined by measuring P_c and L_c during the forced detachment of the cross-bridge ensemble under various experimental conditions in tetanized frog muscle fibers. The relationships between tension developed, critical force and critical length, were determined: (a) during the tetanus rise and at plateau of maximal contractions; (b) at plateau of sub maximal tetanic contractions obtained by using Ringer solution containing 2,3-butanedione monoxime (BDM) at various concentrations; on the rise and at tetanus plateau of tetanic contractions (c) at different medium tonicity and (d) at different temperature.

In principle, changes of the force developed occurring in all these conditions can be due to changes of attached cross-bridge number, changes of the force developed by the individual cross-bridge, or both. P_c and L_c measurements allowed distinguishing amongst these possibilities.

2 Methods

Frogs (*Rana esculenta*) were killed by decapitation followed by destruction of the spinal cord, according to the procedure suggested by the Animal Care and Use Committee of the University of Florence and the Official regulation of the European Community Council (Directive 86/609/EEC). Single intact fibers, dissected from the tibialis anterior muscle (4–6 mm long, 60–120 μm diameter) were mounted by means of aluminum foil clips between the lever arms of a force transducer (natural frequency 30–60 kHz) and a fast electromagnetic motor (minimum stretch time 100 μs) in a thermostatically controlled chamber provided with a glass floor for white and laser light illumination. The experiments were performed at 5 or 14°C (±0.2°C) at about 2.1 μm resting sarcomere length. Stimuli of alternate polarity, 0.5 ms duration and 1.5 times threshold strength, were applied transversely to the fiber by means of platinum-plate electrodes. Frequency of stimulation was adjusted at the minimum frequency necessary to obtain fused tetani at each temperature. Sarcomere length was measured using the striation follower device in a fiber segment (1.2–2.5 mm long) selected for striation uniformity in a region as close as possible to the force transducer. This eliminated the effect of tendon compliance on sarcomere length measurements. Force, fiber length and sarcomere length signals were measured with a digital oscilloscope (4094 Nicolet USA), and transferred to a personal computer for further analysis. Ringer solution had the following composition (mM): NaCl, 115; KCl, 2.5; $CaCl_2$, 1.8; NaH_2PO_4, 0.85; Na_2HPO_4, 2.15. BDM Ringer was obtained by adding the appropriate amount of BDM to Ringer solution to reach the final concentration of 1–5 mM. Hypertonic and hypotonic solutions were obtained by adding or removing NaCl from the Ringer solution in the amount appropriate to reach a tonicity of 1.4 or 0.8 times the Ringer tonicity.

Fast ramp shaped stretches, duration between 0.25 and 0.7 ms; stretching speed, 9.5–90 $l_0\,s^{-1}$ and amplitude ~20 nm per half-sarcomere (hs^{-1}), were applied to one fiber end while force response was measured at the other end. When passive force response to the stretch was measurable, active response was corrected for it. Correction was also made for the contribution due to fiber inertia or viscosity and static stiffness as described previously (Bagni et al. 2005). All these corrections were never greater than 5% of the critical tension.

Fibers developing the maximum tetanic tension (P_0) were easily damaged by the stretches used here. To reduce the damage, in all the conditions studied, most of the stretches were applied when the tension developed was smaller than 0.8 P_0.

3 Results

Figure 1a shows a typical force and sarcomere length response to a fast stretch applied to a single muscle fiber at the tetanus plateau at 14°C. Figure 1b shows a portion of the same response at fast time base. Tension rises steeply and almost linearly during the stretch up to a peak and then falls quickly towards the isometric level in spite of the continued stretching of the sarcomeres. This means that at tension peak fiber stiffness suddenly became negative due to the forced cross-bridge detachment imposed by the stretch. The tension peak represents the critical tension P_c. The sarcomere elongation at tension peak is the critical length L_c. With the stretch amplitude used here, the tension after the stretch, did not fall below the isometric value. This is very likely due to a fast cross-bridge re-attachment which follows the rupture imposed by the stretch (Griffiths et al. 1980).

The mean critical tension measured in ten fibers, was 2.25±0.122 (mean±s.e.m) times greater than the tension developed. This P_c/P ratio is similar to the values reported in literature (Lombardi and Piazzesi 1990; Stienen et al. 1992; Cavagna 1993; Getz et al. 1998) for lower stretching speed.

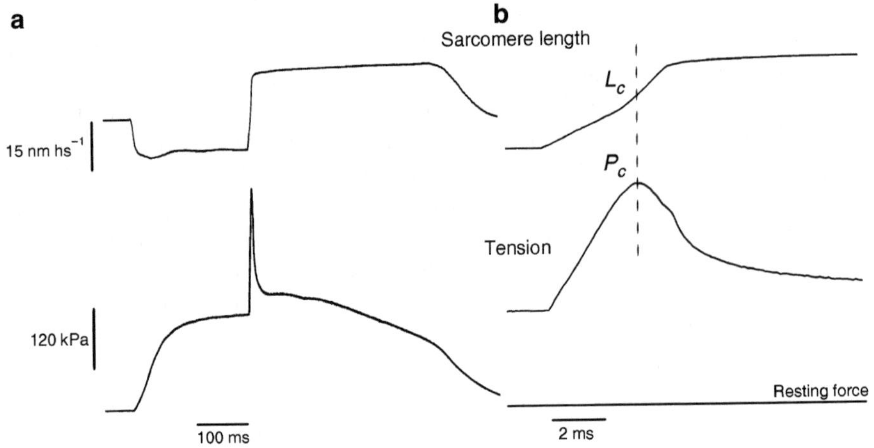

Fig. 1 Force response to a fast ramp stretch applied to a single muscle fiber at tetanus plateau at slow (**a**) and fast time base (**b**). The sharpness of the tension peak at P_c indicates that all cross-bridges are ruptured almost synchronously by the stretch. The *dashed line* in (**b**) indicates the tension peak P_c at which the cross-bridge detachment occurs and the sarcomere elongation at tension peak, L_c. Note that tension starts to fall while sarcomeres are still being stretched. After the attainment of L_c, sarcomere elongation becomes faster, as consequence of the increased sarcomere compliance of the fiber segment monitored by the striation follower due to cross-bridge rupture. Temperature 14°C

3.1 Critical Force and Critical Length on the Tetanus Rise in Normal Ringer and BDM Treated Fiber

As the critical tension is the force needed to forcibly detach the cross-bridges, its value depends, in principle, on both the number of actomyosin bonds in parallel and on the rupture force (F^*) of the individual bond. Therefore the critical force represents a measure of attached cross-bridges number as long as cross-bridges maintain their properties and act independently from each other. To test this point we measured P_c during the tetanus rise, a condition in which it is assumed that tension rises in proportion to the attached cross-bridges. Stretches were applied at different tension levels (P) up to tetanus plateau. Critical tension was normalized by plotting for each fiber the ratio between critical tension on the rise and critical tension at plateau (P_{c0}). Pooled data from ten fibers are shown in Fig. 2a (filled circles). It can be seen that critical tension increases linearly with isometric tension, indicating a direct proportionality with cross-bridge number. This finding is consistent with the idea that P_c measures the cross-bridge number and confirms that tetanus rise is due to a progressive increase of cross-bridges all having the same properties. Similar results (plotted as empty circles, n=3, in Fig. 2a) were obtained when the stretches were applied at plateau of sub-maximal tetanic contractions obtained by using BDM-Ringer at different concentrations. Similarly to the tetanus rise, critical force was directly proportional to tension. This means that, (a) BDM inhibits tension mainly by decreasing cross-bridge number and (b) actomyosin bond rupture force is not affected by BDM.

The critical length L_c was measured in the same experiments reported above and the results obtained on the tetanus rise are shown in Fig. 2b. As expected for the parallel disposition of cross-bridges, L_c was almost independent of the tension

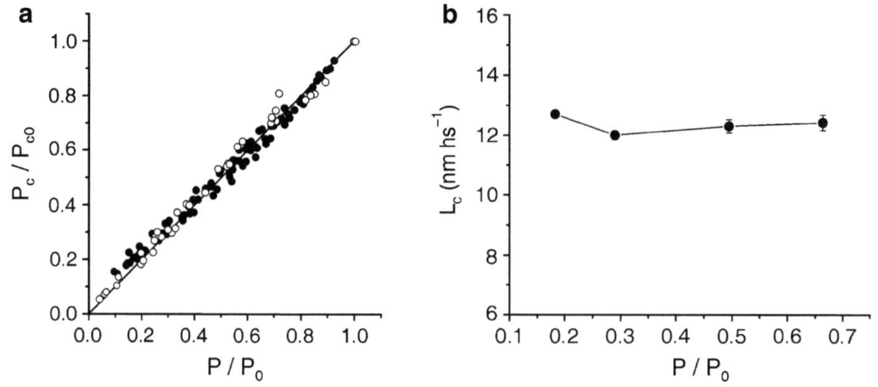

Fig. 2 (**a**) Relative P_c/P_0 relationship during the tetanus rise (*filled circles*) and in presence of various BDM concentrations (*empty circles*). The *line* represents the 1 to 1 proportionality. (**b**) Critical sarcomere length in normal Ringer at various tension during the tetanus rise. Individual points represent the mean values (±s.e.m.) for 0.20 P_c/P_0 class averaging from ten experiments. L_c is almost independent of the isometric tension. Temperature 14°C

developed by the fiber at the time of the stretch. Its mean value was 12.09 ± 0.25 nm hs^{-1}. The same result was found in BDM experiments. L_c was independent of the tension developed by the fiber and its mean value was very similar to Normal Ringer (data not shown).

3.2 *Effects of Tonicity*

Figure 3 shows the well known effects of solution tonicity on tetanic tension (a, b) and the effects of fast stretches applied on the tetanus rise (c, d). Relative to normal Ringer, 1.4T hypertonic solution (a) reduced tetanic tension to 0.82 ± 0.01 (n = 8). The opposite effects (b), a potentiation of 1.09 ± 0.014 (n = 5), was obtained with 0.8T hypotonic solution. These results are in agreement with previous data in literature (Edman and Hwang 1977; Mansson 1989).

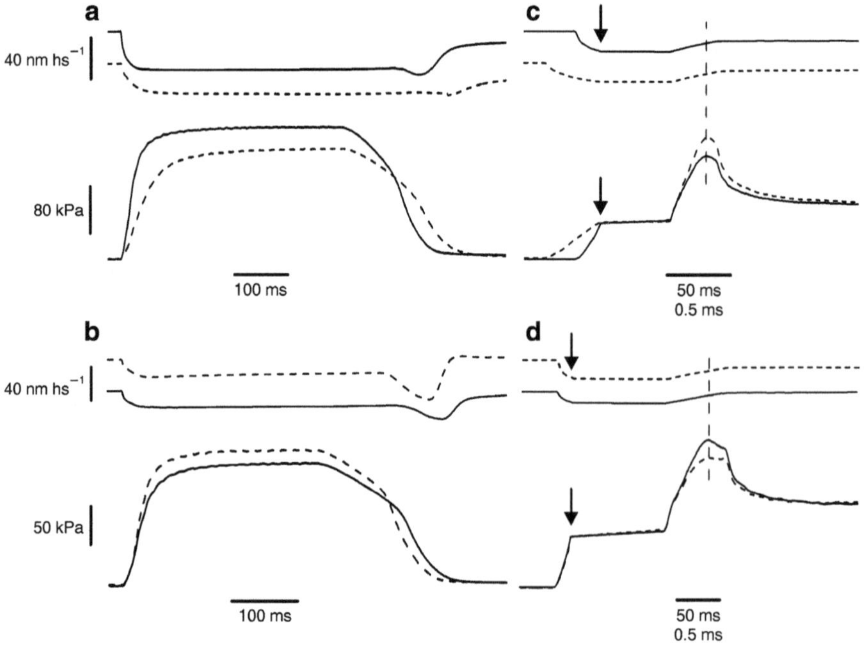

Fig. 3 (**a**) Isometric tetani (*lower traces*) and sarcomere length (*upper traces*) in isotonic (*continuous lines*) and hypertonic solution (*dashed lines*). (**b**) Isometric tetani (*lower traces*) and sarcomere length (*upper traces*) in isotonic (*continuous lines*) and hypotonic solution (*dashed lines*). (**c**) Force response to a fast ramp stretch applied at the same absolute isometric tension during the tetanus rise in isotonic (*continuous lines*) and hypertonic solution (*dashed lines*). (**d**) Force response to a fast ramp stretch applied at the same absolute isometric tension during the tetanus rise in isotonic (*continuous lines*) and hypotonic solution (*dashed lines*). The *arrows* indicate the switch between slow and fast time base recording. The *vertical dashed line* indicates the tension peak. Temperature 14°C

The comparison of force responses to stretches applied at the same absolute tension in normal, hypo- and hypertonic Ringer shows that critical tension increased progressively with the tonicity of solutions, being greater under hypertonic solution and smaller under hypotonic solution. Assuming that the individual cross-bridge rupture force does not change, as in BDM, the greater P_c found in hypertonic solution indicates the presence of a greater number of cross-bridges compared to normal Ringer. Since the tension developed is the same, this means that cross-bridges under hypertonic solution generate a lower mean individual force. Similarly, the smaller P_c found in hypotonic solution shows that under low tonicity solution, cross-bridges develop a greater individual force compared to normal Ringer.

Figure 4a shows the effects of tonicity on the relationship between P_c and isometric tension during the tetanus rise. Both P and P_c were normalized for the plateau tension under the respective solutions. It can be seen that P_c increases linearly with tension in all the solutions, as in normal Ringer, with an intercept not much different from zero. However, the slope of the relationships increases with tonicity. The greatest slope, corresponding to the greatest P_c/P ratio, occurred in hypertonic and the smaller in hypotonic solution. This means that for a given isometric tension, critical tension is always greater in hypertonic solution and smaller in hypotonic solution by the same relative amount compared to isotonic Ringer (Colombini et al. 2007a). The mean P_c/P ratios in isotonic, hypertonic and hypotonic solutions were, 2.25 ± 0.07; 2.70 ± 0.10 and 1.93 ± 0.10 respectively.

Solution tonicity also affected the critical length. This is shown in Fig. 4b. Compared to normal Ringer, L_c increased in hypertonic and decreased in hypotonic solution. Mean values were 12.09 ± 0.25, 13.95 ± 0.21 and 10.25 ± 0.26 nm hs^{-1} in

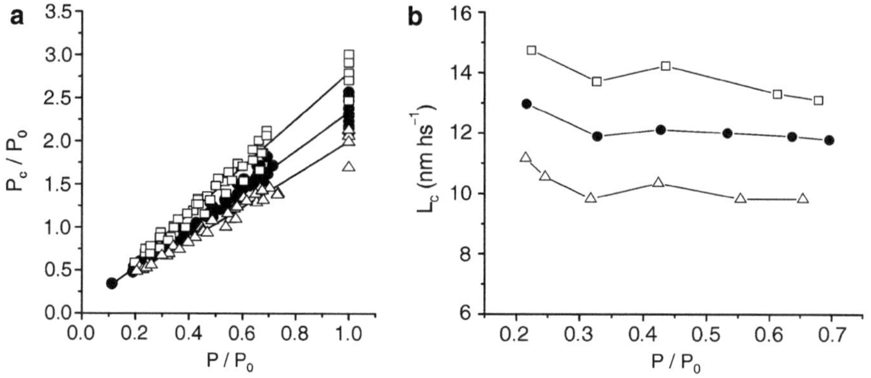

Fig. 4 (**a**) P_c/P_0 as function of P/P_0 during the tetanus rise in normal Ringer (*filled circle*), hypertonic (*empty squares*) and hypotonic (*empty triangles*) solutions. *Continuous lines* are the best fits to the experimental points. The values at $P/P_0 = 1$ are obtained by extrapolation from the linear fitting on the basis of previous data (Bagni et al. 2005) showing the linearity of the relation. Note that the slope of the relation increases with the tonicity of the solutions. (**b**) Critical sarcomere length in normal (*filled circles*) hyper- (*empty squares*) and hypotonic (*empty triangles*) solution at various tension during the tetanus rise in a single fiber. L_c increases in hypertonic and decreases in hypotonic solutions, but in all the solutions it was almost independent of the isometric tension. Temperature 14°C

isotonic, hypertonic and hypotonic solution respectively. Thus, cross-bridge rupture in hypertonic solution requires a greater stretch (by 1.86 nm hs^{-1}) than in normal Ringer, whereas in hypotonic solution a smaller stretch (by 1.84 nm hs^{-1}) is sufficient. This indicates that cross-bridge mean extension depends on solution tonicity, increasing in hypertonic and decreasing in hypotonic solution. These effects are consistent with the effect of tonicity on individual cross-bridge force.

3.3 *Effects of Temperature*

It is well known that tetanic force developed by skeletal muscle fibers is sensitive to temperature. Figure 5a shows that a temperature increase from 5 to 14°C in a single muscle fiber potentiates the tetanic tension by 27%. Similar results were obtained in all the fiber tested (n = 12). For the same temperature increase, tetanic tension increased, by 37 ± 3% in agreement with our previous data (Griffiths et al. 2002).

To examine the effect of temperature on cross-bridges properties, stretches where applied during the tetanus rise at both 5 and 14°C and P_c and L_c were measured. Figure 5b shows the comparison between the force transients obtained when the stretch was applied to the same fiber developing the same absolute tension at high and low temperature. Critical tension is greater at low temperature indicating that, at 5°C, a greater number of cross-bridges is necessary to develop the same isometric tension compared to 14°C. This means that individual cross-bridges develop a smaller force at low temperature, similarly to what occurred in hypertonic solution.

All the P_c data obtained during the tetanus rise at both temperatures are plotted in Fig. 6a as function of the relative tension developed by the fibers. The figure shows that, at both temperatures, critical tension increases linearly with the isometric tension,

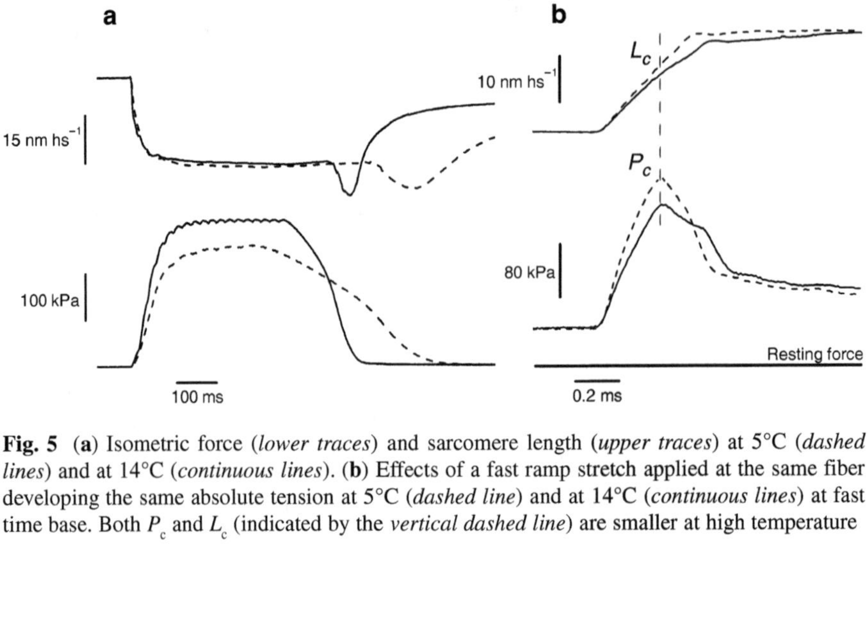

Fig. 5 (**a**) Isometric force (*lower traces*) and sarcomere length (*upper traces*) at 5°C (*dashed lines*) and at 14°C (*continuous lines*). (**b**) Effects of a fast ramp stretch applied at the same fiber developing the same absolute tension at 5°C (*dashed line*) and at 14°C (*continuous lines*) at fast time base. Both P_c and L_c (indicated by the *vertical dashed line*) are smaller at high temperature

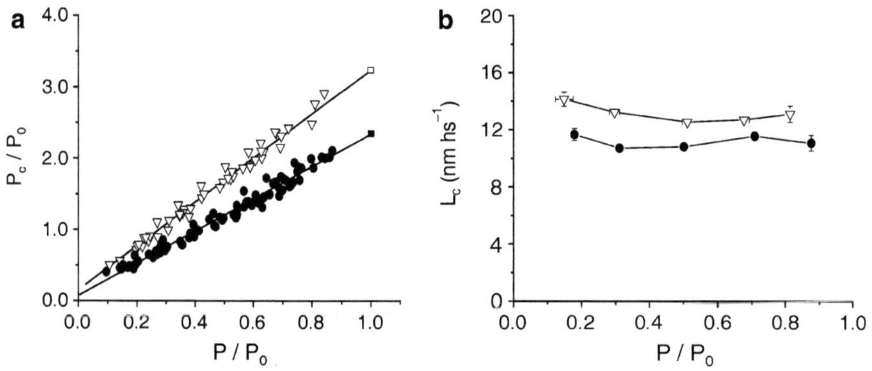

Fig. 6 (**a**) P_c/P_0 as function of P/P_0 at 5°C (*empty triangles*) and at 14°C (*filled circles*) during the tetanus rise. For any given isometric tension, P_c is reduced at high temperature. Values at P_0, indicated by an *empty* and a *filled square*, are obtained by extrapolation from the fitting of the data points at 5 and 14°C (*continuous lines*) respectively. (**b**) L_c measured at various tensions during the tetanus rise at 5°C (*empty triangles*) and 14°C (*filled circles*). Individual points represent the mean values (±s.e.m.) for 0.20 P/P_0 class averaging from ten experiments. L_c depends on temperature but it is almost independent of tension

however the slope of the relationship, representing the P_c/P ratio, is considerably greater at 5°C. The fittings of the experimental data (straight lines) show that P_c/P rises from 2.26±0.04 at 14°C to 3.08±0.09 at 5°C. This means that at any given isometric tension, the force needed to forcibly detach the cross-bridge ensemble is 36% greater at low temperature. This confirms, in all the experiments, the observation above that the number of attached cross-bridges needed to generate a given absolute force is greater at low temperature due to their smaller individual force.

The effect of temperature on L_c is shown in Fig. 6b. L_c is almost independent of the relative tension developed by the fiber at both 5° and 14°C, but depends on temperature. Its mean value decreased by 1.97 nm hs^{-1} going from 13.14±0.28 nm hs^{-1} at 5°C to 11.17±0.19 nm hs^{-1} at 14°C. This suggests that cross-bridges at 14°C are more strained, therefore a smaller elongation would suffice to stretch them up to the rupture force. The greater strain of these bridges also explains the greater force they develop compared to low temperature.

4 Discussion

The experiments reported in this paper were made to investigate cross-bridge properties under various experimental conditions by applying fast stretches to single muscle fibers to induce the forced detachment of cross-bridges. The results show that measurements of force and elongation needed to detach the cross-bridge ensemble provide information on cross-bridge number, cross-bridge mean force and extension.

In all the different experimental conditions examined, the critical tension was always proportional to the tension developed and, therefore, to cross-bridge number. But the degree of proportionality, the P_c/P_0 ratio, depended on the conditions, increasing at lower temperature and at high solution tonicity. In contrast, the critical length was independent of tension but increased at lower temperature and high tonicity.

The direct proportionality between the tension needed to detach the cross-bridges and tension developed during the tetanus rise, is consistent with previous results indicating that P_c was directly proportional to tension when the extent of overlap between myofilaments was varied (Flitney and Hirst 1978a). The independence of L_c from tension developed, or cross-bridge number, is consistent with this view and with the parallel disposition of cross-bridges. Thus P_c represents the tension needed to detach simultaneously all the attached bridges and L_c represents the sarcomere elongation needed to increase the load on the cross-bridges up to their forced rupture.

The values of the P_c/P ratio and L_c found in normal Ringer at 14°C, are in general agreement with those reported on literature with much slower stretches. The P_c/P ratio of 2.25 is slightly greater than the ratio of about 2 reported previously for intact frog preparation and skinned fibers (Getz et al. 1998 and references therein). This is very likely due to the higher stretching speed used here, since P_c/P ratio increases slightly with the stretching speed above $0.5\,l_0\,s^{-1}$ (Flitney and Hirst 1978a; Colombini et al. 2007b). The higher ratio of 3.08 found at 5°C is similar to the 3.2 ratio reported by Getz et al. (1998) in rabbit skinned fibers at the same temperature and it is consistent with previous finding showing that P_c/P ratio increases at low temperature (Chinn et al. 2003).

The increase of critical force in direct proportion to the isometric tension, during the tetanus rise, indicates that tension development is due to a progressive increase of attached cross-bridges all having the same properties. However, this seems in contrast with previous measurements of fiber stiffness, another parameter depending on cross-bridge number, showing that cross-bridges grow more than proportionally during the tetanus rise (Cecchi et al. 1982; Bagni et al. 1999). This discrepancy is mainly due to the effect of myofilament compliance in series with cross-bridges. This compliance is constant whereas cross-bridges compliance decreases in proportion to tension. Therefore, total sarcomere compliance (the sum of cross-bridge and filament compliance) is not inversely proportional to tension, and sarcomere stiffness is not directly proportional to tension. In contrast, it is clear that filament compliance in series has no effect on the tension needed to detach the cross-bridges.

The same result above was found also in BDM Ringer: P_c measured at various tension levels, obtained with different BDM concentrations, was proportional to tension. This finding disagrees with previous reports with BDM Ringer obtained with much lower stretching speed (Rassier and Herzog 2004). The contradiction, however, is only apparent since during slow stretches a significant cross-bridge cycling occurs and this affects both P_c and L_c measurements.

The critical sarcomere length on the tetanus rise and at plateau in normal Ringer was ~12.09 nm hs^{-1}, at 14°C and ~13.14 nm hs^{-1} at 5°C similarly to previously values on single intact or skinned fibers in which length changes were measured at sarcomere level (10–12 nm hs^{-1}, Lombardi and Piazzesi 1990, 8 nm hs^{-1}, Getz et al. 1998). L_c was independent of the tension developed by the fiber and therefore independent of cross-bridge number, in agreement with previous data (Flitney and Hirst 1978a). However, due to filament compliance, L_c was expected to decrease at low tensions. This is because the portion of stretch amplitude absorbed by the filament compliance, which adds to that applied to cross-bridges to give L_c, decreases at low tensions in proportion to P_c. This discrepancy could be due to two possible mechanisms. (a) Filament compliance may not be linear (Bagni et al. 1999 and reference therein) or (b) L_c is affected by the quick force recovery (Bagni et al. 1988). The effect of the quick force recovery is to relief the tension induced on the cross-bridges by the stretch, which would then require a greater L_c to reach the rupture force. With the stretching speed used here, this effect is likely to be negligible at tetanus plateau and at higher tensions. But at low tensions the velocity of the quick recovery becomes much faster (Ford et al. 1986; Bagni et al. 1988) and L_c could increase significantly, compensating the reduction expected from the effect of filament compliance.

Using the stretch technique in bathing solutions of different tonicity we demonstrated that, for a given tension developed, the critical tension increased with solution tonicity. Compared to normal Ringer, the P_c/P_0 ratio increased by about 20% in hypertonic Ringer and decreased by 14% in hypotonic Ringer. If the individual cross-bridge rupture force is insensitive to solution tonicity, these results show that, relative to normal Ringer, at a given tension there are 20% more cross-bridges in hypertonic and 14% less in hypotonic Ringer's solution. This means that cross-bridges develop a 20% lower and a 14% greater individual force in hypertonic and hypotonic solution respectively, in agreement with previous data (Mansson 1989; Piazzesi et al. 1994). This result shows that the depressant effect of hypertonicity on tetanic force (−18%) is entirely accounted for by the reduction of the individual cross-bridge force.

If an individual cross-bridge develops less force, its extension is expected to be smaller, therefore, a greater L_c would be needed to extend it up to the critical length. For this reason, the reduction of the cross-bridge individual force in hypertonic solution, should be accompanied by a greater L_c. This is what we found: L_c at 1.4T was about 15% greater respect to isotonic Ringer. Similar changes of L_c occurred under different experimental conditions (Flitney and Hirst 1978b; Stienen et al. 1992). Is the increase of L_c quantitatively consistent with the individual cross-bridge force reduction? The following calculation answers this question. In isotonic Ringer, L_c was 12.09 nm hs^{-1} and the critical tension was 2.25 times greater than isometric tension. Assuming a linear compliance for cross-bridges and myofilaments this means that total sarcomere extension at P_c is 21.76 nm hs^{-1} (12.09/1.25×2.25) of which 9.67 nm hs^{-1} is the extension at tetanus plateau. In hypertonic saline the relative isometric tension is 0.82, consequently the isometric sarcomere elongation reduces to 7.93 nm hs^{-1} (9.67 × 0.82). Assuming that individual cross-bridge rupture force and total sarcomere extension at P_c are the same as in isotonic Ringer, and considering that P_c/P ratio is 2.70, an external elongation of 13.48 nm hs^{-1} [7.93 × (2.70 − 1)] would be needed to reach the critical tension. This is in good agreement with L_c value of 13.95 nm hs^{-1} found experimentally.

The same analysis above, made for the hypotonic solution data, shows that the calculated L_c was 9.8 nm hs^{-1}, close to the measured value of 10.25 nm hs^{-1}. The good agreement between calculated and measured data in both hyper and hypotonic solutions confirms the validity of the assumptions that cross-bridge rupture force and total sarcomere extension are independent of tonicity.

These mechanical data were confirmed, even quantitatively, by X-ray diffraction experiments, showing that solution tonicity affected the myosin head orientation (Colombini et al. 2007a).

Force potentiation by temperature in the range 0–20°C was generally attributed to the increase of the mean force developed by the individual cross-bridge with no significant changes in cross-bridge number. However recent reports on skinned fibers (Kawai 2003) or single molecule experiments (Kawai et al. 2006) attributed the force potentiation by temperature to an increased cross-bridge number, in opposition to the general view. Our P_c and L_c measurements offer an alternative method to investigate the effects of temperature which helps clarifying this contradiction. P_c/P ratio was 3.08 at 5°C and 2.26 at 14°C, hence for any given tension, P_c, was 27% smaller at 14°C than at 5°C. To calculate precisely the effect of temperature on P_c we need to correct for the effects of temperature on the individual cross-bridge rupture force F^*. It is known that the application of a mechanical force, directed appropriately, increases exponentially the detachment rate constant of a non-covalent chemical bond, such as the actomyosin bond, leading to a rapid rupture of the bond itself. Under these circumstances the rupture force of the single bond is:

$$F^* = (k_B T/x_\beta)\ln(x_\beta / k_0\, k_B T) + (k_B T/x_\beta)\ln(r) \tag{1}$$

where k_B is the Boltzmann constant, k_0 is the detachment rate constant at zero tension, r is the loading rate of the bond and x_β is the distance between the stable state to the transition state. In a previous paper (Colombini et al. 2007b) we showed that (1), can be applied to the cross-bridge ensemble in the overlap region of the half-sarcomere and this allowed to calculate the total rupture force of the parallel cross-bridge ensemble with the (2)

$$P_c = F^* \times N \tag{2}$$

where N is the total number of cross-bridges. Thus, by correcting for the effect of absolute temperature on F^* it is possible to calculate cross-bridge number from critical tension measurements. Whit this correction P_c/P ratio at 14°C reduced to 2.19 and cross-bridge number was 0.71 times the number at 5°C. Thus, the development of the same isometric tension at the higher temperature, requires 29% less bridges than at 5°C which means that at 14°C cross-bridges develop a 40% greater individual force. For the same temperature range, tetanic force increased on average by 37%. Thus, we can say that, within the experimental error, the tetanic force enhancement by temperature is entirely accounted for by the increase of the mean cross-bridge force (Colombini et al. 2008).

As in the tonicity experiments, at higher temperature it is expected that cross-bridges developing a greater individual force are more strained and have a mean position shifted further along the power stroke compared to bridges at low temperature. Then, to raise the tension of these cross-bridges up to the rupture force it is necessary a smaller external elongation and therefore L_c should be reduced at higher temperature. This is what we found: L_c decreased by 1.97 nm hs^{-1}, as the temperature rose from 5 to 14°C.

The following calculation shows that the increase of the average cross-bridge force justifies the whole force potentiation if both cross-bridge and myofilament compliances are assumed to be linear. At 5°C P_c is 3.08 times P and L_c is 13.14 nm hs^{-1}. From these values it can be calculated that the half-sarcomere extension (y_0) at isometric plateau is 13.14/2.08 is 6.32 nm hs^{-1}. To account for the 37% potentiation occurring when temperature raises from 5 to 14°C, with the sole change in individual cross-bridge force, y_0 should also increase by 37%. Therefore, at 14°C, y_0 should be $6.32 \times 1.37 = 8.66$ nm hs^{-1}, 2.34 nm hs^{-1} greater than at 5°C. This predicted value can be compared with the value calculated directly: at 14°C, P_c/P ratio is 2.26 and L_c is 11.17 nm hs^{-1} therefore, $y_0 = 11.17/1.26 = 8.86$ nm hs^{-1}, close to the expected value. It should be pointed out, however, that this predictions is subjected to some uncertainties since critical length is influenced by other factors as the speed of the quick recovery and the speed of the stretch.

The value of 6.32 nm hs^{-1} for cross-bridge mean extension (y_0) at 5°C is relatively high compared with previous data in literature. This depends essentially on the relatively slows stretching speed and on the procedure used to calculate y_0.

At higher temperature the cross-bridge strain increase is in good quantitative agreement with the shift of the myosin head position of 0.73 nm towards the end of the power stroke, measured for 28% of force potentiation with X-ray diffraction technique (Griffiths et al. 2002) and supports the idea that both force generation and temperature potentiation occur through the same molecular mechanism. A similar myosin head shift is also associated to the force potentiation induced by hypotonic solutions (Colombini et al. 2007a).

Our observation that the absolute value of critical tension is independent of temperature is consistent with the result that force potentiation by a temperature jump is strongly reduced when the t-jump is applied to a fiber during stretching (Ranatunga et al. 2007).

5 Conclusion

In conclusion, our data show that by measuring the force needed to detach the cross-bridges and the corresponding sarcomere elongation it is possible to obtain information about the cross-bridge properties. The main findings obtained with this technique are (a) tension development during the tetanus rise is due to the progressive formation of cross-bridges all having the same properties; (b) BDM produces its depressant effect on tetanic tension by altering in a dose dependent way cross-bridge number; (c) tetanic force potentiation induced by a temperature increase from 5 to 14°C is entirely accounted for by the increase of the individual cross-bridge force without a significant change in cross-bridge number; (d) the same mechanism also accounts for the force depression and potentiation occurring by bathing the fiber in hypertonic and hypotonic solution respectively; (e) the insensitivity of individual rupture force of the cross-bridges to the tonicity of the bathing solution, suggests a non ionic nature of the actomyosin bond.

Acknowledgments The authors wish to thank the University of Florence and the Ministero della Ricerca Scientifica (PRIN) for financial support.

References

Bagni MA, Cecchi G, Colomo F, Tesi C (1988) The mechanical characteristics of the contractile machinery at different levels of activation in intact single muscle fibers of the frog. Adv Exp Med Biol 226:473–487

Bagni MA, Cecchi G, Colombini B, Colomo F (1999) Sarcomere tension-stiffness relation during the tetanus rise in single frog muscle fibres. J Muscle Res Cell Motil 20:469–476

Bagni MA, Cecchi G, Colombini B (2005) Crossbridges properties investigated by fast ramp stretching of activated frog muscle fibres. J Physiol 565(1):261–268

Cavagna GA (1993) Effect of temperature and velocity of stretching on stress relaxation of contracting frog muscle fibres. J Physiol 462:161–173

Cecchi G, Griffiths PJ, Taylor SR (1982) Muscular contraction: kinetics of crossbridge attachment studied by high-frequency stiffness measurements. Science 217:70–72

Chinn M, Getz EB, Cooke R, Lehman SL (2003) Force enhancement by PEG during ramp stretches of skeletal muscle. J Muscle Res Cell Motil 24:571–578

Colombini B, Bagni MA, Cecchi G, Griffiths PJ (2007a) Effect of solution tonicity on crossbridge properties and myosin lever arm disposition in intact frog muscle fibres. J Physiol 578(1):337–346

Colombini B, Bagni MA, Romano G, Cecchi G (2007b) Characterization of actomyosin bond properties in intact skeletal muscle by force spectroscopy. Proc Natl Acad Sci U S A 104(22): 9284–9289

Colombini B, Nocella M, Benelli G, Cecchi G, Bagni MA (2008) Effect of temperature on cross-bridge properties in intact frog muscle fibers. Am J Physiol Cell Physiol 294:C1113–C1117

Edman KA, Hwang JC (1977) The force-velocity relationship in vertebrate muscle fibres at varied tonicity of the extracellular medium. J Physiol 269:255–272

Edman KAP, Elzinga G, Noble MIM (1978) Enhancement of mechanical performance by stretch during tetanic contractions of vertebrate skeletal muscle fibres. J Physiol 281:139–155

Flitney FW, Hirst DG (1978a) Cross-bridge detachment and sarcomere "give" during stretch of active frog's muscle. J Physiol 276:449–465

Flitney FW, Hirst DG (1978b) Filament sliding and energy absorbed by the crossbridges in active muscle subjected to cyclical length changes. J Physiol 276:467–479

Ford LE, Huxley A, Simmons RM (1986) Tension transients during the rise of tetanic tension frog muscle fibres. J Physiol 372:595–609

Getz EB, Cooke R, Lehman SL (1998) Phase transition in force during Ramp stretches of skeletal muscle. Biophys J 75:2971–2983

Griffiths PJ, Guth K, Kuhn HJ, Ruegg JC (1980) Crossbridge slippage in skinned frog muscle fibres. Biophys Struct Mech 7(2):107–124

Griffiths PJ, Bagni MA, Colombini B, Amenitsch H, Bernstorff S, Ashley CC, Cecchi G (2002) Changes in myosin S1 orientation and force induced by a temperature increase. Proc Natl Acad Sci U S A 99:5384–5389

Julian FJ, Morgan DL (1981) Variation of muscle stiffness with tension during tension transients and constant velocity shortening in the frog. J Physiol 319:193–203

Katz B (1939) The relation between force and speed in muscular contraction. J Physiol 96(1):45–64

Kawai M (2003) What do we learn by studying the temperature effect on isometric tension and tension transients in mammalian striated muscle fibres? J Muscle Res Cell Motil 24:127–138

Kawai M, Kido T, Vogel M, Fink RHA, Ishiwata S (2006) Temperature changes does not affect force between regulated actin filaments and heavy meromyosin in single-molecule experiments. J Physiol 574(3):877–887

Lombardi V, Piazzesi G (1990) The contractile response during steady lengthening of stimulated frog muscle fibres. J Physiol 431:41–171

Mansson A (1989) Changes in force and stiffness during stretch of skeletal muscle fibers, effects of hypertonicity. Biophys J 56(2):429–433

Piazzesi G, Linari M, Lombardi V (1994) The effect of hypertonicity on force generation in tetanized single fibres from frog skeletal muscle. J Physiol 476:531–546

Ranatunga KW, Coupland ME, Pinniger GJ, Roots H, Offer GW (2007) Force generation examined by laser temperature-jumps in shortening and lengthening mammalian (rabbit psoas) muscle fibres. J Physiol 585:263–277

Rassier DE, Herzog W (2004) Active force inhibition and stretch-induced force enhancement in frog muscle treated with BDM. J Appl Physiol 97(4):1395–1400

Stienen GJM, Versteeg PGA, Papp Z, Elzinga G (1992) Mechanical properties of skinned rabbit psoas and soleus muscle fibres during lengthening: effects of phosphate and Ca^{2+}. J Physiol 451:503–523

Crossbridge and Non-crossbridge Contributions to Force in Shortening and Lengthening Muscle

K.W. Ranatunga, H. Roots, G.J. Pinniger, and G.W. Offer

Abstract Analysis of tension responses to ramp length changes in muscle can provide important information about the crossbridge cycle. During a ramp length change, the force response of an active muscle shows an early change in slope (the P_1 transition) followed by a later, gradual change in slope (the P_2 transition). Modeling shows that the first transition reflects the tension change associated with the crossbridge power stroke in shortening and with its reversal in lengthening; the reduction in slope at the second transition occurs when most of the crossbridges (myosin heads) that were attached at the start of the ramp become detached; the steady tension during shortening is borne mainly by post-stroke heads whereas tension during lengthening is borne mostly by pre-stroke heads. After the P_2 transition, the tension reaches a steady level in the model whereas in the experiments the tension continues to increase during lengthening or to decrease during shortening; this tension change is seen at a wide range of sarcomere lengths and even when active force is reduced by a myosin inhibitor. It appears that some non-crossbridge components in muscle fibers stiffen upon activation and contribute to the continued tension rise during lengthening; release of such tension leads to tension decline during shortening. Thus, non-crossbridge visco-elasticity in sarcomeres may also contribute to energy storage and release during in situ muscle function.

Keywords Muscle force • Crossbridge force • Non-crossbridge force • Shortening • Lengthening

K.W. Ranatunga (✉)
Muscle Contraction Group, Department of Physiology and Pharmacology, School of Medical Sciences, University of Bristol, Bristol BS8 1TD, UK
e-mail: k.w.ranatunga@bris.ac.uk

D.E. Rassier (ed.), *Muscle Biophysics: From Molecules to Cells*, Advances in Experimental Medicine and Biology 682,
DOI 10.1007/978-1-4419-6366-6_12,

1 Introduction

1.1 Background

The force–velocity (F–V) relation of active muscle remains a cornerstone in understanding both the in situ function of muscle and the underlying molecular basis of the contractile process. The force that an active muscle develops declines with increase of shortening velocity reaching zero at the maximum velocity (V_{max}); the maximal mechanical power output occurs at some intermediate velocity (0.2–0.4 V_{max}) (Jewell and Wilkie 1958; Edman et al. 1976). Force increases sharply to ~2·P_0 (isometric force), and a muscle stores energy, as lengthening velocity is increased (Edman et al. 1978, 1981, 1982; Flitney and Hirst 1978; Lombardi and Piazzesi 1990; Månsson 1994; Piazzesi et al. 1992; Stienen et al. 1992). The shortening and lengthening limbs of the F–V relation are not symmetrical and their underlying bases are different; thus, both energy liberation and rate of ATP-breakdown in muscle are increased during shortening but depressed during lengthening (Fenn 1924; Curtin and Davies 1973; Smith et al. 2005).

It was found in some recent experiments on intact rat muscle fibers (Pinniger et al. 2006; Roots et al. 2007) that the force response during a ramp shortening or lengthening shows an early change in slope (the P_1 transition) followed by a more gradual change in slope (the P2 transition); both transitions could be attributed to crossbridge characteristics. The tension level at the P_2 transition could be used to determine the F–V relation and the velocity dependence of the time to the transitions and the length change at the transitions showed asymmetries between shortening and lengthening. These features could be simulated using a mechano-kinetic model of the crossbridge cycle based on a Lymn and Taylor (1971) scheme. After the P_2 transition, the tension reached a steady level in the model but in the experiments continued to decline during a ramp shortening or continued to rise during a ramp lengthening. Therefore, it was proposed that, during shortening and lengthening of active muscle, non-crossbridge elements contribute to muscle tension.

2 Aim

The primary aim of this review is to outline the features of the force response during a ramp (isovelocity) length change in maximally activated intact rat muscle fibers with a view to understanding the underlying mechanisms; evidence of crossbridge and non-crossbridge contributions to force in active muscle is considered.

3 Materials and Methods

All the experimental details have been published before (see Coupland and Ranatunga 2003; Roots et al. 2007). Experiments were performed on small bundles of intact muscle fibers isolated from the flexor hallucis brevis (FHB) muscle of adult male rats (250–300 g, body mass) that were humanely killed with an intra-peritoneal injection of an overdose (>200 mg kg^{-1} body mass) of sodium pentobarbitone (Euthatal, Rhône Mérieux). FHB contains a predominance of fast (~90% type 2) fibers (Coupland and Ranatunga 2003). Bundles of five to ten intact excitable fibers (bundle width ~150 μm) were dissected from the muscle under dark-field illumination and aluminum foil T-clips were attached to the tendons within 0.2 mm of the fiber-ends. The preparation was mounted horizontally between a force transducer (AE 801 element; AME, Horten, Norway) and a servomotor in a flow-through stainless steel chamber (volume ~2 ml); the natural resonant frequency of the force transducer was >5 kHz. The fiber bundle was perfused at 0.5 ml min^{-1} with a physiological saline solution containing (mM): NaCl, 109; KCl, 5; $MgCl_2$, 1; $CaCl_2$, 4; $NaHCO_3$, 24; NaH_2PO_4, 1; sodium pyruvate, 10 and 200 mg l^{-1} of bovine fetal serum; the solution was continuously bubbled with 95% O_2 and 5% CO_2. In experiments using myosin inhibitor N-benzyl-*p*-toluene sulphonamide (BTS), a small volume (<1 ml per 100 ml) of BTS dissolved in DMSO was added to the physiological saline solution to obtain 5–10 μM BTS concentrations; the normal saline flow through the chamber was then replaced with BTS saline flow to expose the preparation to BTS.

A preparation was set to an initial length where isometric tetanic tension was maximal; the sarcomere length using He–Ne laser was ~2.5 μm. The fiber length (L_0) of a bundle was ~2.2 mm (total number of sarcomeres, ~900). Following an equilibration period of at least 30 min, control isometric tetanic contractions were recorded at L_0. Fiber bundles were stimulated with appropriate frequency and duration to give a steady tetanic tension and ramp shortening or stretches of 5–20% L_0 were applied on the tension plateau at different velocities. In some preparations the change in the relative position of two markers, placed on the surface of a fiber bundle, was monitored using a position-sensitive detector (Hamamatsu Photonics, S3932); the two markers (small aluminum foil pieces or human hair) were placed 0.5–0.7 mm apart on the force transducer half of the bundle, so that the relative displacement of their magnified images provided a signal for the segment-length change. Segment length change followed the applied length change to fiber end (see Pinniger et al. 2006; Roots et al. 2007).

4 Tension Response During a Ramp Length Change

4.1 General Features

The superimposed tension traces shown in Fig. 1a illustrate the general features of the tension responses to ramp shortening and lengthening in active muscle (At resting sarcomere length of 2.5 μm, the passive tension in relaxed muscle and its change by ramps were minimal). During a 5% L_0 ramp lengthening, the tension rises rapidly initially and then more slowly until the end of the ramp; the point at which a change in slope occurs is referred to as the P_2 transition. After the end of the ramp, the tension recovers incompletely towards P_0 so that there is a residual incremental force (P_3) at the stretched length. Strikingly, an inverse

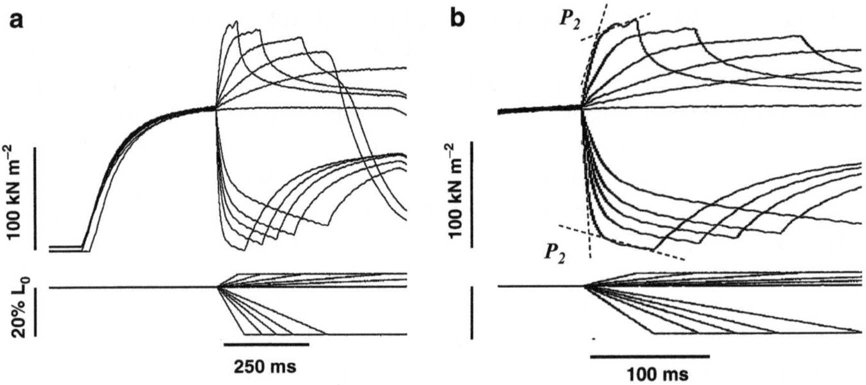

Fig. 1 Tension responses to ramp length changes. (**a**) An isometric contraction and a selection of sample recordings from one preparation at L_0 and 20°C illustrating the tension responses (*upper traces*) to ramp length steps (*lower traces*) applied on the plateau of a tetanic contraction. The duration of tetanic stimulation was shorter in two of the contractions. The tension change to a ramp lengthening and ramp shortening consists of an initial fast phase followed by a later slow phase. (**b**) The same tension traces are shown at an expanded time scale. The tension at the change in slope (the P_2 transition) was estimated from the intersection between two straight lines fitted to the tension records (as shown in two records) (adapted from Roots et al. (2007))

of this tension response is seen during ramp shortening: the tension falls rapidly initially and undergoes a transition (P_2) after which tension declines more slowly till the end of the shortening ramp. After completion of the 20% L_0 ramp shortening, the tension rises to a value less than P_0 indicating a residual force deficit. No detailed analyses were made on the residual force enhancement and the residual force deficit; our particular interest here is to characterize features of the tension response during a ramp length change. The tension level at the P_2 transition was estimated from the point of intersection between two linear regressions fitted to the tension record on either side of the P_2 transition (see Fig. 1b).

4.2 P_2 Transition and the Force–Velocity (F–V) Relation

Assuming that the tension at the P_2 transition represents the crossbridge tension (see modelling below) during a constant velocity length change, a F–V relation can be constructed. Figure 2a shows such a F–V relation at 20°C from one preparation. The lengthening limb of the F–V relation is similar to that previously reported (Pinniger et al. 2006) from the same muscle; the tension increases to a plateau of ~1.6 P_0 as the velocity is increased above ~1–2 L_0/s. The shortening limb of the F–V relation (F–V_s relation) is similar to that reported from rat fast (extensor digitorum longus) muscle at different temperatures, using the isotonic

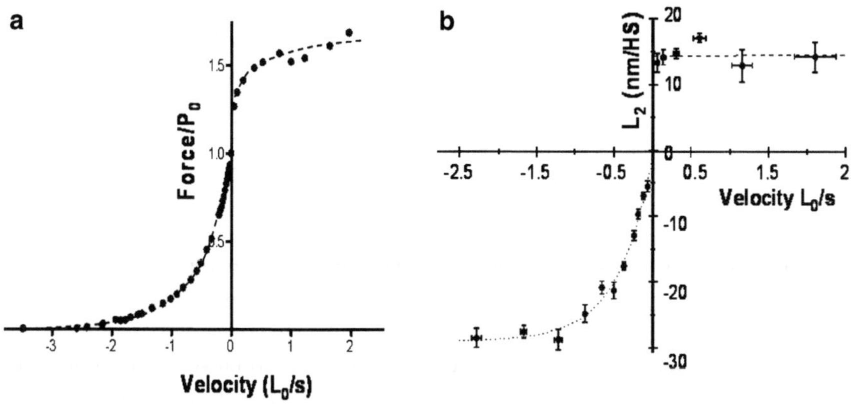

Fig. 2 The P_2 transition and the force–velocity relation. (**a**) The tension at the P_2 transition was normalised to P_0 and is plotted against the shortening (negative) velocity and lengthening (positive) velocity. (**b**) L_2, the length change at the P_2 transition, was determined using the ramp length record and is plotted (in nm per half-sarcomere, HS, taken as 1.2 μm) against the velocity. The *lines* are fitted by eye to the pooled data (mean ± SEM, n = 5–15) from ten fiber bundles. Note that during lengthening L_2 is ~15 nm/HS (<1.5% L_0) and it is relatively insensitive to velocity; the magnitude of L_2 increases with shortening velocity to reach a steady (negative) value of ~25–30 nm/HS (<2.5% L_0) at velocities above ~1 L_0/s (adapted from Roots et al. (2007))

release method (Ranatunga 1984). In those isotonic experiments, however, the data were limited to force levels <0.6 P_0; it can be seen from Fig. 2a that, using the isovelocity approach, a fuller F–V_s relation could be obtained in the present experiments. For comparison with the previous study, such F–V_s data at 20°C and at 30°C, were analysed using A.V. Hill's (1938) hyperbolic equation using a non-linear curve fitting to the force and velocity values. Although the absolute values for V_{max} and a/P_0 (governing the curvature of the plot) were somewhat different (see Roots et al. 2007, for a detailed discussion), the temperature-dependent changes (i.e. a higher V_{max} and a lower curvature at 30°C than at 20°C) were similar between the two studies.

Figure 2b shows the velocity dependence of the length change L_2 at which the P_2 transition occurs. The data shows that L_2 during lengthening, sometimes referred to as the "critical stretch", is insensitive to velocity – as reported in many previous studies (see references in Pinniger et al. 2006). Interestingly, Bagni et al. (2005) have shown that, even with fast (>10 L_0/s) ramp stretches where the force declined at or after P_2 during a stretch, L_2 in frog fibers was independent of stretching velocity; L_2 was ~11 nm/HS and it correlated with the mean crossbridge extension. In our rat fiber experiments at 20°C, L_2 is ~1.2% L_0 (~14 nm/HS) for lengthening. Figure 2b shows that the magnitude of L_2 increases with shortening velocity, up to about 1 L_0/s, and reaches an approximately steady value of ~28 nm/HS as the velocity is increased further to ~3 L_0/s. The L_2 value for shortening velocities >1 L_0/s is thus much higher than that for lengthening.

4.2.1 The P_1 Transition

Figure 3 shows the tension response to a ramp shortening, displayed at different scales, to illustrate that the P_2 transition is preceded by an earlier transition (P_1, indicated by an arrow in a). Close inspection revealed that P_1 transition was an inflection (see Fig. 3b); Ford et al. (1977) and Bressler (1985) clearly commented on this feature in frog muscle fiber experiments and, also, we characterised in rat fiber experiments the inverse of this tension change during ramp lengthening (Pinniger et al. 2006). The basic characteristics of this transition are the following (see Roots et al. 2007). (1) The P_1 tension amplitude was proportional to velocity during shortening as well as during lengthening; F–V relation for P_1 was ~linear. (2) The time to the P_1 transition was uncorrelated with lengthening velocity but showed a significant negative correlation with shortening velocity. (3) The length change L_1 at which the P_1 transition (measured as in Fig. 3b) occurs was correlated with velocity both during shortening and lengthening although it was proportional to velocity in lengthening but indicated a non-linear distribution at different shortening velocities. (4) The average stiffness at the P_1 transition was similar for shortening and lengthening, ~50 P_0/L_0, and within the range as the slope of the T_1 tension versus length step amplitude (~65 P_0/L_0) in this preparation at 20°C (Pinniger et al. 2005), although such stiffness values may be under-estimates because of end-compliance effects and some tension recovery at P_1. Thus, both P_1 and P_2 show asymmetries between shortening and lengthening.

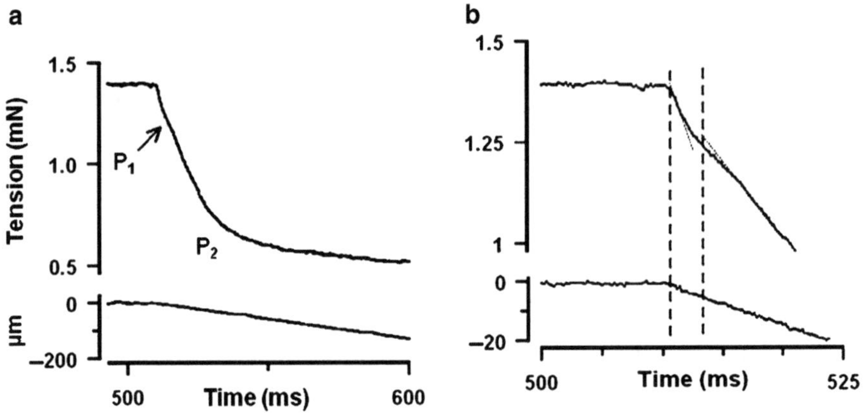

Fig. 3 The P_1 transition: (**a**) A tension trace (*upper trace*) during a ramp shortening (*lower trace*) from a tetanus plateau; the velocity is ~0.7 L_0/s. An early change in slope (P_1 transition, denoted by an *arrow*) is seen on the tension decline. (**b**) Records in **a** are displayed at higher expansion. The P_1 transition is an inflection, where the slope of tension change is initially high, then it decreases and increases again (*superimposed dotted lines*). The P_1 tension was measured near the midpoint, as indicated by the *two vertical dashed lines*; the corresponding length change (L_1) was determined from the length record (adapted from Roots et al. (2007))

5 Crossbridge Modelling

To investigate the molecular processes underlying the P_1 and P_2 transitions we performed simulations of the changes in crossbridge tension during ramp length changes. By using a strain-dependent mechano-kinetic model based on the Lymn-Taylor scheme (1971) for the crossbridge cycle (Scheme 1), the basic features of the tension response during a ramp length change could be simulated; moreover, analyses of the occupancies of pre- and post-stroke attached states provided some insight into the underlying mechanisms (for details, see Pinniger et al. 2006). Figure 4a shows the model simulation of the tension response to ramp shortening. Here, the early inflection marks the P_1 transition and it is due to pre-stroke heads undergoing the power stroke (see Fig. 4c). The P_2 transition occurs when the half-sarcomere shortening (L_2) is ~17–18 nm and, at the P_2 transition, the switchover from original to newly attached crossbridges is essentially complete (Fig. 4a). The simulated force response to lengthening (Fig. 4d) shows that reversal of the power stroke signalled the P_1 transition. The tension rose monotonically after the P_1 inflection and then levelled off at a half-sarcomere length increase (L_2) of ~14 nm, representing the P_2 transition. As in shortening, the P_2 transition occurs when the switchover from original to newly attached crossbridges is essentially complete (see Lombardi and Piazzesi 1990). Although the absolute values are different, a longer L_2 in shortening than in the simulated lengthening tension response is found experimentally.

Scheme 1 (step 1 is attachment/detachment, step 2 is power stroke/reversal, step 3 is the irreversible step involving Pi-release, ATP binding and detachment and step 4 is ATP cleavage on detached heads).

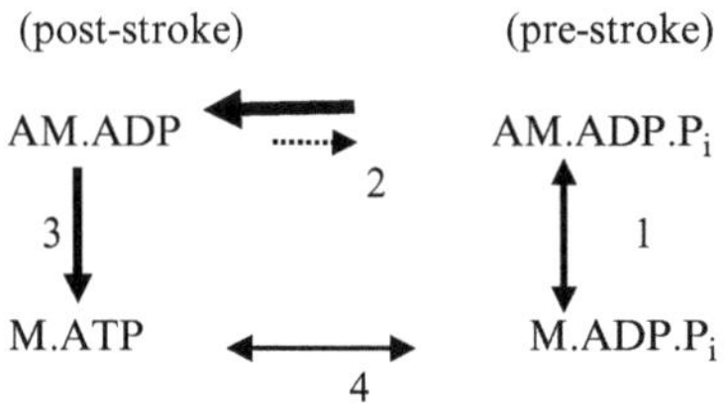

(During shortening the rate of the power stroke in step 2 is enhanced and the full cycle operates faster than in isometric).

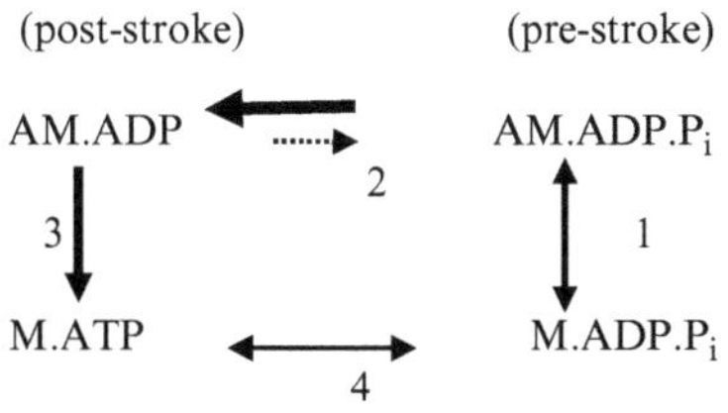

(During lengthening the rate of the reverse power stroke in step 2 is enhanced and the steps indicated by the dotted lines are slowed and hence the cycle becomes truncated to step 1).

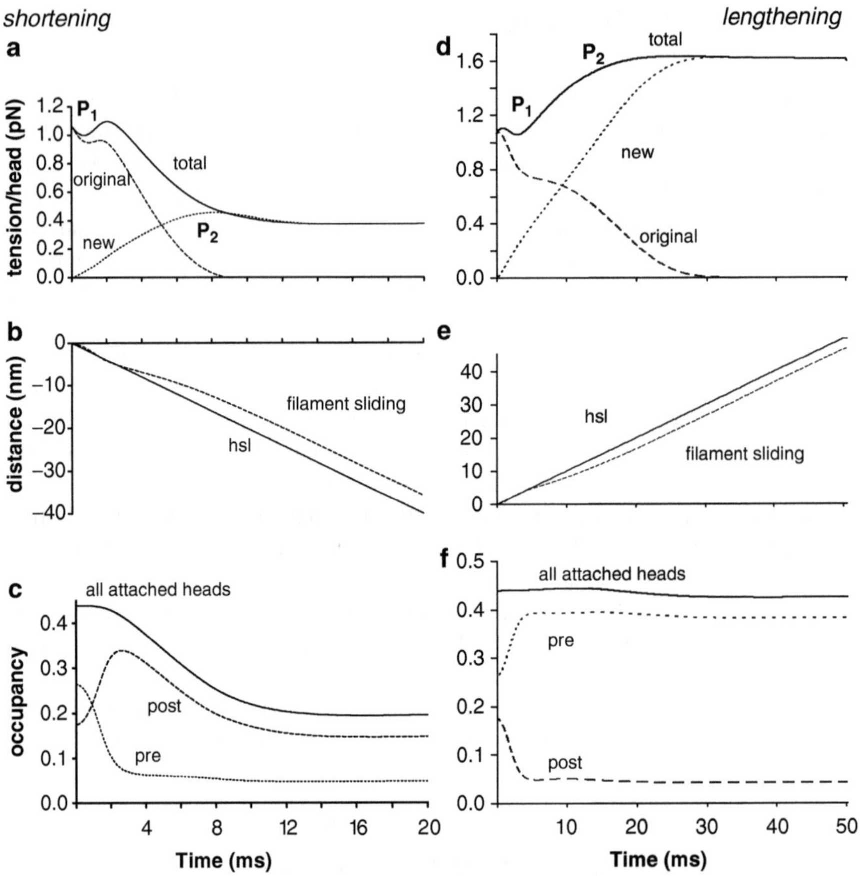

Fig. 4 Simulation of the tension response and other features during a ramp shortening (*left*) and ramp lengthening (*right*): The simulations were based on the Lymn-Taylor kinetic scheme for actomyosin ATPase in solution; the rate constants and their dependence on strain were obtained by refining the model against both the shortening and lengthening limbs of the force–velocity relation of fast mammalian skeletal muscle at 20°C (see Pinniger et al. 2006). (**a, d**) Time-course of changes in tension expressed in pN per myosin head. *Continuous line* is the tension contribution from all attached heads (total); *dashed line*, contribution from heads that were attached at the beginning of the ramp (original) and *dotted line*, the contribution from heads that become attached during the course of the ramp (new). (**b, e**) Time-course of change (nm) in half sarcomere length (hsl) and distance of filament sliding (filament sliding). At the beginning of a ramp, there is no net detachment of heads (crossbridges) and tension is near isometric and, hence, the filament sliding is expected to lag behind hsl. Velocities were ~2 L_0/s (shortening) and ~1 L_0/s (lengthening). (**c, f**) Time-course of changes in fractional occupancy of attached heads in the two conformations pre + post (all attached heads); pre-stroke conformation (pre) and post-stroke conformation (post). Adapted from Pinniger et al. (2006) and Roots et al. (2007)

In experiments, P_1 and P_2 transitions both showed some quantitative asymmetry between the shortening and lengthening limbs. For example, the time to the P_1 transition decreased with shortening velocity, but showed little change with lengthening velocity. Similarly, in Fig. 2b, there is a marked difference between the shortening and lengthening limbs in the dependence of L_2 on velocity; L_2 increases with increasing shortening velocity to a plateau, whereas L_2 is constant in the lengthening limb. Modelling showed that such asymmetries arise because different molecular events dominate during lengthening and shortening and because these events have different strain dependencies.

At an early time during a shortening ramp, the P_1 transition occurs when pre-stroke heads execute the forward power stroke whereas during a lengthening ramp it occurs when some post-stroke heads undergo the reverse power stroke. At later times during shortening, the P_2 transition occurs when the originally attached heads have largely detached by step 3, while during lengthening it occurs when the originally attached heads have largely detached by the reverse of step 1 (see scheme 1). Since the rate constants for these four processes and their sensitivities to strain are different, the asymmetry of slope between shortening and lengthening limbs is understandable; moreover, modeling shows that post-stroke heads carry most of tension during steady shortening (Fig. 4c) whereas during lengthening pre-stroke heads carry most of tension (Fig. 4f).

6 Non-crossbridge Contribution

6.1 Experimental Findings

The simulated tension responses produced by the model for both shortening and lengthening showed the basic features of the tension responses, except for the continued tension change seen after the P_2 transition in the experimental records; the tension reaches a steady level in the model whereas in the experiments the tension continues to decrease during shortening or to increase during lengthening (see Fig. 1).

A continued tension rise during muscle fiber lengthening has been reported in several previous studies (Edman and Tsuchiya 1996; see references in Pinniger et al. 2006). The tension records during ramp shortening reported in mammalian muscle studies (e.g. Joyce et al. 1969; Asmussen and Marechal 1989) clearly show a continued tension decline after an initial fast decrease; Asmussen and Marechal (1989) indeed used an initial, small, length-release to unload series elasticity, but the tension decline was still seen during ramp shortening. Moreover, Joyce and Rack (1969) showed that, in the converse load-clamp experiment, the isotonic shortening velocity decreased significantly with time and that it could not be explained by changes in filament overlap.

From our analyses, the main features of the continuing tension change during a ramp are the following. Firstly, at rest length L_0 (sarcomere length of ~2.5 μm), the tension continues to rise during lengthening or to decline during shortening at a rate that is velocity-sensitive (see Fig. 5a); in shortening, the slope of tension decline reaches a steady value at velocities of 0.5 to ~2 L_0/s. The stiffness(slope/velocity)

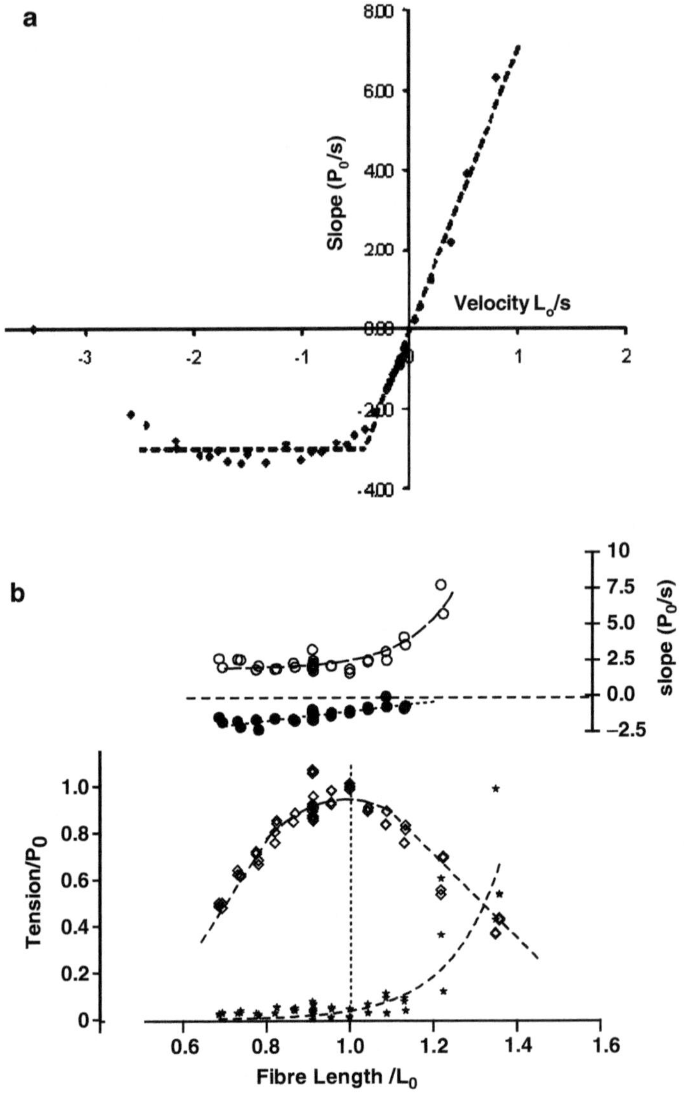

Fig. 5 Tension change after the P_2 transition. (**a**) The rate of tension change after the P_2 transition was calculated from the slope of a linear regression fitted to the post-P_2 tension trace (see Fig. 1b) and is plotted as P_0/s against velocity (20°C). Excluding values at higher shortening velocities (>2 L_0/s) where the determination of the post-P_2 slope was difficult, two phases are seen in the data distribution; phase (i), a proportionate dependence on velocity for low shortening velocities that extends to lengthening and (ii), a velocity-independent phase at intermediate shortening velocities. (**b**) The tension decline or rise after the P_2 transition at different fiber lengths. Pooled data from two fiber bundles showing the tetanic tension (*open diamonds*) and the resting tension (*stars*) at a range of fiber lengths. Values for the tension slope during a standard ramp length step (velocity ~0.7 L_0/s, amplitude 10% L_0) are plotted as P_0/s, *open circles* for lengthening and filled circles for shortening; since velocity was constant, the tension slope relates to "stiffness". All *lines* through data points were fitted by eye. Note that the slope (stiffness) is greater for lengthening than for shortening and it increases at longer fiber length (adapted from Roots et al. (2007))

associated with the continuing tension is ~5–10 P_0/L_0 during lengthening and slow shortening, which is ~10% of the (rapid) crossbridge stiffness for this muscle at this temperature (see Pinniger et al. 2005). Secondly, the continuing tension decline during shortening and tension rise during lengthening are found at the optimal sarcomere length range, and at lengths shorter and longer than optimal (Fig. 5b). Thirdly, the continued tension decline during shortening was not abolished by caffeine indicating that it is not primarily due to Ca-sensitive thin filament de-activation. Fourthly, the continued tension decline during a shortening is reduced, but not abolished, when the active force is depressed by myosin inhibitor BTS; the continued tension rise during lengthening is also reduced with BTS, but only at low force levels, and the stiffness remains much higher than in resting muscle.

6.2 *Mechanism(s) for the Continued Tension Change*

Some mechanisms that have been proposed as contributing to a continuous tension change during ramp shortening and/or lengthening are the following.

6.2.1 Sarcomere Instability

The continuing tension rise during a ramp stretch and the residual tension that persists afterwards have been attributed to increased sarcomere non-uniformity in muscle fibers and "sarcomere popping", i.e. lengthening of weak sarcomeres to beyond filament overlap, the tension in "popped sarcomeres" being then held by non-crossbridge elements (Morgan 1994). However, the latter idea has not received experimental support from recent experiments monitoring sarcomere and half sarcomere length changes during ramp stretch of myofibrils (Rassier et al. 2003; Telley et al. 2006; Rassier 2008); whereas the tension rise and residual tension were observed during and after a ramp stretch, as in seen in muscle fiber recording, there was no evidence of sarcomere popping. Moreover, although sarcomere heterogeneity was apparent, sarcomere heterogeneity did not increase during stretch; thus, while there is evidence for some sarcomere inhomogeneity during stretch, there appear to be mechanisms resisting the anticipated popping. Additionally, in experiments on intact mammalian muscle fibers, the tetanic tensions measured in end-held isometric contractions over a wide range of initial lengths accord with the expected tension versus sarcomere-length relation (see Figs. 3 and 4 in Elmubarak and Ranatunga 1984 and Fig. 6 in Roots et al. 2007); the findings support the notion that, at least in intact mammalian muscle fibers, there may be mechanisms that stabilise sarcomeres. Nevertheless, the possibility that segmental non-uniformities and series compliance may make some contribution to the tension behaviour during lengthening and shortening in intact fibers can not be entirely ruled out (see Julian and Morgan 1979; Edman and Reggiani 1984; Curtin and Edman 1989; Sugi and Tsuchiya 1988; Telley et al. 2003).

6.2.2 Thin Filament Deactivation

Our caffeine experiments (see Roots et al. 2007) suggested that changes in Ca-sensitive thin filament activation/deactivation is not the primary cause for the continued tension decline during shortening. On the other hand, Edman (1975) showed the occurrence of shortening deactivation and Colomo et al. (1986) showed a velocity-sensitive deactivation during muscle fiber shortening that may be Ca-insensitive; although the exact mechanism(s) remain unclear, the general implication was that changes in crossbridge cycling may interfere with thin filament activation/deactivation. Such mechanism(s) in combination may account for the finding that the slope of tension decline during ramp shortening was decreased with depression of active force (crossbridge attachments) by the myosin inhibitor BTS in our experiments (see Roots et al. 2007).

The increased rate of tension decline during shortening, particularly at shorter sarcomere lengths, could be due to decrease in filament overlap. At longer sarcomere length, the continuing tension rise during a stretch was enhanced whereas the rate of tension decline during shortening was decreased (Fig. 5); this might arise from the hysteresis in the force–extension relation and visco-elasticity that is also evident in resting muscle.

6.2.3 Stiffening of Non-crossbridge Elements

Our modelling of the crossbridge cycle showed that, after the P_2 transition, the tension reaches a steady level and no further change in the distribution of strain in attached crossbridges occurs; this suggests that any further tension change observed after the P_2 transition has a non-crossbridge origin. Other basic experimental findings are the following. (1) Edman and Tsuchiya (1996) showed that during stretch of frog fibers the tension rise from the P_2 transition to the end of the ramp and the residual force enhancement after stretch are associated with strain building up in viscoelastic elements other than cross-bridges. (2) Bagni et al. (2002) demonstrated the existence of an increased non-crossbridge stiffness in active muscle fibers, although its magnitude seems small. (3) The visco-elastic tension responses to stretch in resting muscle fibers are qualitatively similar to those in active muscle (Mutungi and Ranatunga 1996; Ranatunga 2001), but their amplitudes ~10-times smaller. The difficulty in identifying the non-crossbridge structural elements responsible may be caused by adhering to the classical assumption that the parallel sarcomeric elastic elements, present in relaxed muscle fibers, are unchanged on activation. A plausible explanation is that non-crossbridge contribution arises from an increase upon activation of titin stiffness (see Labeit et al. 2003) and, possibly, from reversible binding of C-protein (Offer et al. 1973) to actin in a calcium-dependent manner (Moos 1981). Such mechanisms, of course, will not alter the isometric active force versus sarcomere length relationship but they would contribute to tension changes when the sarcomere length is changing. In principle, stiffening of non-crossbridge, visco-elastic elements can account for the continuing tension rise during stretch and tension decline during slow shortening at L_0 (see Fig. 5), their occurrence at a range of sarcomere lengths and, also, it may stabilise sarcomeres and prevent sarcomere popping during lengthening.

6.3 *Implications*

It appears that, during shortening and lengthening, force in active muscle is contributed to (1) by crossbridge mechanisms that are different for shortening and lengthening and (2), as indicated by the continued tension change, by other (non-crossbridge) mechanisms. The continued tension rise during ramp lengthening may be due to stretch of non-crossbridge visco-elastic structures that stiffen on activation; also, release of tension in such structures may be significant in slow ramp shortening. The tension decline during moderately fast ramp shortening, on the other hand, may in part be due to a form of Ca-insensitive de-activation of thin filament. The changes in sarcomere heterogeneity might contribute and modulate to varying extents these processes. The continued tension change is a characteristic feature during lengthening and shortening in intact mammalian muscle and it provides a time (history)-dependent feature to the force–velocity relation; in so far that - within the force–velocity relation – the low velocity length changes may be more physiologically relevant, the stiffening of non-crossbridge elements would contribute significantly to energy storage (during lengthening) and release (during shortening) in active muscle in situ. Linari et al. (2003) indeed commented that two-thirds of the total energy absorption by muscle in a ramp stretch could not be accounted for by cross-bridge (and thick and thin filament) mechanisms; so the non-crossbridge components (Cavagna 1993) may play a major role in animal locomotion by storing energy during lengthening which in the short term could be used in a subsequent shortening.

Acknowledgements We thank the Wellcome Trust Foundation for financial support of our research, and Blackwell Publishing and Springer for permission to include data we had published in the Journal of Physiology and the Journal of Muscle Research and Cell Motility.

References

Asmussen G, Marechal G (1989) Maximal shortening velocities, isomyosins and fibre types in soleus muscle of mice, rats and guinea-pigs. J Physiol 416:245–254

Bagni MA, Cecchi G, Colombini B, Colomo F (2002) A non-cross-bridge stiffness in activated frog muscle fibers. Biophys J 82:3118–3127

Bagni AM, Cecchi G, Colombini B (2005) Crossbridge properties investigated by fast ramp stretching of activated frog muscle fibers. J Physiol 565:261–268

Bressler BH (1985) Tension responses of frog skeletal muscle to ramp and step length changes. Can J Physiol Pharmacol 63:1617–1620

Cavagna GA (1993) Effect of temperature and velocity of stretching on stress relaxation of contracting frog muscle fibres. J Physiol 462:161–173

Colomo F, Lombardi V, Piazzesi G (1986) A velocity-dependent shortening depression in the development of the force–velocity relation in frog muscle fibres. J Physiol 380:227–238

Coupland ME, Ranatunga KW (2003) Force generation induced by rapid temperature jumps in intact mammalian (rat) skeletal muscle fibres. J Physiol 548:439–449

Curtin NA, Davies RE (1973) Chemical and mechanical changes during stretching of activated frog skeletal muscle. Cold Spring Harb Symp Quant Biol 37:619–626

Curtin NA, Edman KAP (1989) Effects of fatigue and reduced intracellular pH on segment dynamics in isometric relaxation of frog-muscle fibres. J Physiol 413:159–174

Edman KAP (1975) Mechanical deactivation induced by active shortening in isolated muscle fibres of the frog. J Physiol 246:255–275

Edman KAP, Reggiani C (1984) Redistribution of sarcomere length during isometric contraction of frog muscle fibres and its relation to tension creep. J Physiol 381:169–198

Edman KAP, Tsuchiya T (1996) Strain of passive elements during force enhancement by stretch in frog muscle fibres. J Physiol 490:191–205

Edman KAP, Mulieri LA, Scuborn-Mulieri B (1976) Non-hyperbolic force–velocity relationship in single muscle fibres. Acta Physiol Scand 98:143–156

Edman KAP, Elzinga G, Noble MI (1978) Enhancement of mechanical performance by stretch during tetanic contractions of vertebrate skeletal muscle fibres. J Physiol 281:139–155

Edman KAP, Elzinga G, Noble MI (1981) Critical sarcomere extension required to recruit a decaying component of extra force during stretch in tetanic contractions of frog skeletal muscle fibres. J Gen Physiol 78:365–382

Edman KAP, Elzinga G, Noble MI (1982) Residual force enhancement after stretch of contracting frog single muscle fibres. J Gen Physiol 80:769–784

Elmubarak MH, Ranatunga KW (1984) Temperature sensitivity of tension development in a fast-twitch muscle of the rat. Muscle Nerve 7:298–303

Fenn WO (1924) The relationship between the work performed and the energy liberated in muscular contraction. J Physiol 59:373–395

Flitney FW, Hirst DG (1978) Crossbridge detachment and sarcomere 'give' during stretch of active frog's muscle. J Physiol 276:449–465

Ford LE, Huxley AF, Simmons RM (1977) Tension responses to sudden length change in stimulated frog muscle fibres near slack length. J Physiol 269:441–515

Hill AV (1938) The heat of shortening and the dynamic constants of muscle. Proc R Soc Lond B 126:136–195

Jewell BR, Wilkie DR (1958) An analysis of the mechanical components in frog's striated muscle. J Physiol 143:5151–5540

Joyce GC, Rack PM (1969) Isotonic lengthening and shortening movements of cat soleus muscle. J Physiol 204:475–491

Joyce GC, Rack PM, Westbury DR (1969) The mechanical properties of cat soleus muscle during controlled lengthening and shortening movements. J Physiol 204:461–474

Julian F, Morgan D (1979) The effect on tension of non-uniform distribution of length changes applied to frog muscle fibres. J Physiol 293:379–392

Labeit D, Watanabe K, Witt C, Fujita H, Wu Y, Lahmers S, Funck T, Labeit S, Granzier H (2003) Calcium-dependent molecular spring elements in the giant protein titin. Proc Natl Acad Sci USA 100:13716–13721

Linari M, Woledge RC, Curtin NA (2003) Energy storage during stretch of active single fibres from frog skeletal muscle. J Physiol 548:461–474

Lombardi V, Piazzesi G (1990) The contractile response during steady lengthening of stimulated frog muscle fibres. J Physiol 431:141–171

Lymn RW, Taylor EW (1971) Mechanism of adenosine triphosphate hydrolysis by actomyosin. Biochemistry 10:4617–4624

Månsson A (1994) The tension response to stretch of intact skeletal muscle fibres of the frog at varied tonicity of the extracellular medium. J Muscle Res Cell Motil 15:145–157

Moos C (1981) Fluorescence microscope study of the binding of added C protein to skeletal muscle myofibrils. J Cell Biol 90:25–31

Morgan DL (1994) An explanation for residual increased tension in striated muscle after stretch during contraction. Exp Physiol 79:831–838

Mutungi G, Ranatunga KW (1996) The visco-elasticity of resting intact mammalian (rat) fast muscle fibres. J Muscle Res Cell Motil 17:357–364

Offer G, Moos C, Starr R (1973) A new protein of the thick filaments of vertebrate skeletal myofibrils. Extraction, purification and characterization. J Mol Biol 74:653–676

Piazzesi G, Francini F, Linari M, Lombardi V (1992) Tension transients during steady lengthening of tetanized muscle fibres of the frog. J Physiol 445:659–711

Pinniger GJ, Bruton JD, Westerblad H, Ranatunga KW (2005) Effects of a myosin-II inhibitor (N-benzyl-p-toluene sulphonamide, BTS) on contractile characteristics of intact fast-twitch mammalian muscle fibers. J Muscle Res Cell Motil 26:135–141

Pinniger GJ, Ranatunga KW, Offer GW (2006) Crossbridge and non-crossbridge contributions to tension in lengthening muscle: force-induced reversal of the power stroke. J Physiol 573:627–643

Ranatunga KW (1984) The force–velocity relation of rat fast- and slow-twitch muscles examined at different temperatures. J Physiol 351:517–529

Ranatunga KW (2001) Sarcomeric visco-elasticity of chemically skinned skeletal muscle fibres of the rabbit at rest. J Muscle Res Cell Motil 22:399–414

Rassier DE. (2008) Pre-power stroke cross bridges contribute to force during stretch of skeletal muscle myofibrils. Proc R Soc B 275:2577–2586

Rassier DE, Herzog W, Pollack GH (2003) Stretch-induced force enhancement and stability of skeletal muscle myofibrils. Adv Exp Med Biol 538:501–515

Roots HR, Offer GW, Ranatunga KW (2007) Comparison of the tension response to ramp shortening and lengthening in intact mammalian muscle fibers: crossbridge and non-crossbridge contributions. J Muscle Res Cell Motil 28:123–139

Smith NP, Barclay CJ, Loiselle DS (2005) The efficiency of muscle contraction. Prog Biophys Mol Biol 88:1–58

Stienen G, Versteeg P, Papp Z, Elzinga G (1992) Mechanical properties of skinned rabbit psoas and soleus muscle fibres during lengthening: effects of phosphate and Ca^{2+}. J Physiol 451:503–523

Sugi H, Tsuchiya T (1988) Stiffness changes during enhancement and deficit of isometric force by slow length changes in frog skeletal muscle fibres. J Physiol 407:215–229

Telley A, Denoth J, Ranatunga KW (2003) Inter-sarcomere dynamics in muscle fibers: a neglected subject? Adv Exp Med Biol 538:481–500

Telley A, Stehle R, Ranatunga KW, Pfitzer G, Stüssi E, Denoth J (2006) Dynamic behaviour of half-sarcomeres during and after stretch in activated psoas myofibrils: sarcomere asymmetry but no "sarcomere popping". J Physiol 573:173–185

Short-Range Mechanical Properties of Skeletal and Cardiac Muscles

Kenneth S. Campbell

Abstract Striated muscles are disproportionately stiff for small movements. This facet of their behavior can be demonstrated by measuring the force produced when the muscle is stretched more than about 1% of its initial length. When this is done, it can be seen that force rises rapidly during the initial phases of the movement and much less rapidly during the latter stages of the stretch. Experiments performed using chemically permeabilized skeletal and cardiac muscles show that the initial stiffness of the preparations increases in proportion with isometric force as the free Ca^{2+} concentration in the bathing solution is raised from a minimal to a saturating value. This is strong evidence that the short-range mechanical properties of activated muscle result from stretching myosin cross-bridges that are attached between the thick and thin filaments. Relaxed intact muscles also exhibit short-range mechanical properties but the molecular mechanisms underlying this behavior are less clear. This chapter summarizes some of the interesting features of short-range mechanical properties in different types of muscle preparation, describes some of the likely underlying mechanisms and discusses the potential physiological significance of the behavior.

Keywords Muscle stiffness • Muscle mechanics • Myosin • SREC

1 Introduction

When a skeletal or cardiac muscle is stretched further than ~1% of its initial length, force rises more quickly at the beginning of the movement than it does thereafter. This behavior has been recognized since at least 1924 (Liddell and Sherrington 1924)

K.S. Campbell (✉)
Department of Physiology & Center for Muscle Biology, University of Kentucky, Lexington, KY 40503, USA
e-mail: k.s.campbell@uky.edu

D.E. Rassier (ed.), *Muscle Biophysics: From Molecules to Cells*, Advances in Experimental Medicine and Biology 682,
DOI 10.1007/978-1-4419-6366-6_13,

and is a general property of both relaxed and activated living striated muscle. The mechanical non-linearity is also evident when chemically permeabilized skeletal and cardiac muscle preparations are stretched in Ca^{2+}-activating solutions but is much less obvious when the permeabilized samples are immersed in solutions with nanomolar free Ca^{2+} concentrations. The initial phase of the non-linear force response has a complex velocity-dependence but is traditionally regarded as a predominantly elastic effect (force proportional to length) as opposed to a predominantly viscous phenomenon (force proportional to stretch velocity).

Two of the most influential papers describing the non-linear behavior were written by D.K. Hill (1968) and Rack and Westbury (1974). Hill used relaxed frog muscles in his experiments and attributed the non-linear force response produced by the relaxed tissue to its "Short Range Elastic Component" (SREC). Rack and Westbury measured the non-linearity in tetanized cat soleus and lateral gastrocnemius muscles and attributed it to the "short range stiffness" of the activated tissue. This chapter will discuss the mechanical properties of both relaxed and activated muscles. The initial phase of the non-linear force response observed in both types of tissue will be referred to in general terms as a short-range mechanical property. The goals of the chapter are to summarize the basic features of the non-linear force response in different types of muscle preparation, consider some of the likely underlying mechanisms and finally discuss the potential physiological significance of the behavior.

2 Short-Range Mechanical Properties are a General Feature of Muscle

To the Author's knowledge, short-range mechanical properties have been observed in every type of muscle preparation that has been tested. He personally, for example, has published experimental records of short-range properties obtained using living frog fibers (Campbell and Lakie 1998), chemically permeabilized mammalian fast skeletal muscle fibers (Campbell and Moss 2000), chemically permeabilized mammalian slow skeletal muscle fibers (Campbell and Moss 2002) and chemically permeabilized mammalian cardiac muscle (Campbell et al. 2003). Figure 7 shows an unpublished experimental record obtained in his laboratory that demonstrates short-range mechanical behavior in an electrically-stimulated rat cardiac trabecula.

Short-range properties have also been measured in smooth muscle preparations. The experimental records published by Gunst (1983) and Meiss (1987), for example, were obtained using canine airway and rabbit mesotubarium superius and overian ligament smooth muscles, and with the exception of the relatively slow time-scale, appear very similar to those measured using striated muscles. It therefore seems possible that the molecular mechanism(s) underlying short-range molecular properties are similar in all muscle types but the behavior in smooth muscle will not be considered further in this work.

One of the many interesting features of short-range mechanical properties is that they are relatively easy to observe in experiments performed on human subjects. Lakie et al. (1984), for example, demonstrated that relaxed fore-arm muscles were disproportionately stiff for small movements. Loram et al. (2007) have recently used ultrasound measurements to show that the short-range stiffness of relaxed human calf muscles resides within the muscle fibers themselves. The short-range stiffness cannot therefore be attributed to non-linear mechanical properties of tendons.

A particularly interesting study was published by Axelson and Hagbarth in 2001. Their measurements showed that the short-range mechanical properties of the fore-arm muscles influenced wrist movements in healthy humans. More importantly, at least for those who are interested in how people control their own movements, the experimental results suggested that the central nervous system "… takes account of and compensates for the … short-range stiffness of the muscles antagonistic to the prime movers." (Axelson and Hagbarth 2001).

3 Molecular Mechanism

3.1 Evidence Supporting a Cross-Bridge Mechanism

Short-range force responses are elicited when a muscle is stretched from any length in its normal operating range. The short-range properties are therefore not defined by the overall length of the muscle but rather self-reset whenever the muscle is held at a fixed length (Campbell and Lakie 1998). This, in itself, suggests that that the short-range mechanical properties of muscle are unlikely to reflect the behavior of a non-linear elastic component in parallel with the contractile apparatus but more quantitative information can be obtained from experiments performed using chemically permeabilized muscle preparations. These studies are particularly enlightening because they allow the short-range properties to be measured in muscles that are differentially activated in solutions with distinct free Ca^{2+} concentrations.

3.2 Calcium Dependence

Figure 1 shows force, sarcomere and fiber length records obtained using a single chemically permeabilized rat soleus fiber immersed in solutions with different free Ca^{2+} concentrations. The records were obtained using a computer-controlled negative feedback system (SLControl: Campbell and Moss 2003) to ensure that the sarcomere length change (measured using laser diffraction) in the center of the muscle fiber was the same at every level of Ca^{2+}-activation. This approach was designed to reduce potential artifacts that could have arisen due to Ca^{2+}-dependent compliance in the ends of the fiber preparation.

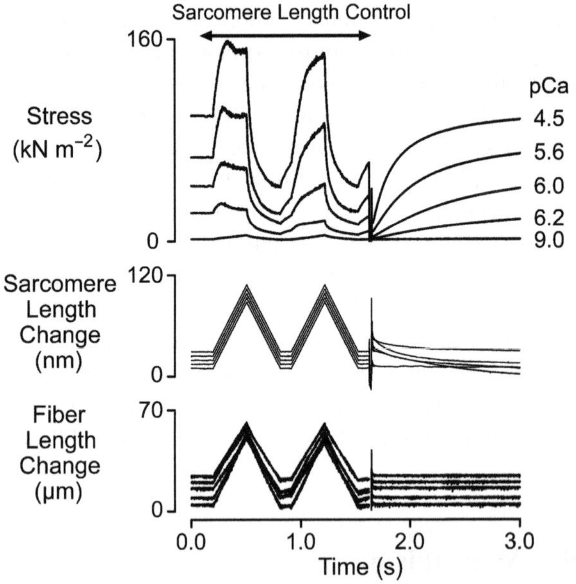

Fig. 1 Ca^{2+} dependence. Superposed experimental records for a chemically permeabilized rat soleus muscle fiber subjected to paired ramp stretches and releases (0.03 l_0, ±0.12 l_0^{-1}, where l_0 is the length of the relaxed muscle fiber at a mean sarcomere length of ~2.60 μm) followed by a large shortening/re-stretch perturbation (0.2 l_0, 20 ms duration) in solutions with different free Ca^{2+} concentrations. The sarcomere and fiber length records are offset vertically from each other for clarity. The sarcomere and fiber length signals are 'clipped' during the large shortening/re-stretch perturbation in this figure so that the quality of the sarcomere length control during the ramp stretch/shortening cycles can be assessed (reproduced with permission from Campbell and Moss 2002)

It is clear from Fig. 1 that the magnitude of the short-range response increased when the fiber was developing high levels of active force. More detailed analysis demonstrated that the stiffness of the initial phase of the response (defined in this example as the slope of a regression line fitted to the first 25 ms of an xy-plot of force against sarcomere length) increased in proportion with the steady-state isometric force (Campbell and Moss 2002). This result is consistent with the hypothesis that the initial phase of the tension response results from stretching cross-bridges bound between the actin and myosin filaments if complications due to thick and thin filament compliance (Huxley et al. 1994; Wakabayashi et al. 1994) are ignored.

3.3 *Myosin ATP-ase Inhibitors*

Further support for the cross-bridge basis for short-range mechanical properties comes from experiments using myosin ATP-ase inhibitors. Figure 2 shows short-range tension responses of a murine diaphragm fiber immersed in solutions with pCa (=-$\log_{10}[Ca^{2+}]$) 9.0, pCa 5.7, and pCa 5.7 + 16 mM 2,3-butanedione-2-monoxime (BDM). Force increased approximately linearly throughout the movement when the

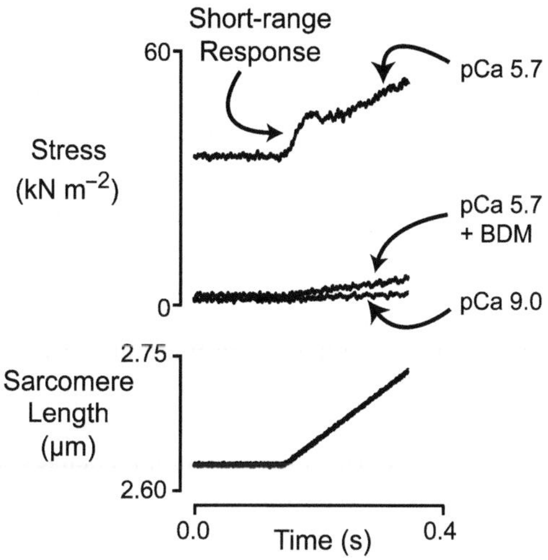

Fig. 2 Short-range response of a single murine diaphragm fiber. Superposed force and sarcomere length records for a chemically permeabilized murine diaphragm fiber subjected to a single ramp stretch (0.04 l_0, 0.20 l_0 $^{-1}$) imposed under sarcomere length control in three different experimental solutions. The fiber developed a steady-state isometric force equal to ~0.44 of the maximum isometric value measured in pCa 4.5 solution when it was immersed in the pCa 5.7 solution lacking BDM. Experiments were performed at 15°C (unpublished records)

fiber was stretched in pCa 9.0 solution (minimal Ca^{2+}-concentration) but there was a clear short-range force response when the trial was repeated using pCa 5.7 solution. When BDM was added to the pCa 5.7 solution, steady-state isometric force was reduced from its initial value of ~45% of the maximal isometric value measured in saturating pCa 4.5 solution to a level very near that measured in pCa 9.0 solution. The short-range force response to the linear stretch was also eliminated.

These records suggest that the short-range force response can only be observed in a muscle in which the myosin cross-bridges are actively cycling. Additional measurements performed in the Author's laboratory show that another myosin ATP-ase inhibitor, blebbistatin, also reduces the short-range force response of activated muscle fibers in a concentration-dependent manner (Mitov and Campbell unpublished data). (The magnitude of the reduction observed in these measurements also depended on the time that the muscle was incubated with blebbistatin (Farman et al. 2008)). This adds further support to the idea that the initial phase of the short-range response represents the stretch of myosin heads that are bound to the thin filament.

4 Dependence on Mechanical History

Up until this point, this chapter has only discussed the non-linear force response measured during a single ramp stretch. However, one of the most intriguing features of the short-range mechanical properties of muscle is the fact that they depend on the history of movement. The effect is clearly seen in Fig. 1. The tension response to the second lengthening movement is different from that measured during the first movement in every activating solution.

This behavior was demonstrated by Denny-Brown in 1929 in experiments in which one end of a cat limb muscle was freed from the skeleton and connected to the ceiling via a length of wire. The force response was then measured as the muscle was stretched by lowering the operating table (mounted on an oil-filled piston) towards the floor. Denny-Brown noted the short-range force response measured in the experiments and commented that it "lessens … upon rapid repetition of the stretch but recovers with rest at a stationary length" (Denny-Brown 1929).

A very similar effect, demonstrated using more modern technology, is shown in Fig. 3. The two upper panels show superposed force and sarcomere length change records for a sub-maximally activated chemically permeabilized rat soleus muscle fiber subjected to two identical length changes separated by different inter-stretch intervals in different trials. The graph in the bottom panel shows

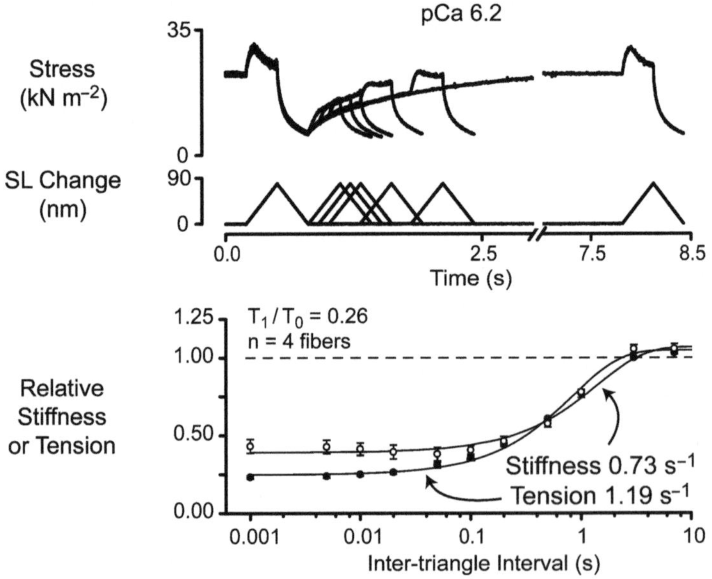

Fig. 3 Recovery time-courses of relative stiffness and relative tension. The *top two panels* show superposed force and sarcomere length records for a single chemically permeabilized rat soleus fiber subjected to paired ramp stretch/shortening cycles separated by time intervals ranging from 1 ms to 7 s. The fiber was immersed in pCa 6.2 solution and developed a steady-state isometric force equal to ~0.26 of the maximum value measured in pCa 4.5 solution throughout the duration of the experiment. The *bottom panel* shows the relative initial stiffness and the relative prevailing tension measured at the beginning of the second length change plotted as functions of the time interval between the two stretch/shortening cycles. The *continuous lines* are best-fits of the form $y = [a - b.\exp(-c.\Delta t)]$ where a and b are constants greater than or equal to zero and Δt is the time interval between the length changes. The labels show the calculated values of the rate constant c for the recovery time-courses of relative stiffness and relative tension. Experiments were performed at 15°C (reproduced with permission from Campbell and Moss 2002)

the relative prevailing force at the beginning of the second movement and the relative short-range stiffness during the second stretch as functions of the time interval between the two movements. Both parameters recover towards unity with a time-course that can be modeled as an exponential recovery process (Campbell and Moss 2002).

The temporary reduction in muscle stiffness that follows movement in a relaxed intact muscle was attributed to a thixotropic effect by Buchthal and Kaiser (1951). The term thixotropy is still used in the biomedical literature (for example, Bianco et al. 2007) but it is not clear that different researchers always use the term in the same way. Another problem is that thixotropy in the more general contact of rheology (the study of the flow of matter) means a "decrease in the apparent viscosity under shear stress, followed by a gradual recovery when the stress is removed." (Harris 1977). The distinction between a thixotropic and a shear-thinning effect therefore seems to be rather arbitrary, perhaps depending in a poorly-defined manner on the time-scale of the observation (Blair 1969).

The term thixotropy also has the unfortunate consequence of implying that the history-dependence of the short-range mechanical properties reflects non-linear viscous effects. This is probably not the case because short-range mechanical properties do not scale linearly with the stretch velocity (Hill 1968; Rack and Westbury 1974). The Author has suggested that muscle's thixotropic behavior would be more accurately described as history-dependent stiffness (Campbell and Moss 2000) but he is not aware of a single word that describes this type of effect.

4.1 Recovery Rate

The fact that the short-range mechanical properties of a muscle depend on its history of movement seems to have been acknowledged for relaxed tissue before it was formally recognized in activated preparations. A particularly interesting set of experiments was performed by Herbst (1976). He used paired length changes similar to those shown in Fig. 3 to measure the recovery time-course of short-range mechanical properties in relaxed frog muscles and suggested that the asymptotic recovery reflected redevelopment of a population of bound cross-bridges that were perturbed by the first movement.

If myosin heads cycled only between a single detached and a single attached state, the rate of stiffness recovery following an imposed perturbation would, in the simplest kinetic models, be equal to the rate of tension recovery (Brenner 1988). The results published by Campbell and Moss in 2002 show that, at least in chemically permeabilized rat soleus fibers, the real behavior is more complicated. The rate of stiffness recovery in these measurements did vary with the level of Ca^{2+}-activation but it was always lower than the rate of tension recovery between the two length changes and k_{tr}, the rate of tension recovery following a much larger mechanical perturbation (Brenner 1988). Although these results seem at first to argue against Herbst's cross-bridge theory, computer simulations presented by Campbell and Moss (2002) showed that stiffness and tension can recover at different rates if myosin heads can be bound to the thin filament in more than one state. It is also now known that filament compliance effects will affect the measured rate of tension recovery (Campbell 2006).

5 The 'Range' of the Short-Range Response

An experimenter who wishes to demonstrate the existence of the short-range mechanical properties of muscle has to stretch their preparation far enough to observe a reduction in the rate of rise of force during the movement. The length change required to elicit the reduction is a key experimental parameter because it defines the 'range' of the short-range response. The parameter also provides important information about the potential underlying mechanisms. If, for example, the observed range was 100 nm half-sarcomere^{-1} (many times the physical size of an S1 myosin head), it is unlikely that the biphasic force response observed experimentally could reflect the stretch and subsequent cyclic reformation of actin-myosin links.

5.1 *Difficulty of Analysis*

Although there is now no doubt that the length change required to demonstrate the short-range response is not greater than the distance that actin-myosin links could be stretched, it is difficult to be substantially more precise. One of the issues is that the range of the response varies with the experimental conditions so that it is impossible to come up with a single definitive value (Campbell and Holbrook 2006; Stienen et al. 1992). A more subtle point is that the transition between the 'short' and 'long' range portions of the force response often occurs as a gradual change in the measured force gradient rather than abruptly at a well-defined instant in time. Consequently it is difficult to envisage a method that can identify a single value for the range without at least some element of uncertainty.

Different researchers have used different techniques to calculate the experimental parameter. For example, Getz et al. (1998) defined L_c, the 'critical strain' of chemically permeabilized rabbit psoas fibers, as the length change at the first time point during the stretch at which the first derivative of force with respect to time became zero. Their results showed that L_c was about 7.7 nm half-sarcomere^{-1} and independent of the stretch velocity for stretch speeds varying from 0.5 to 3.0 muscle lengths per second. Lombardi and Piazzesi (1990) measured l_p in tetanized frog fibers as the "lengthening necessary to attain the peak of force". They obtained a value of ~15 nm half-sarcomere^{-1} for stretch velocities of ~0.4 muscle lengths per second and ~11 nm half-sarcomere^{-1} for stretch velocities of ~1.0 muscle lengths per second. Previous measurements (for example, Edman et al. 1981) performed without sarcomere length control have produced slightly larger values for the 'range' which suggests that experimental estimates of the parameter can be compromised by series compliance in the ends of the muscle preparation.

As noted above, even when experiments are (as in the case of Getz et al. and Lombardi et al.) confined to maximally activated fibers, there is probably not a single best method for measuring the range of the initial phase. Getz et al. could not calculate L_c values for stretch velocities less than or equal to 0.5 muscle lengths per second, for example, because the first derivative of force with respect to length did not drop below zero. Lombardi et al.'s method (noting the length change at the peak of force) would not have worked for the experimental records obtained by Getz et al. where (unlike the case with frog fibers) force continued to rise (albeit at a slower rate) beyond the first phase.

The situation is even more difficult when the experiments involve fibers activated in solutions with different free Ca^{2+}-concentrations. In Fig. 2, for example, it seems clear that the fiber exhibits a short-range response in the pCa 5.7 condition (without BDM) and not when it is immersed in the pCa 9.0 solution. At some point between these conditions, there must be a Ca^{2+}-concentration which is 'just sufficient' to elicit a short-range response. Is there a way of unambiguously defining the threshold level?

One such technique was described in a recent publication by Mitov et al. (2009) that summarized measurements of short-range mechanical properties in chemically permeabilized rat myocardial preparations. The authors in this case assumed that the slower rise in force beyond the short-range response was a linearly increasing tension resulting from the increased strain in passive mechanical components such as titin, collagen and elastin molecules. The authors estimated the stiffness of these structures from the rate of rise of force during the latter stages of the movement and used this parameter to deduce the force due to non-cross-bridge components at different time-points in the lengthening response. The force due to attached cross-bridges was then estimated at each time-point in the record by subtracting the calculated passive mechanical response from the experimental force signal.

These procedures allowed short-range responses to be detected using automated computer programs. A short-range response was defined as having occurred when the peak value in the force due to the attached cross-bridges exceeded a threshold level defined as the root mean square of the random noise in the measured force signal multiplied by a constant factor. When an event was detected, the 'elastic limit' of the short-range response was accordingly defined as the length change required to produce the peak cross-bridge force.

The main advantage of this technique was that it provided an unambiguous way of defining whether experimental records obtained in a wide range of experimental conditions exhibited short-range responses or not. It could also be implemented to run in automated computer programs without requiring continual user intervention. This allowed many hundreds of records to be processed in a very efficient manner. Visual inspection of selected traces showed that the calculated elastic limit values were generally slightly smaller than the 'critical strain' equivalents defined in the manner of Getz et al. as the length change at the time-point at which the first derivative of force with respect to time fell to zero.

Several other techniques have also been used to analyze short-range force responses. Stienen et al. (1992), for example, estimated the transition point between the ‘short’ and the ‘long’ range portions of the force response of rabbit psoas fibers using piece-wise linear regression (Vieth 1989). This technique is probably about as good as any when the force transition is very gradual but would underestimate the force at the end of the first phase in records where the first derivative of force is transiently negative. This Author, and probably many others preceding him, has published records (Campbell and Lakie 1998; Campbell and Moss 2000) in which the transition point was simply identified ‘by eye’.

In this Author’s opinion, ‘fitting by eye’, while not easy to implement as a computer algorithm, is probably not much worse at identifying plausible transition points than techniques that are considerably more mathematically sophisticated. The simple fact is that it is not easy to develop a single algorithm that all researchers will agree identifies the most appropriate transition point in every type of mechanical record showing a short-range response.

6 Velocity-Dependence

One of the reasons that so many researchers have tried to identify a single transition point at the end of the short-range response is that investigating how the point varies with the velocity of the imposed stretch can provide insights into the underlying biophysical mechanism(s). For example, the fact that the short-range force did not increase linearly with stretch velocity when Getz et al. (1998) stretched psoas fibers at rates ranging from 0.5 to 3.0 muscle lengths per second indicates that viscous drag of the myofilaments (a force that would be directly proportional to stretch velocity) was probably not important in these experiments.

There are however conditions in which the short-range response does vary with the velocity of the imposed stretch. One example from the Author’s previous work is shown in Fig. 4. Panel a in this diagram shows records of force, sarcomere length and preparation length for a sample of canine myocardium developing ~3% of its maximal isometric force and stretched at five different velocities ranging from 0.04 to 0.87 muscle lengths per second. The magnitude of the short-range force response is clearly increased in the faster stretches. An interesting feature of these records is that the short-range force peaks at an earlier time in each record as the stretch velocity increases. This is important information because it implies that the short-range response, even at this low level of activation, cannot be attributed to passive visco-elastic components unless the relaxation time of the visco-elastic mechanism is itself influenced by stretch velocity.

Plotting the records from panel a on different axes reveals more information. When force records are plotted not against time but against sarcomere length, as in panel b, the initial phases of the force records for the different stretch velocities now superpose. This analysis indicates that the faster stretch velocities do not produce greater force changes by producing a greater force increment per unit length change, but rather by stretching the muscle further along a fixed force/length trajectory before the transition to a slower rate of rise of force occurs.

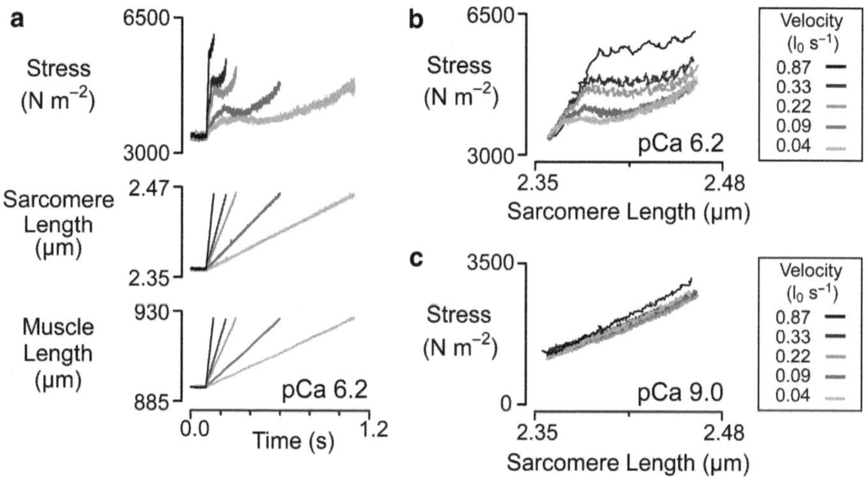

Fig. 4 Velocity dependence. Experimental records for a chemically permeabilized sample of canine myocardium subjected to ramp stretches with velocities ranging from 0.04 to 0.87 l_0 s^{-1}. *Panel* (**a**) shows experimental records plotted against time. *Panels* (**b** and **c**) show force records plotted against sarcomere length. *Panels* (**a** and **b**) show records obtained when the preparation was immersed in pCa 6.2 solution and developing ~0.03 of its maximal Ca^{2+}-activated tension. *Panel* (**c**) shows the response of the same preparation when it was immersed in pCa 9.0 solution with minimal free Ca^{2+}. Experiments were performed at 15°C (reproduced with permission from Campbell et al. 2003)

This behavior is not consistent with simple visco-elastic mechanisms but could be explained by a cross-bridge mechanism if in these experiments fast stretches displaced attached myosin heads further before they detached than slow stretches. This mechanism would still be consistent with Getz et al.'s measurement of a critical strain in rabbit psoas fibers that was insensitive to stretch velocities for movements faster than 0.5 muscle lengths per second if there is a physical limit to the distance that an actin-myosin bond can be extended before it breaks. Forced detachment might, for example, occur when the strain energy stored in the actin-myosin link exceeds the binding strength of the actin-myosin interface.

Before leaving the discussion of velocity-dependence it is interesting to note that while many current-day experiments are carried out using very fast stretches (much faster than those encountered by a muscle in vivo), some of the earlier measurements were performed using very slow length changes. Hill (1968) measured short-range force responses in relaxed frog fibers using stretch velocities down to 1×10^{-5} muscle lengths per second. (This corresponds to a stretch that takes more than an hour to lengthen a muscle by 4%). It seems exceptionally unlikely that the short-range responses measured in this type of experiment are due to pure viscous components (elements that produce a force directly proportional to the stretch velocity) unless the viscosity 'disappears' during contraction (Ford et al. 1977). Any viscous resistance that was large enough to be detected in relaxed muscles during ultra-slow stretches and persisted during a contraction would drastically impede the muscle during normal physiological movements.

7 Second Component

Although this chapter focuses on the first phase of the non-linear tension response to an imposed stretch, the second phase of the response is also worthy of attention. This Author's view of the "Second Component" (Campbell and Lakie 1995) has been heavily influenced by a model developed by Haugen and Sten-Knudsen (1981) to help analyze short-range force responses measured in relaxed frog muscle fibers stretched from different initial sarcomere lengths.

Haugen and Sten-Knudsen hypothesized that the gradual increase in force that occurred in their records after the short-range response resulted from the extension of a passive linear spring (Haugen and Sten-Knudsen 1981). This spring (the Parallel Elastic Component) was arranged in parallel with the "Short Range Elastic Component" which, in Haugen and Sten-Knuden's work on relaxed muscle fibers, reflected the mechanical contribution from a relatively small number of myosin heads that continued to cycle between attached and detached states under resting conditions. The mean strain in the cross-bridge links was hypothesized to increase during interfilamentary movements so that the Short-Range Elastic Component produced a force that rose to a constant level during the initial stages of the length change and was maintained at that force plateau throughout the remainder of the stretch. The experimentally measured force response was the arithmetic sum of the Parallel and Short-Range Elastic Components at each time-point in the record.

The Author has since used this simple framework in several of his own publications (Campbell and Lakie 1998; Campbell and Moss 2000, 2002; Mitov et al. 2009). One of the advantages of the approach is that it is readily applicable to experimental data obtained using chemically permeabilized preparations. Mechanical responses obtained using muscles immersed in solutions with different free Ca^{2+}-concentrations can be simulated simply by varying the number of cycling myosin heads in the "cross-bridge component" (a more general form of Haugen and Sten-Knudsen's Short Range Elastic Component). The stiffness of the Parallel Elastic Component in a given sample can be estimated by fitting a (potentially non-linear) mathematical function to the latter stages of the measured force response. This should be appropriate for any cross-bridge mechanism because it can be shown mathematically that the force due to a fixed population of cycling cross-bridges eventually reaches a steady-state level during an imposed movement, irrespective of the number of bound states and/or the strain-dependence of the actin-myosin kinetics (Campbell and Moss 2000).

Although Haugen and Sten-Knudsen (1981) did not specify which proteins were likely to contribute to the Parallel Elastic Component, it now seems likely that the major contribution in single skeletal muscle fibers comes from titin molecules

(Granzier and Labeit 2002). (Note that although these molecules are arranged in series between the actin and myosin filaments, they act as if they are mechanically in parallel with force-generating cross-bridges as long as minor complications due to actin and myosin filament extensibility (Huxley et al. 1994; Wakabayashi et al. 1994) are ignored.) In multicellular cardiac muscle samples, there are probably additional contributions from extracellular collagen components and intracellular tubulin and intermediate filaments (Granzier and Irving 1995).

All of the molecular components that can contribute to the Parallel Elastic Component were initially thought to have mechanical properties that are independent of the prevailing free Ca^{2+}-concentration. It is therefore interesting to note that there are some experiments that suggest that the stiffness of the Parallel Elastic Component increases with the free Ca^{2+}-concentration. One of these is shown in Fig. 2. The pCa 5.7+BDM condition is an experimental situation in which the muscle fiber is immersed in a solution with a free Ca^{2+}-concentration that is sufficient to activate significant cross-bridge cycling but in which the myosin ATP-ase activity is inhibited by BDM. There are two separate indications that the myosin inhibition is almost complete. First, the isometric force is reduced to a value very near that measured in the pCa 9.0 condition. Second, there is no obvious short-range response in the experimental record.

The interesting result demonstrated here is that the stiffness of the muscle (proportional to the slope of the force response) is increased relative to that measured in pCa 9.0 solution in the pCa 5.7 solution containing BDM. This would be the expected result if the Parallel Elastic Component had a stiffness that increased with the free Ca^{2+}-concentration. Control experiments (not shown) demonstrated that BDM did not change the stiffness of the diaphragm muscle fiber when it was immersed in pCa 9.0 solution.

In a skeletal muscle fiber, the Parallel Elastic Component probably consists primarily of titin molecules, so the simplest interpretation is that titin molecules have Ca^{2+}-dependent mechanical properties. Experiments performed by Fujita et al. (2004) using bovine atrial tissue also suggest that some titin isoforms have a stiffness that increases with the free Ca^{2+} concentration but the effect shown in Fig. 2 seems to be quite a lot larger than that described in their experiments. Similar effects were observed by Bagni et al. (2002, 2004) in frog muscle fibers subjected to very fast (~40 muscle lengths per second) ramp stretches. They suggested that there was a "static stiffness component" which had an elastic modulus which increased with the free Ca^{2+} concentration but was positioned "outside the cross-bridges". The molecular identity of the 'static' structure remains unclear but there is no evidence to suggest that it is not titin.

Other published short-range force responses support the conjecture that titin molecules may exhibit Ca^{2+}-dependent mechanical properties but the evidence from these measurements is less direct. Campbell and Moss measured short-range force responses in rabbit psoas (2000) and rat soleus (2002) muscles fibers. They then tried to fit their experimental records using a mathematical model consisting of a fixed population of cycling cross-bridges arranged in parallel with a non-linear elastic spring. In both cases, the best-fit to the force responses measured in fibers immersed in maximally-activating pCa 4.5 solution was obtained with a parallel elastic stiffness that was ~3 times greater than that measured in the same fibers when they were immersed in pCa 9.0 solutions. One interpretation of these findings is clearly that titin filaments have a stiffness that depends on the free Ca^{2+}-concentration.

Although this remains a genuine possibility, computer simulations currently being developed by the Author suggest a potential alternative explanation. Work presented in abstract form at the 2009 Biophysical Society meeting (Campbell 2009) shows that the slope of the force response during the Second Component is increased at high levels of Ca^{2+}-activation if adjacent half-sarcomeres contain different numbers of myosin heads. This is because the half-sarcomeres with fewer than average bound cross-bridges can only bear the force generated by half-sarcomeres with a greater than average number of bound cross-bridges by continually 'creeping' towards longer lengths. This leads to dynamic inter-half-sarcomere interactions and an apparent Second Component stiffness that is greater than that of the Parallel Elastic Component in any one half-sarcomere (Morgan 1990).

A similar effect might explain why muscle stiffness is apparently increased in the pCa 5.7 + BDM condition shown in Fig. 2. It could be that the BDM reduces cross-bridge cycling in most half-sarcomeres but leaves a small number of units (perhaps in the ends of the preparation) where a few half-sarcomeres continue to develop force. This could create a situation in which sarcomere length heterogeneities developed during an imposed stretch which lead to over-estimates of the real stiffness of the Parallel Elastic Component. Simulations to test this hypothesis are currently underway.

8 Relaxed Muscle

Although there seems little doubt that short-range mechanical properties measured in activated muscle fibers primarily reflect the effects of interfilamentary movement on populations of cycling cross-bridges, the situation in relaxed muscle fibers is more controversial (Bianco et al. 2007; Campbell and Lakie 2008; Kellermayer et al. 2008; Mutungi and Ranatunga 2000; Proske and Morgan 1999; Proske et al. 1993). There is good evidence to support Hill's conjecture (1968) that a small number of cross-bridges (of order 1% of the number involved in a maximal isometric contraction) continue to cycle in a relaxed frog muscle but new results increase the probability that short-range mechanical properties could reflect other mechanisms. For example, elegant experiments performed using optical tweezers by Bianco et al. (2007) show that fragments of the PEVK domain of titin molecules can bind to F-actin. If these interactions occur in vivo as well as in vitro, they probably exhibit all of the features required to produce a short-range response (Campbell and Lakie 2008).

It is possible of course that short-range mechanical properties can reflect more than one molecular mechanism. Some of the discrepancies in the existing literature might be resolved, for example, if the non-linear force response in relaxed frog fibers reflects actin-titin interactions while the behavior in mammalian muscle results from a small number of cycling cross-bridges. The converse situation is also possible. Finally, both mechanisms (cycling cross-bridges and actin-titin interactions) and potentially additional molecular effects could contribute simultaneously to the observed behavior.

The idea that cross-bridges underlie the short-range mechanical properties of relaxed muscles has been questioned for many different reasons since Hill's original paper (1968). For example, Sandow (1970) was quite critical of Hill's cross-bridge hypothesis because the stiffness of the short-range response remained fairly constant when the muscle was subjected to small stretches imposed in separate trials from starting sarcomere lengths estimated as ranging from 2.0 to 3.0 μm. Sandow argued that filament overlap would have decreased by ~40% over this experimental range and a short-range response due to bound cross-bridges should have shown a proportional change. Given the fact that all of Hill's experiments were performed using whole amphibian muscles, there must now be some doubt as to whether or not the sarcomere length was the same in all regions in all the fibers in his preparations. Moreover, Hill did not correct his experimental results for the expected increase in Parallel Elastic Component stiffness at long sarcomere lengths. This could have masked a potential reduction in the stiffness of the short-range response at longer overall muscle lengths.

When Haugen and Sten-Knudsen (1981) repeated Hill's experiments using single frog fibers (to improve the measurements of sarcomere length) and their Parallel Elastic Component framework (to correct for length-dependent stiffness of passive structures) they showed that the stiffness of the Short Range Elastic Component extrapolated to zero at a sarcomere length near 3.7 μm – the point at which thick and thin filaments no longer overlapped. However, as often seems to be the case with short-range mechanical properties, Haugen and Sten-Knuden's results were not as simple to interpret as they might have hoped. The experimental data showed that the Short Range Elastic Component stiffness actually peaked at a sarcomere length of 3.0 μm rather than at full filament overlap. With the benefit of hindsight, this might reflect an effect of sarcomere length heterogeneity (Morgan 1990) but Haugen and Sten-Knudsen proposed that the increased stiffness of the short-range response at a sarcomere length of 3.0 μm could result from length dependent activation of their isolated frog fibers (Endo 1973). Both of these effects seem to be genuine possibilities. For example, the idea that the contractile apparatus in frog fibers can be activated by stretch is supported by the Feng effect – the increase in metabolic rate observed when a resting frog muscle is stretched to more than ~120% of its normal in vivo length (Feng 1932).

The sequence of experiments described above is an interesting example of the sort of difficulties that researchers have encountered when they have set out to measure short-range mechanical properties in relaxed muscles. The results have rarely been clear-cut and it often seems that researchers end up trying to compare results obtained in separate labs using different experimental preparations under dissimilar experimental conditions. For example, it is not completely clear that the very slight discontinuity observed when chemically permeabilized rat soleus fibers are stretched in pCa 9.0 solution (Campbell and Moss 2002) results from exactly the same mechanism as the Short Range Elastic Component response observed in relaxed intact frog fibers. The biochemical environments around the myofilaments in the two preparations are probably quite different which may emphasize distinct molecular mechanisms.

In this Author's view (Campbell and Lakie 2008), the balance of evidence still suggests that at least a portion of the short-range mechanical properties of relaxed muscles is due to a small proportion of cross-bridges that continue to cycle in the resting tissue. Put simply, the contractile apparatus may not completely switch off when the muscle relaxes (Campbell and Lakie 1998).

The best evidence to support this view probably comes from experiments performed using chemically permeabilized muscle fibers. All of the experimental results obtained to date show that the stiffness of the short-range response increases linearly with isometric force as the free Ca^{2+} concentration in the bathing solution is raised from minimal to saturating values (Campbell and Moss 2000, 2002; Campbell et al. 2003). This is the expected result if the short-range response reflects the displacement of attached cross-bridges and complications due to filament compliance (which are probably quite small at very low levels of Ca^{2+}-activation) are ignored. As shown, for example, in Fig. 2, chemicals that inhibit myosin ATP-ase also reduce the stiffness of the short-range response in permeabilized fibers though it must be mentioned that in relaxed intact muscle fibers, BDM appears have no effect (Mutungi and Ranatunga 1996). The interpretation of mechanical records will also become more complicated if additional experiments support a recently hypothesized interaction between myosin binding protein-C and actin filaments (Harris et al. 2009).

9 Cardiac Muscle

Figure 5 shows the force responses that were measured when chemically permeabilized multicellular myocardial preparations were subjected to three paired ramp stretches and releases followed by a rapid shortening/re-stretch perturbation. Trials were performed at different levels of Ca^{2+}-activation and superposed. The records in panel a were obtained using canine myocardial samples (Campbell et al. 2003). The records in panel b were obtained using rat preparations (Mitov et al. 2009).

Although there are some features of the force records that look quite different in the two muscle types, it is the Author's view that both sets of data can be explained by the same general mechanism. As described in several of the earlier sections, the force response can be decomposed into two separate signals: (a) a force representing the passive mechanical properties of the preparation and (b) a force representing the mechanical effects of a variable number of cycling cross-bridges. When described in this manner, it becomes obvious that the records from the different preparations look different because of two significant molecular effects. First, the Parallel Elastic Component is stiffer in the rat myocardium than

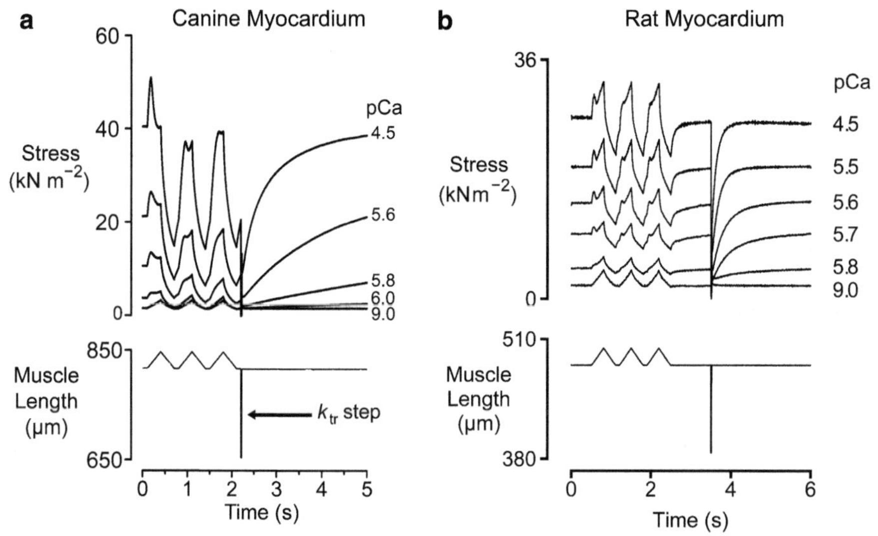

Fig. 5 Mechanical properties of different types of myocardium. *Panel* (**a**) shows superposed force and muscle length records for a sample of chemically permeabilized canine myocardium subjected to repeated stretch/shortening cycles in solutions with different free Ca^{2+} concentrations. *Panel* (**b**) shows very similar experiments performed using samples of chemically permeabilized rat myocardium. Both sets of records were obtained at 15°C (*panel* **a** reproduced with permission from Campbell et al. 2003; *panel* **b** reproduced with permission from Mitov et al. 2009)

it is in the canine samples. This probably reflects increased expression of the shorter, stiffer N2B titin isoform in the rodent samples (Wu et al. 2000). Second, the cross-bridges cycle more quickly in the rat myocardium than they do in the canine myocardium. This is consistent with the different rates of tension recovery measured in the two preparations. The k_{tr} values (Brenner 1988) in maximally-activating pCa 4.5 solution were about 1.3 s^{-1} and 11.2 s^{-1} for the canine and rat preparations respectively.

The reason that the rate of cross-bridge cycling makes such a big difference to the qualitative appearance of the records was explained by Hill (1968). He pointed out that if cross-bridge heads are attaching to the thin filament very quickly, any myosin head that is detached as a result of the imposed movement will be rapidly replaced by a new actin-myosin bond. The force response to a ramp stretch will therefore consist of a sustained, elevated, force plateau because the mean actin-myosin bond length will be stretched (relative to the isometric value) because of the sustained interfilamentary movement. If, on the other hand, cross-bridges are attaching quite slowly, not all the myosin heads that are detached by the imposed movement will be replaced. The force response to a ramp stretch will therefore consist of a large transient followed by a decline to a new steady-state level. This level will be substantially lower than the steady-state isometric force if the increase in mean cross-bridge strain is not sufficient to compensate for the reduction in the number of attached cross-bridges.

Another consequence of the different myosin cycling rates is that the mechanical properties of the two preparations have different recovery time-courses. In the canine samples, the responses to the second stretch are very different to those measured during the first stretch at every level of Ca^{2+}-activation. This is because the cross-bridge populations were significantly perturbed by the first movement and have not had time to recover to their steady-state profiles in the time interval between the stretches. This contrasts with the behavior seen in the rodent samples where the faster cross-bridge cycling means that the measured forces are very similar for all three stretches, irrespective of the free Ca^{2+} concentration in the bathing solution.

9.1 Myosin Heavy Chain Effects

Rats express two cardiac myosin heavy chains and the proportion of the total myosin occurring as the slower β-isoform increases with age (for example Carnes et al. 2004). Figure 6 shows the results of experiments in which myocardial samples were isolated from Fischer 344 rats of different ages and subjected to experimental trials similar to those shown in Fig. 5. When the mechanical experiments were completed, the individual preparations (approximately three or four cells in size) were collected, processed, and analyzed by gel electrophoresis (Tikunov et al. 2001).

Figure 6a shows representative optical density profiles obtained from the SDS-PAGE gels. Figure 6b–d show selected parameters calculated from the measured force responses plotted against the proportion of slow β-myosin heavy chain in the experimental preparations. The initial stiffness of the muscle and the relative force produced during the short-range response both correlated with the proportional β-myosin heavy chain content. This was a particularly exciting result when it was published in 2009 because, to the Author's knowledge, it was the first time that the short-range mechanical properties of cardiac muscle had been directly associated with the differential expression of gene products.

10 Physiological Significance

It is possible that the short-range mechanical properties that seem to be a feature of all types of muscle are an inevitable consequence of using cross-bridges (thought of in this context as short elastic springs arranged in linear arrays) to generate force. It is also possible that this arrangement confers important advantages that mean that cross-bridges are a more effective way of powering animal movements than alternative mechanisms, for example flagellar whips.

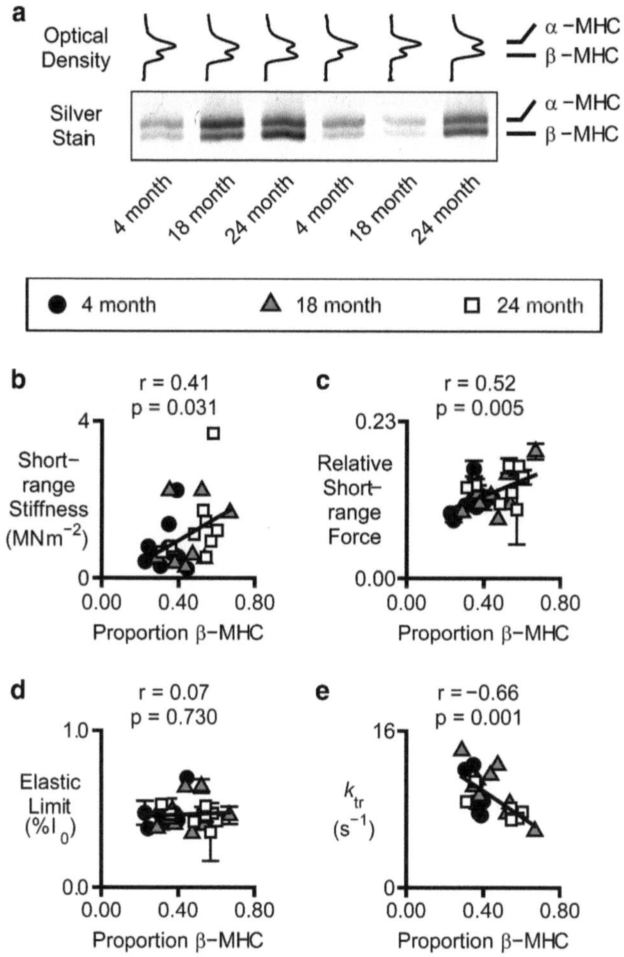

Fig. 6 Myosin heavy chain effects. *Panel* (**a**) shows optical density profiles and a representative gel showing electrophoretic separation of the α and β myosin heavy chain isoforms in samples of chemically permeabilized rat myocardium that had been previously used for mechanical experiments. *Panels* (**b–e**) show various mechanical properties plotted as functions of the relative β myosin heavy chain content in the different experimental preparations. The initial stiffness and the maximum force produced during the short-range response increased linearly with the relative β myosin heavy chain content (reproduced with permission from Mitov et al. 2009)

It has been argued, for example, that short-range mechanical properties may help to stabilize sarcomere lengths in a working muscle and therefore may play an important role in minimizing sarcomere length inhomogeneity (Allinger et al. 1996). This theory is controversial (Zahalak 1997) but preliminary simulations currently under development by the Author support the general idea (Campbell 2009).

10.1 Skeletal Muscle

Campbell and Lakie (1998) pointed out that if short-range mechanical properties in relaxed muscles reflect a low level of persistent cross-bridge cycling, this would have significant metabolic implications. Muscle accounts for about 40% of an adult human's weight. If the tissue is not completely deactivated when it relaxes, cycling myosin heads will continue to use ATP and to produce heat. The sheer amount of muscle in the body means that this could add up to a significant energy expenditure. Indeed, A.V. Hill, extrapolating from data obtained using isolated frog muscles, suggested that relaxed muscle might account for as much as 25% of the basal metabolic rate in humans (Hill 1965). Small variations in the magnitude of this resting activity might explain why some individuals have higher basal metabolic rates than others.

Irrespective of the molecular mechanism underlying short-range mechanical properties in relaxed muscle, the fact that they are history-dependent means that the neuromuscular control system has to allow for them if it is to control movements with precision. It is particularly interesting, for example, to learn that the integrated surface electromyogram signals recorded from the agonist muscles are different in repeated movements when healthy humans are asked to track a repeating signal (Axelson and Hagbarth 2003).

10.2 Cardiac Muscle

The stiffness of myocardium is important because it influences the amount of blood that enters into the ventricles during diastole. When the chambers are compliant, blood enters through the atrio-ventricular valves with little hindrance and the chambers distend fully at normal filling pressures. Conversely, when the walls are stiff, pressure rises sharply when blood enters and adequate ventricular filling is prevented. This can lead to a condition known as diastolic dysfunction (Zile et al. 2004).

The experimental records shown in Fig. 5 show that short-range mechanical properties clearly affect the stiffness of cardiac muscle. Whether these measurements obtained using chemically permeabilized samples are directly relevant to *in vivo* function has sometimes been regarded as less clear. The free Ca^{2+}-concentration required to significantly augment the passive myocardial stiffness in the chemically permeabilized canine preparations was pCa 6.2 (630 nM) (Campbell et al. 2003). This is substantially higher than the basal free Ca^{2+}-concentration (75–200 nM) normally assumed for working hearts (Bers 1991). The Ca^{2+}-threshold for significant stiffness augmentation in chemically permeabilized rat myocardium is probably higher still (Mitov et al. 2009).

It has been argued (Campbell et al. 2003; Mitov et al. 2009) that it is not the precise threshold for Ca^{2+}-activation that matters in these experiments but rather the relative level of Ca^{2+}-activated force that the tissue is developing when the short-range

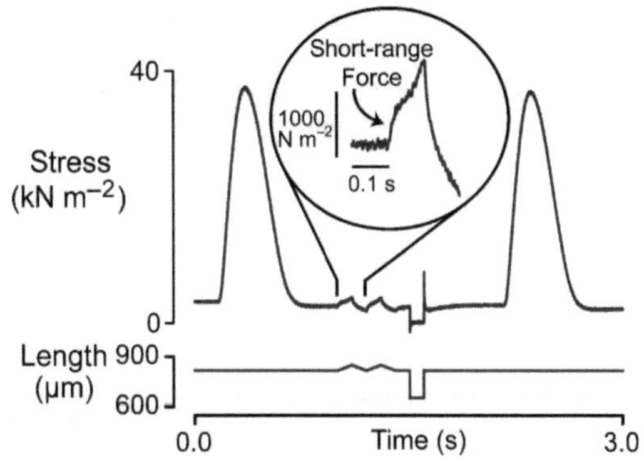

Fig. 7 Intact rat trabecula. Two triangular length changes were imposed during the 2 s interval between electrically-excited twitch contractions. The relaxed trabecula exhibited a clear short-range force response. The experiment was performed at 22°C (unpublished records)

mechanical properties become significant. This is because there is evidence to suggest that the tension-pCa relationship in cardiac muscle is significantly left-shifted in intact cardiac muscle compared to that measured in permeabilized preparations (Gao et al. 1994). Nevertheless, it would be interesting to see whether intact cardiac muscle exhibits short-range mechanical properties during the diastolic interval.

The necessary experiments are technically difficult because it is not easy to rule out the possibility that the mechanical measurements are compromised by series compliance artifacts. However, preliminary measurements obtained in the Author's laboratory with twitching rat trabeculae (Fig. 7) show a clear short-range elastic response to an imposed rat stretch. This result suggests that there may be sufficient myosin heads bound to actin filaments in relaxed intact cardiac muscle to influence ventricular filling. Data obtained with single intact murine myocytes presented in abstract form at the 2009 Biophysical Society Annual Meeting by King et al. (2009) support a similar conclusion.

Acknowledgements This work was supported by American Heart Association Scientist Development Grant 0630079N, NIH AG021862, NIH HL090749 and the University of Kentucky Research Challenge Trust Fund.

References

Allinger TL, Epstein M, Herzog W (1996) Stability of muscle fibers on the descending limb of the force-length relation. A theoretical consideration. J Biomech 29:627–633

Axelson HW, Hagbarth KE (2001) Human motor control consequences of thixotropic changes in muscular short-range stiffness. J Physiol 535(1):279–288

Axelson HW, Hagbarth KE (2003) Human motor compensations for thixotropy-dependent changes in resting wrist joint position after large joint movements. Acta Physiol Scand 179:389–398

Bagni MA, Cecchi G, Colombini B, Colomo F (2002) A non-cross-bridge stiffness in activated frog muscle fibers. Biophys J 82:3118–3127

Bagni MA, Colombini B, Geiger P, Berlinguer Palmini R, Cecchi G (2004) A non cross-bridge calcium-dependent stiffness in frog muscle fibers. Am J Physiol Cell Physiol 286:C1353–C1357

Bers DM (1991) Excitation-contraction coupling and cardiac contractile force. Kluwer, Dordrecht

Bianco P, Nagy A, Kengyel A, Szatmári D, Mártonfalvi Z, Huber T, Kellermayer MSZ (2007) Interaction forces between F-actin and titin PEVK domain measured with optical tweezers. Biophys J 93:2102–2109

Blair GWS (1969) Elementary rheology, Academic, London

Brenner B (1988) Effect of Ca2+ on cross-bridge turnover kinetics in skinned single rabbit psoas fibers: Implications for regulation of muscle contraction. Proc Natl Acad Sci USA 85:3265–3269

Buchthal F, Kaiser E (1951) The rheology of the cross striated muscle fibre with particular reference to isotonic conditions. Vol. 21.7. Det Kongelige Danske Videnskabernes Selskab Biologiske Meddelser, Copenhagen

Campbell KS (2006) Filament compliance effects can explain tension overshoots during force development. Biophys J 91:4102–4109

Campbell KS (2009) Short-range mechanical properties simulated with a mechanical mdoel incorporating multiple half-sarcomeres. Biophys J 96:615a

Campbell KS, Holbrook AM (2006) Myocardial stiffness in experimental conditions that mimic ischemia. Biophys J 90:1270A

Campbell KS, Lakie M (1995) Tension responses to imposed length changes in isolated relaxed muscle fibre bundles from *Rana temporaria.* J Physiol 487:155–156P

Campbell KS, Lakie M (1998) A cross-bridge mechanism can explain the thixotropic short-range elastic component of relaxed frog skeletal muscle. J Physiol 510.3:941–962

Campbell KS, Lakie M (2008) Response to Bianco et al.: Interaction forces between F-actin and titin PEVK domain measured with optical tweezers. Biophys J 94:327–328; 329–330

Campbell KS, Moss RL (2000) A thixotropic effect in contracting rabbit psoas muscle: prior movement reduces the initial tension response to stretch. J Physiol 525(2):531–548

Campbell KS, Moss RL (2002) History-dependent mechanical properties of permeabilized rat soleus muscle fibers. Biophys J 82:929–943

Campbell KS, Moss RL (2003) SLControl: PC-based data acquisition and analysis for muscle mechanics. Am J Physiol Heart Circ Physiol 285:H2857–H2864

Campbell KS, Patel JR, Moss RL (2003) Cycling cross-bridges increase myocardial stiffness at sub-maximal levels of Ca2+ activation. Biophys J 84:3807–3815

Carnes CA, Geisbuhler TP, Reiser PJ (2004) Age-dependent changes in contraction and regional myocardial myosin heavy chain isoform expression in rats. J Appl Physiol 97:446–453

Denny-Brown D (1929) On the nature of postural reflexes. Proc R Soc Lond B Biol Sci 104:252–301

Edman KA, Elzinga G, Noble MI (1981) Critical sarcomere extension required to recruit a decaying component of extra force during stretch in tetanic contractions of frog skeletal muscle fibers. J Gen Physiol 78:365–382

Endo M (1973) Length dependence of activation of skinned muscle fibres by calcium. Cold Spring Harb Symp Quant Biol 37:505–510

Farman GP, Tachampa K, Mateja R, Cazorla O, Lacampagne A, de Tombe PP (2008) Blebbistatin: use as inhibitor of muscle contraction. Pflugers Arch 455:995–1005

Feng TP (1932) The thermo-elastic properties of muscle. J Physiol 74:455–470

Ford LE, Huxley AF, Simmons RM (1977) Tension responses to sudden length change in stimulated frog muscle fibres near slack length. J Physiol 269:441–515

Fujita H, Labeit D, Gerull B, Labeit S, Granzier HL (2004) Titin isoform-dependent effect of calcium on passive myocardial tension. Am J Physiol Heart Circ Physiol 287:H2528–H2534

Gao WD, Backx PH, Azan-Backz M, Marban E (1994) Myofilament Ca2+ sensitivity in intact versus skinned rat ventricular muscle. Circ Res 74:408–415

Getz EB, Cooke R, Lehman SL (1998) Phase transition in force during ramp stretches of skeletal muscle. Biophys J 75:2971–2983
Granzier HL, Irving TC (1995) Passive tension in cardiac muscles: contribution of collagen, titin, microtubules, and intermediate filaments. Biophys J 68:1027–1044
Granzier H, Labeit S (2002) Cardiac titin: an adjustable multi-functional spring. J Physiol 541:335–342
Gunst SJ (1983) Contractile force of canine airway smooth muscle during cyclical length changes. J Appl Physiol 55:759–769
Harris J (1977) Rheology and non-Newtonian flow. Longman, London and New York
Harris SP, Shaffer JF, Bezold KL, Kensler RW (2009) Switching gears with myosin binding protein-C. Biophys J 96:4A
Haugen P, Sten-Knudsen O (1981) The dependence of the short-range elasticity on sarcomere length in resting isolated frog muscle fibres. Acta Physiol Scand 112:113–120
Herbst M (1976) Studies on the relation between latency relaxation and resting cross-bridges of frog skeletal muscle. Pflugers Arch 364:71–76
Hill AV (1965) Trails and trials in physiology. Edward Arnold, London
Hill DK (1968) Tension due to interaction between the sliding filaments in resting striated muscle. The effect of stimulation. J Physiol 199:637–684
Huxley HE, Stewart A, Sosa H, Irving T (1994) X-ray diffraction measurements of the extensibility of actin and myosin filaments in contracting muscle. Biophys J 67:2411–2421
Kellermayer MS, Bianco P, Martonfalvi Z, Nagy A, Kengyel A, Szatmari D, Huber T, Linari M, Caremani M, Lombardi V (2008) Muscle thixotropy: more than just cross-bridges? Biophys J 94:329–330
King NMP, Helmes M, Granzier H (2009) A direct method to measure the restoring force and slack sarcomere length of intact cardiomyocytes. Biophys J 96:498a
Lakie M, Walsh EG, Wright GW (1984) Resonance at the wrist demonstrated by the use of a torque motor: an instrumental analysis of muscle tone in man. J Physiol 353:265–285
Liddell EGT, Sherrington C (1924) Reflexes in response to stretch (myotatic reflexes). Proc R Soc Lond B Biol Sci 96:212–242
Lombardi V, Piazzesi G (1990) The contractile response during steady lengthening of stimulated frog muscle fibres. J Physiol 431:141–171
Loram ID, Maganaris CN, Lakie M (2007) The passive, human calf muscles in relation to standing: the short range stiffness lies in the contractile component. J Physiol 584:677–692
Meiss RA (1987) Stiffness of active smooth muscle during forced elongation. Am J Physiol Cell Physiol 253:C484–C493
Mitov MI, Holbrook AM, Campbell KS (2009) Myocardial short-range force responses increase with age in F344 rats. J Mol Cell Cardiol 46:39–46
Morgan DL (1990) New insights into the behavior of muscle during active lengthening. Biophys J 57:209–221
Mutungi G, Ranatunga KW (1996) The viscous, viscoelastic and elastic characteristics of resting fast and slow mammalian (rat) muscle fibres. J Physiol 496.3:827–836
Mutungi G, Ranatunga KW (2000) Do cross-bridges contribute to the tension response during stretch of passive muscle? A response. J Muscle Res Cell Motil 21:301–302
Proske U, Morgan DL (1999) Do cross-bridges contribute to the tension during stretch of passive muscle? J Muscle Res Cell Motil 20:433–442
Proske U, Morgan DL, Gregory JE (1993) Thixotropy in skeletal muscle and in muscle spindles: a review. Prog Neurobiol 41:705–721
Rack PMH, Westbury DR (1974) The short range stiffness of active mammalian muscle and its effect on mechanical properties. J Physiol 240:331–350
Sandow A (1970) Skeletal muscle. Ann Rev Physiol 32:87–138
Stienen GJ, Versteeg PG, Papp Z, Elzinga G (1992) Mechanical properties of skinned rabbit psoas and soleus muscle fibres during lengthening: effects of phosphate and Ca2+. J Physiol 451:503–523
Tikunov BA, Sweeney HL, Rome LC (2001) Quantitative electrophoretic analysis of myosin heavy chains in single muscle fibers. J Appl Physiol 90:1927–1935

Vieth E (1989) Fitting piecewise linear regression functions to biological responses. J Appl Physiol 67:390–396

Wakabayashi K, Sugimoto Y, Tanaka H, Ueno Y, Takezawa Y, Amemiya Y (1994) X-ray diffraction evidence for the extensibility of actin and myosin filaments during muscle contraction. Biophys J 67:2422–2435

Wu Y, Cazorla O, Labeit D, Labeit S, Granzier H (2000) Changes in titin and collagen underlie diastolic stiffness diversity of cardiac muscle. J Mol Cell Cardiol 32:2151–2162

Zahalak GI (1997) Can muscle fibers be stable on the descending limbs of their sarcomere length-tension relations? J Biomech 30:1179–1182

Zile MR, Baicu CF, Gaasch WH (2004) Diastolic heart failure – abnormalities in active relaxation and passive stiffness of the left ventricle. N Engl J Med 350:1953–1959

Crossbridge Mechanism(s) Examined by Temperature Perturbation Studies on Muscle

K.W. Ranatunga and M.E. Coupland

Abstract An overall view of the contractile process that has emerged from temperature-studies on active muscle is outlined. In isometric muscle, a small rapid temperature-jump (T-jump) enhances an early, pre-phosphate release, step in the acto-myosin (crossbridge) ATPase cycle and induces a characteristic rise in force indicating that crossbridge force generation is endothermic (force rises when heat is absorbed). Sigmoidal temperature dependence of steady force is largely due to the endothermic nature of force generation. During shortening, when muscle force is decreased, the T-jump force generation is enhanced; conversely, when a muscle is lengthening and its force increased, the T-jump force generation is inhibited. Taking T-jump force generation as a signature of the crossbridge – ATPase cycle, the results suggest that during lengthening the ATPase cycle is truncated before endothermic force generation, whereas during shortening this step and the ATPase cycle, are accelerated; this readily provides a molecular basis for the Fenn effect.

Keywords Temperature perturbation • Force generation • Endothermic force • Muscle force

1 Introduction

The process that underlies muscle contraction is an ATP-driven, cyclic interaction of crossbridges (myosin heads) between thick (M = myosin) and thin (A = actin) filaments in a sarcomere (Huxley 1957, 1969); during a cycle, a crossbridge (myosin head) attaches to actin, undergoes a conformational change generating muscle force, and power, and then detaches. It has been known for over half-century (Hadju 1951) that muscle force is sensitive to temperature; in mammalian muscle, maximal active force

K.W. Ranatunga (✉)
Muscle Contraction Group, Department of Physiology and Pharmacology, School of Medical Sciences, University of Bristol, Bristol BS8 1TD, UK
e-mail: k.w.ranatunga@bris.ac.uk

D.E. Rassier (ed.), *Muscle Biophysics: From Molecules to Cells*, Advances in Experimental Medicine and Biology 682,
DOI 10.1007/978-1-4419-6366-6_14,

increases by ~2-fold (Ranatunga and Wylie 1983), shortening velocity increases by ~6-fold (Ranatunga 1984) and power increases by >10-fold (Ranatunga 1998) in warming from 10°C to more physiological temperatures (>30°C). Thus, temperature-sensitivity determination and temperature perturbation (T-jump) are obvious techniques to examine the mechanisms of force generation and crossbridge cycle in muscle.

The purpose of this short review is to summarise some of the main findings from such temperature-studies on mammalian muscle fibers with emphasis on our T-jump experiments. It will be concluded that crossbridge force generation is endothermic (is perturbed by T-jump) and is coupled to a defined kinetic step prior to the release of inorganic phosphate (Pi) in the cycle; the temperature dependence of steady active force is largely due to the endothermic nature of the force generation step; the strain-sensitivity of this step can provide a basis for the well-known Fenn-effect in muscle contraction (Fenn 1924); structural mechanism of endothermic crossbridge force generation remains unclear.

2 Materials and Methods

This account refers to complementary findings from two different fast muscle fiber experimental preparations, namely electrically activated intact fiber bundles and chemically activated skinned fibers. Only some essential features of materials and methods, which would be of interest to temperature studies, are summarized below; comprehensive details are published elsewhere.

2.1 *Mechanical Recording System*

The trough system, mounted on an optical microscope stage, consisted of three ~50 μl troughs milled in a titanium block and a front experimental trough with glass windows in the front and bottom (Ranatunga 1996). The temperature of the trough system was kept <5°C but, by Peltier modules, the front trough temperature could be independently clamped at a suitable temperature. In order to reduce the temperature sensitivity of the force (=tension) recording, the transducer was built using two AE 801 elements (Akers, Norway): one cut-beam element was connected to the muscle fiber (natural resonant frequency ~14 kHz) and the other set close-by acted as a dummy, forming a full bridge (Ranatunga 1999a). A moving-coil motor was used to produce ramp length changes in muscle fibers.

2.2 *Temperature-Jump Technique*

A near infra-red ($\lambda = 1.32$ μm) pulse of 200 μs duration (maximum power = 2 J per pulse) from a Nd-YAG laser (Schwartz Electro-Optics Inc., Florida, USA) was used to induce a T-jump in the front trough. The energy absorption by water at this wavelength (~50%) was such that the laser pulse that entered the trough through the front window was reflected back by the aluminium foil in the back wall and raised the temperature of the 50 μl aqueous medium in the trough, and the fiber immersed in it, by 3–5° in 200 μs. The raised solution temperature remained constant for ~500 ms, but the duration could be increased by using the Peltier T-clamping (see Ranatunga 1996).

2.3 *Intact Fiber Experiments*

Intact fiber experiments referred to here were performed on bundles of fibers isolated from fast flexor hallucis brevis muscle of adult male rats (Coupland and Ranatunga 2003); animals were killed with an intra-peritoneal injection of an overdose (~150 mg kg^{-1} body weight) of Sodium Pentobarbitone (Euthatal, Rhône Mérieux). Small bundles of ~5–10 intact excitable fibers were dissected from the mid-belly of the whole muscle and aluminium foil T-clips were fixed on the tendons within 0.2 mm of the fiber-ends; resting fiber length was ~2 mm, sarcomere length ~2.5 μm and the bundle width was 100–200 μm.

A preparation was super-fused with physiological saline solution containing (mM) NaCl, 109; KCl, 5; $MgCl_2$, 1; $CaCl_2$, 4; $NaHCO_3$, 24; NaH_2PO_4, 1; sodium pyruvate, 10 and 200 mg l^{-1} of bovine foetal serum. The solution was bubbled with a mixture of 95% O_2 and 5% CO_2 and was directly stimulated with supra-maximal voltage pulses (<0.5 ms duration) applied to two platinum plate electrodes placed on either side of a bundle (see Ranatunga 1984).

2.4 *Skinned Fiber Experiments*

Bundles of fibers from psoas muscle of adult male rabbits (killed by an intravenous injection of an overdose of sodium pentobarbitone) were chemically skinned using 0.5% Brij 58. A segment of a single fiber (2–4 mm in length) was mounted (using nitro-cellulose glue) between two hooks, one attached to the force transducer and the other to the motor. The sarcomere length in a 0.5 mm region of the fiber near the tension transducer was monitored using He–Ne laser diffraction (see Ranatunga et al. 2002); the sarcomere length was ~2.5 μm.

The buffer solutions contained 10 mM glycerol-2-phosphate as a low temperature-sensitive pH buffer and also 4% Dextran (mol. wt. ~500 kDa) to compress the filament lattice spacing in the fibers to normal dimensions; major anion was acetate, ionic strength = 200 mM, pH = 7.1. For other specific details of the solution compositions, see Coupland et al. (2001, 2005).

2.5 Some General Considerations

When dealing with tension responses induced by different perturbations, some difficulty is encountered in identifying the homologous components. In keeping with the nomenclature used in our previous papers (see Vawda et al. 1999; Coupland et al. 2001; Ranatunga et al. 2002), the following components, or phases, may be recognised in the force transients.

Phase 1: The force change that occurs concomitant with a perturbation, where the extreme force reached is referred to as T_1, as used in the original length perturbation experiments (Huxley and Simmons 1971). An instantaneous drop in force (phase 1) is also seen in T-jump (pressure-release = P-jump) experiments, probably induced by expansion in series elasticity. A T-jump or a P-jump induces a drop in force in rigor muscle fibers and this provides support for the expansion in some elasticity (see Goldman et al. 1987; Ranatunga et al. 1990).

Phase 2: Following phase 1, the force recovers quickly to T_2 force level after a length perturbation. This partial force recovery consists of two exponential components (Davis and Harrington 1993) – referred to as phase 2a (fast) and phase 2b (intermediate) (Ranatunga et al. 2002). The force generation, i.e. force rise above the pre-perturbation level, induced by a T-jump (and a P-jump) corresponds to phase 2b. In T-jump and P-jump experiments where a prominent phase 1 was seen (see Ranatunga 1999b; Vawda et al. 1999), a quicker force recovery corresponding to phase 2a was also seen; this phase tends to partially recover the force to pre-perturbation level as in length perturbation. Both phase 1 and phase 2a are not prominent in the small amplitude T-jumps used here.

Phase 3: The slower exponential component of the force transient induced by a T-jump (and a P-jump) in isometric state.

In summary, only phases 2b and 3 are readily seen in the isometric T-jump transients; in shortening muscle phase 3 is not seen. Hence, particular emphasis is phase 2b (endothermic force generation).

3 Temperature Dependence of Isometric Force

3.1 Intact Muscle Fibers

Figure 1a shows isometric tetanic contractions from an intact (fast) muscle fiber bundle; the steady active force (=tension) is higher at higher temperatures. Figure 1b shows normalised tetanic tension data from several experiments, obtained at a range of temperatures in warming and cooling; the data shows that the relation between force and (reciprocal) temperature is sigmoidal with a half-maximal tension at ~10°C. Figure 2 shows the tension response induced by a T-jump, applied on the plateau of tetanus. As found in Ca-activated skinned fibers (see Davis and

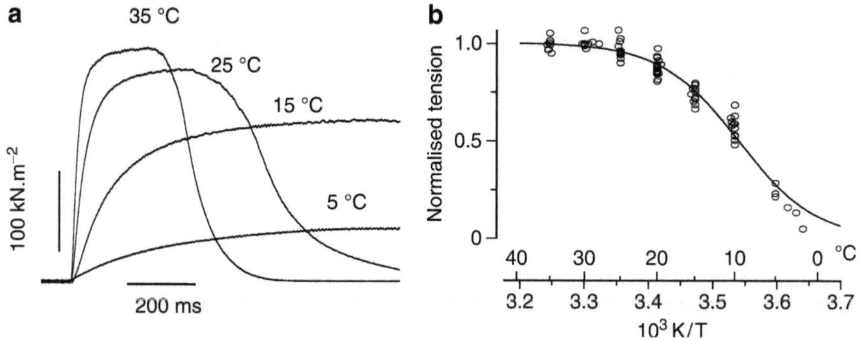

Fig. 1 (**a**) Tetanic contractions recorded from one intact fiber bundle at four temperatures, using appropriate stimulation frequencies and durations. (**b**) Tetanic tension data from eight bundles are plotted as a ratio of that at 35°C; abscissa is reciprocal absolute temperature. The relationship between tetanic tension and temperature is approximately sigmoidal with a half-maximal tension occurring at ~10°C (adapted from Coupland and Ranatunga 2003)

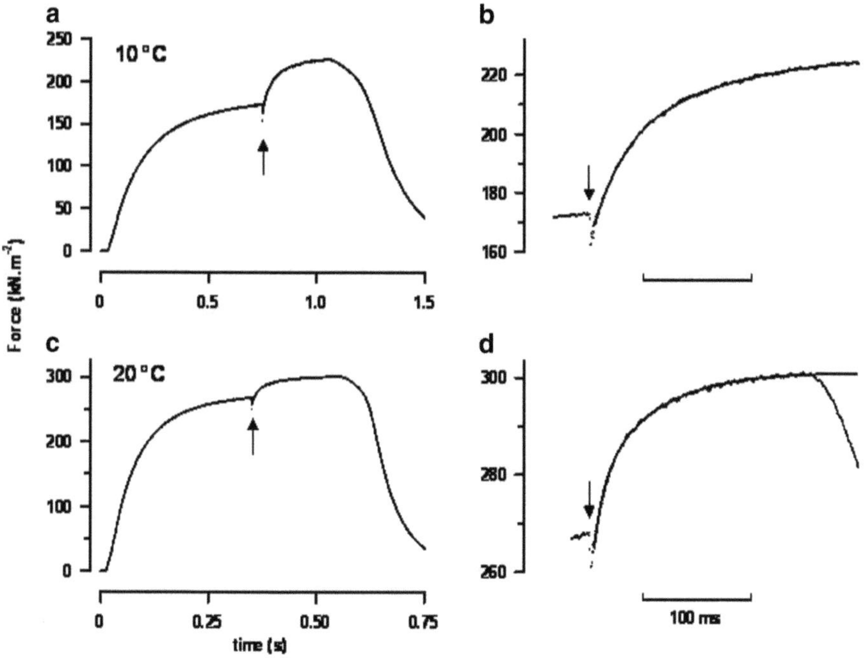

Fig. 2 T-jump tension responses. (**a**) A T-jump of ~4°C applied during the tetanus plateau, at 10°C, induces a small instantaneous drop in tension (phase 1) followed by a rise to a new steady level. (**b**) The tension rise on an expanded time scale is fitted with a bi-exponential curve and the rates for phases 2b and 3 were 37 and 9 s^{-1} respectively. (**c, d**) A corresponding pair of records from the same bundle at 20°C. Note that the tetanic tension is higher but the tension increment induced by the T-jump is smaller but faster (from Coupland and Ranatunga 2003)

Harrington 1987; Goldman et al. 1987; Bershitsky and Tsaturyan 1992; Ranatunga 1996), a T-jump induces a biphasic tension rise that reaches a new steady level, both at 10 and 20°C; for a given T-jump, the initial tension rise (phase 2b) is faster but the amplitude is less at the higher temperature. Phase 2b is thought to represent endothermic force generation in attached crossbridges. Studies have shown that stiffness remains unchanged (see ref in Roots and Ranatunga 2008) or tension/stiffness ratio is increased (Galler and Hilber 1998) with increase of temperature; additionally, single-molecule experiments of Kawai et al. (2006) showed that the force each crossbridge generates is independent of temperature. On the basis of such findings, it may be argued that steady active force in isometric muscle may be simplified to and treated as a two state system – pre-force generating (low-force) and force generating (high-force) states and, while the total number of attached crossbridges remains the same, higher temperature favours (endothermic) force generation. In principle, such a simplistic scheme can account for the sigmoidal temperature dependence as being due to the shift in the equilibrium from low- to high-force states (Davis 1998; Roots and Ranatunga 2008).

3.2 *Skinned Fibers*

Although recording individual contractions from a preparation in warming and cooling and at a range of temperatures (as in intact fibers) was not feasible with the more fragile skinned fiber preparations, use of skinned fibers enables one to readily change the chemical composition of the intracellular medium. Thus, endothermic force generation to a T-jump and sigmoidal temperature dependence of force are not seen in rigor fibers (depleted of ATP and crossbridges attached but not cycling), nor in relaxed fibers (crossbridges detached); due to thermal expansion (Goldman et al. 1987; Ranatunga 1996), rigor force dropped instantly with a T-jump (phase 1) and showed no recovery and the steady rigor force decreased linearly with increase of temperature (Ranatunga 1994). Hence, endothermic force arises from attached crossbridges when actively cycling.

Under control conditions, a sigmoidal temperature dependence of maximal active force, similar to intact fibers, is obtained in maximally Ca-activated skinned fibers (Ranatunga 1994); Fig. 3 from two studies (Coupland et al. 2001, 2005) show that the position of the sigmoidal relation, with respect to temperature, is sensitive to the levels of inorganic phosphate (Pi) and MgADP, two products released during crossbridge (acto-myosin ATPase) cycling; Pi is released earlier and ADP later in the cycle (see Goldman et al. 1984; Hibberd et al. 1985; He et al. 1997, 1999). Figure 3a shows that active force is depressed by Pi so that the curve is shifted to higher temperatures, whereas Fig. 3b shows that MgADP potentiates force and the curve is shifted to lower temperatures. It is also seen that, because of the endothermic nature of force generation, the relative effects on tension at a given level of Pi or ADP is less at higher temperature; for instance, the force depression by 25 mM Pi is ~50% at ~10°C where as it is ~20% at physiological temperatures of ~30°C (Fig. 3a).

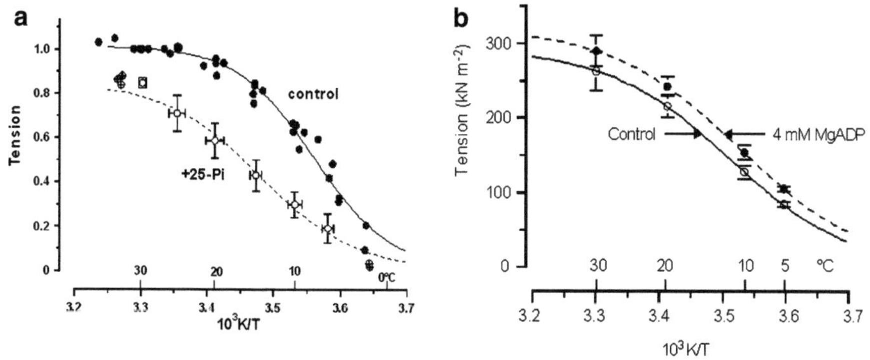

Fig. 3 (**a**) Effect of Pi. Pooled data from five fibers in each of which tension data (with maximal Ca-activation) were collected at different temperatures (range ~5–30°C). A fiber was activated at ~5°C, and temperature raised by laser T-jumps and/or Peltier. Tensions recorded in each contraction were normalized to control at 30°C, and plotted against reciprocal absolute temperature. *Filled circles* and *solid curve* are from activation in control solution (no added Pi). Open symbols are the means (±SD) pooled tensions in the presence of 25 mM Pi from two series (i.e. before and after control). Pi depresses tension, the relative Pi-induced tension depression is less at the higher temperatures and the curve shifted to left (From Coupland et al. 2001). (**b**) Effect of ADP: Pooled data from 18 fibers where tension was measured in control solution and with 4 mM added MgADP at one or more temperatures. Mean (±SEM) specific tension (in kN m^{-2}) are shown for control (*open symbols*) and for 4 mM MgADP (*filled symbols*); with ADP tension is potentiated, relative potentiation is less at higher temperatures and the curve is shifted to right (From Coupland et al. 2005)

4 Tension Responses to Temperature-Jump (T-Jump)

4.1 Effects of Pi and ADP

To determine the molecular step in the ATPase cycle that underlies endothermic force generation, force transients induced by a standard T-jump and at ~9–10°C were examined in control and in the presence of Pi or MgADP. Figure 4a shows a force transient induced by a T-jump in control and Fig. 4b shows the force transient from the same fiber when it was subsequently reactivated in the presence of 12.5 mM added Pi. It is seen that, compared to the control, the steady force before and after the T-jump is lower in the presence of Pi but the initial force rise (phase 2b or endothermic force generation) is clearly faster with Pi. Figure 5a shows T-jump force transient from a fiber in control and Fig. 5b the force transient in the presence of 4 mM MgADP. Compared to the control, the steady force is higher but the T-jump force rise is clearly slower in the presence of ADP.

The contrasting effects of Pi and ADP on the time course of T-jump force transient are summarised in Fig. 6; the amplitude of tension rise remained similar (not illustrated) and phase 3 was not much sensitive to Pi or ADP. On the other hand, endothermic force generation (phase 2b) becomes faster with increase of Pi (Fig. 6a) whereas it becomes slower with increase of ADP (Fig. 6b); in both cases the time course change saturates at higher levels.

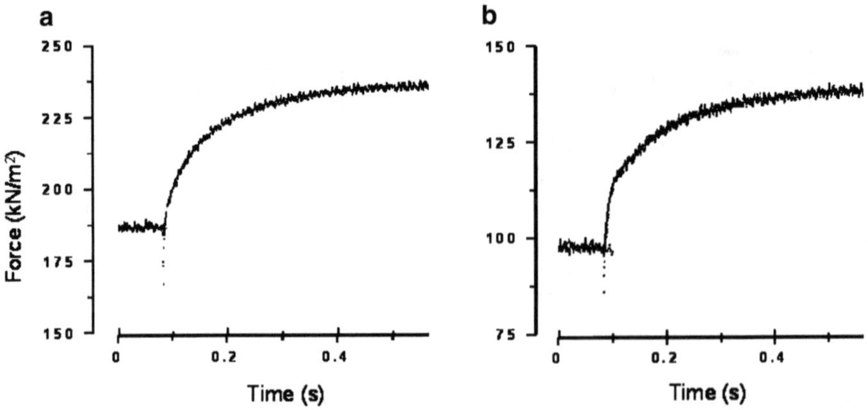

Fig. 4 Effect of Pi. Tension response to a T-jump of ~3°C after the fiber was maximally Ca-activated to steady state at ~9°C. (**a**) On control activation (no added Pi) (**b**) On reactivation in a medium containing 12.5 mM added Pi. A bi-exponential curve is fitted to each transient. With Pi the steady force is depressed, but the initial T-jump transient (phase 2b) is faster; the amplitude of the tension rise is similar to control (adapted from Ranatunga 1999a)

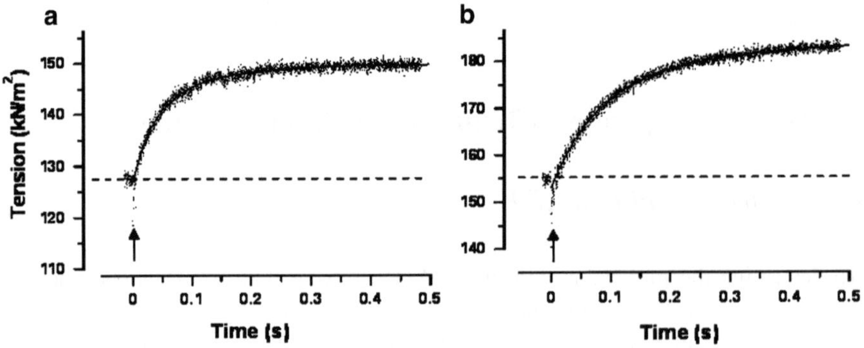

Fig. 5 Effect of MgADP. (**a**) Control (no added MgADP), protocol similar to Fig. 4. (**b**) An identical T-jump induced in the same fiber after reactivation in the presence of 4 mM MgADP. Note that the pre-T-jump tension is higher but the T-jump induced tension rise is slower in the presence of MgADP; the amplitude of the tension rise is similar (from Coupland et al. 2005)

Pi (inorganic phosphate) is released earlier in the crossbridge cycle and the steady muscle force is decreased with added Pi (Cooke and Pate 1985; Hibberd et al. 1985) but the kinetics of the approach to the new steady state were enhanced; this has been shown studies using different techniques, such as hydrostatic pressure-release (P-jump, Fortune et al. 1991), sinusoidal length oscillation (Kawai and Halvorson 1991) and Pi-jump (Dantzig et al. 1992; Tesi et al. 2000) and Pi-measurement (He et al. 1997). The unified thesis that arose from such different

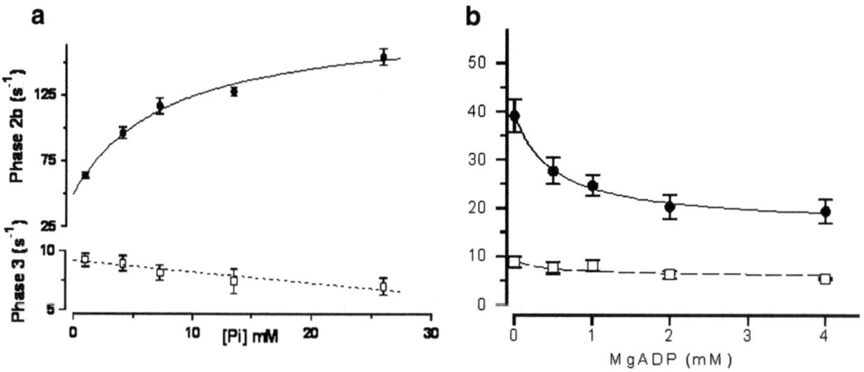

Fig. 6 From experiments as in Figs. 4 and 5 (T-jump from ~9 to 12°C), the mean (±SEM) rate constants for phase 2b (*filled symbols*) and phase 3 (*open symbols*) are shown. (**a**) Pi-dependence: Phase 2b rate increases with Pi, exhibits saturation at high P_i levels and the relation is hyperbolic (the curve fitted to the data). Phase 3 shows minimal sensitivity to Pi and may represent contribution to force rise of crossbridges going through a slower step in the cycle after Pi-release (adapted from Ranatunga 1999b). (**b**) ADP dependence: Phase 2b rate decreases with ADP, exhibits saturation at high ADP levels and the relation is hyperbolic. Phase 3 shows minimal sensitivity to ADP (adapted from Coupland et al. 2005)

studies was that Pi-release in active muscle occurs in two steps, both steps are reversible and, kinetically, the crossbridge force generation precedes Pi-release. Findings from T-jump experiments are consistent with that thesis (Figs. 4 and 6a) and also show that this force generation is endothermic.

The steady active tension was potentiated when [Mg.ADP] is increased; the binding of MgADP to nucleotide-free crossbridges (AM) leading to accumulation of force-bearing AM-ADP states (Cooke and Pate 1985; Dantzig et al. 1991; Lu et al. 1993, 2001; Seow and Ford 1997), in general, may underlie the tension increase. When [MgADP] is increased, the tension rise induced by a T-jump was slower (Figs. 5 and 6b) indicating that the approach to the new steady state at the post-T-jump temperature is slower.

4.2 *Effect of Strain (During Shortening and Lengthening)*

Temperature perturbation experiments referred to above were on isometric muscle fibers, whereas it is known that the force that a muscle develops varies with velocity of filament sliding in their sarcomeres, i.e. during steady muscle shortening and lengthening. With increase of shortening velocity, force declines below isometric force (P_0) and reaches zero at the maximum velocity; conversely, an active muscle develops ~$2\times P_0$ as lengthening velocity is increased to 1–2 L_0 (muscle fiber length)/s (Hill 1938; Katz 1939; Lombardi and Piazzesi 1990). Moreover, the energy production and the acto-myosin ATPase rate in muscle are increased with shortening and decreased with lengthening (Fenn 1924; Curtin and Davies 1973;

Getz et al. 1998; He et al. 1999; Linari et al. 2003). Hence, in a recent study, in addition to recording under isometric condition as before, we also examined the tension response induced by a T-jump in maximally active muscle fibers, when their force was decreased to a steady level by ramp shortening or increased to a steady level by ramp lengthening. The range of velocities used was 0–0.2 L_0/s and the unloaded (maximum) shortening velocity at this temperature was ~1 L_0/s; force decreased to $<0.5 \times P_0$ when shortening at 0.2 L_0/s and increased to $2–3 \times P_0$ when lengthening at >0.05 L_0/s (for details, see Ranatunga et al. 2007).

Superimposed tension responses from a fiber in Fig. 7a illustrate the T-jump tension responses (middle panel) in the different mechanical states. During ramp lengthening (top tension traces), the tension rises towards a level of about twice the isometric tension (P_0) and a T-jump produces no further tension rise; a small instantaneous drop in tension was sometimes seen, indicative of thermal expansion in the fiber (see Ranatunga 1996). During steady shortening (lower traces), the tension decreases to a level lower than P_0, and a T-jump produces a marked (mono-phasic) tension rise. As given above, a T-jump induces a biphasic rise in tension when the fiber is held in isometric (middle trace). Figure 7b shows that, plotted as a ratio of the post-T-jump tension, the amplitude of the T-jump tension rise in shortening fibers is higher than isometric and is correlated with velocity (filled circles); the amplitude of tension change during lengthening is not different from zero. Figure 7c shows that the rate of T-jump-induced tension rise during steady shortening increases linearly with velocity. For completeness, the data for the post-T-jump tension rise during steady lengthening are also shown (open circles). Since the amplitude is negligible (Fig. 7b), this is only an apparent rate and it increases slightly with velocity; the rates of tension rise determined by curve fit to the late part of pre-T-jump tension trace (crosses) were not significantly different from the post-T-jump rates (Students t-test, $P > 0.05$). As given above, the T-jump-induced tension rise in the isometric case contained two components; their rates (40–50 and 5–10 s^{-1}) are shown by open squares on the ordinates. Thus, the data in Fig. 7 shows that the tension during steady lengthening is not changed by a T-jump, whereas the tension in shortening is particularly enhanced by a T-jump.

4.3 Kinetic Simulation of Basic Findings

Incorporation of the above features to an adapted Lymn and Taylor (1971) cycle leads to a minimal, five-step, kinetic scheme for the crossbridge/AM-ATPase cycle (Scheme 1) that can qualitatively simulate some of the findings above (see Coupland et al. 2005; Ranatunga et al. 2007).

Briefly, the forward rate constant (k_{+1}) of Step 1 is endothermic force generation, step 2 is Pi release/binding (similar to Dantzig et al. 1992); steps 3 and 4 represent the slow, two-step, ADP release: k_{+3}/k_{-3} was forward biased (Dantzig et al. 1991). Step 5 is irreversible and includes all the necessary steps after ADP release to reprime a crossbridge for the next cycle. The overall rate in this route is low

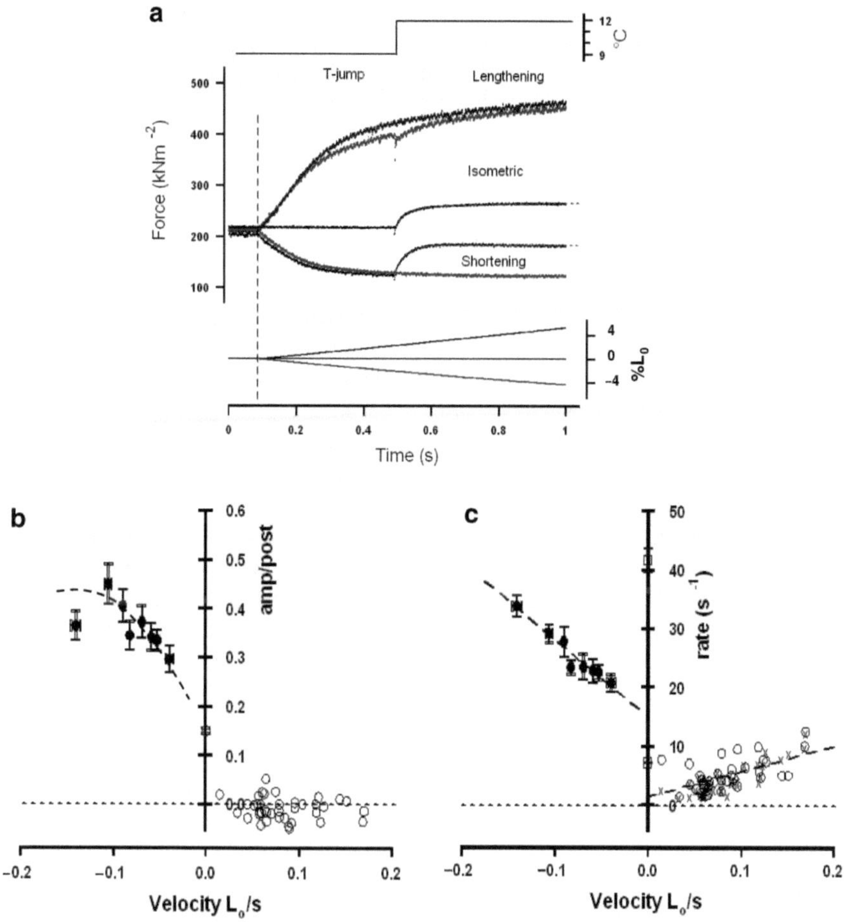

Fig. 7 (**a**) A fiber held isometric was maximally Ca-activated at ~9°C and, during the tension plateau (P_0), a T-jump of ~3°C applied (*top panel* – schematic) to obtain the "isometric" tension trace; *bottom panel* shows length records. Temperature was clamped again at ~9°C and fiber stretched at a constant velocity to obtain the "lengthening" tension traces (one without and the other with a T-jump); during lengthening, the tension rises to ~2.2 P_0 and T-jump does not lead to a net increase of tension, but induces an obvious the instantaneous tension drop (phase 1). The procedure was then repeated with fiber shortening to obtain "shortening" tension traces; tension decreases to ~0.5 P_0, but T-jump induces a pronounced tension rise; the post T-jump tension trace (300 ms) is fitted with a single exponential function. (**b**, **c**): Data from six fibers, velocity is plotted as L_0/s on the abscissa, negative for shortening and positive for lengthening. (**b**) The net tension change after a T-jump is plotted as a ratio of the post-T-jump tension. During shortening (*filled circles*) T-jump tension rise is correlated with velocity ($P<0.05$); each symbol is a mean (±SEM, n=7, 8) and curve fitted by eye. During lengthening, the tension change (open symbols, individual data) is not significantly different from zero. Mean (SEM, n=30) for isometric state is plotted on the ordinate. (**c**) With shortening, rate of tension rise (mean ± SEM) is correlated velocity ($P<0.001$). During lengthening, rate of tension rise from curve fit to the late part of pre-T-jump tension trace (*crosses*) is not significantly different ($P>0.1$) from the post-T-jump rate (*open circles*). Two isometric rates are on the ordinate (adapted from Ranatunga et al. 2007)

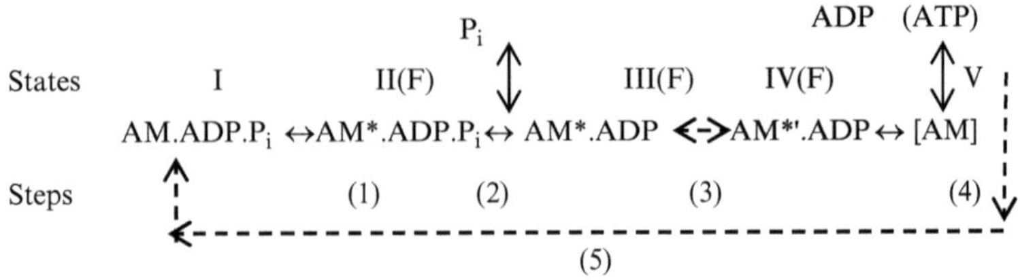

Scheme 1 The crossbridge/acto-myosin ATPase cycle used for simulations

(k_{+5}, ~10 s^{-1}), probably limited by the M.ATP↔M.ADP.P_i cleavage step after detachment (see He et al. 2000). The linear kinetic Scheme 1 above was solved by the matrix method using Mathcad 2000 software (Mathsoft) as described previously (Gutfreund and Ranatunga 1999). According to the scheme, states II, III and IV (AM*.ADP.P_i, AM*ADP and AM*'.ADP) are equal-force bearing states (F) and the sum of their fractional occupancy is taken as force.

As shown in Fig. 8a, temperature effect was simulated by increase of the rate constant k_{+1} (force generation, Q_{10} of 4) and a small increase in step k_{+4} (Q_{10} of 1.32), since there is evidence that ADP release itself is temperature-sensitive (Siemankowski et al. 1985). The sum of the occupancies of states II, III and IV at ~1 s was taken as steady tension. Figure 8b filled circles show the control steady state tension and its distribution approximates a sigmoidal relation with half-maximal at ~10–12°C. Within a temperature range of ~0–40°C, the relation is shifted upwards along the tension axis (potentiation) and slightly to the right with 4 mM added ADP and shifted down (depression) and to the left (to higher temperatures) with Pi. Figure 8c and d show simulated tension responses induced by a standard T-jump at ~10°C under control conditions (continuous trace) and with Pi (Fig. 8c) or MgADP (Fig. 8d); in each case, the two traces are vertically shifted for superimposition to show that they have similar amplitudes; the tension rise is faster with Pi but slower with ADP. Figure 9a shows that, during simulated shortening, a T-jump tension response is faster at higher velocity and Fig. 9b shows that, at a given velocity, T-jump tension response becomes faster at higher temperatures.

The model is simplistic, used a minimal kinetic scheme for actin-myosin-ATPase pathway and did not have strain-sensitive features; hence, it could only qualitatively account for endothermic force in isometric and shortening muscle.

5 General Discussion

Findings from various temperature-studies on intact and skinned muscle fibers, briefly referred to in this review, provide a useful overall picture for the process and mechanism of muscle contraction. Thus, a fundamental characteristic of active

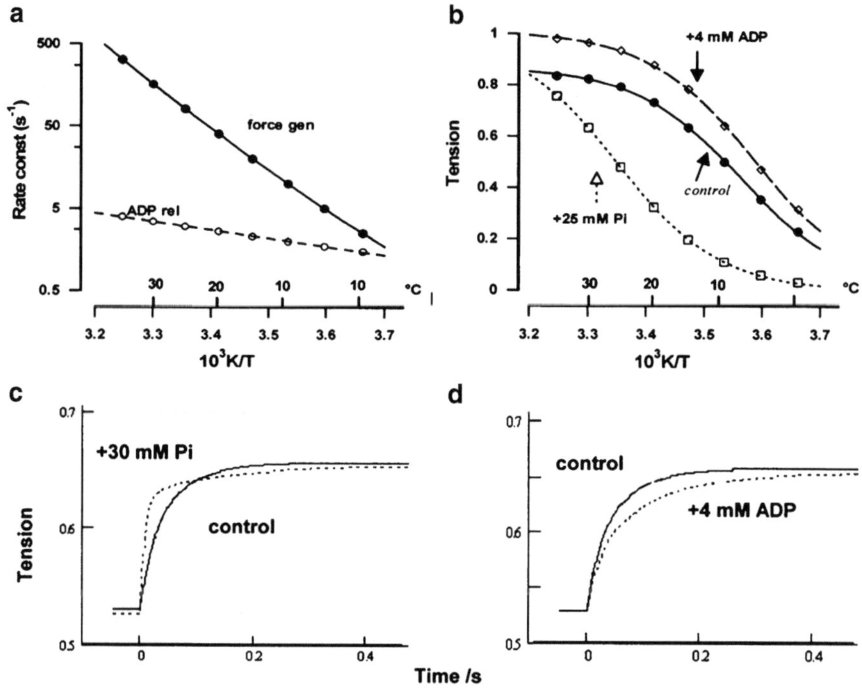

Fig. 8 Model simulations – isometric. (**a**) Arrhenius plots of the two rate constants of the model (see Scheme 1) that were changed to simulate temperature effects; the rate of the force generation step (k+1) increased markedly (see Zhao and Kawai 1994; Q10 ~4) and that of ADP-release step (k+4) slightly (Q10 of 1.3). (**b**) Filled circles show simulated, steady state tension (sum of the fractional occupancies of states II, III and IV) under control conditions (0.5 mM Pi and 10 μM ADP) at 5°C intervals. Note the approximate sigmoidal relation with half-maximal at ~10–12°C. *Open symbols* with *dotted curves* show that with simulated 25 mM Pi and 4 mM [MgADP], the sigmoidal relation is shifted to the left and right, respectively. (**c**) Simulated tension transients to T-jump at ~10°C in control (*solid curve*) and in the presence of 30 mM Pi (*dotted curve*); tension rise is faster with Pi. (**d**) Tension transients, in control (*solid curve*) and in the presence of MgADP (*dotted curve*); the tension rise is slower but amplitudes are similar with MgADP. (In both **c** and **d**, the tension records with Pi or ADP were shifted vertically for superimposition) – adapted from Coupland et al. (2005)

muscle is that its active force is endothermic (force rises on absorption of heat); this is largely due to the force generation by an attached crossbridge itself being temperature-sensitive. The particular molecular step is identified as being before phosphate release in a linear acto-myosin ATPase pathway, it is enhanced during muscle shortening and depressed during muscle lengthening and hence strain-sensitive; as summarised below, this overall thesis that has emerged from temperature studies is broadly consistent with most findings reported from other kinetic and mechanics studies on muscle.

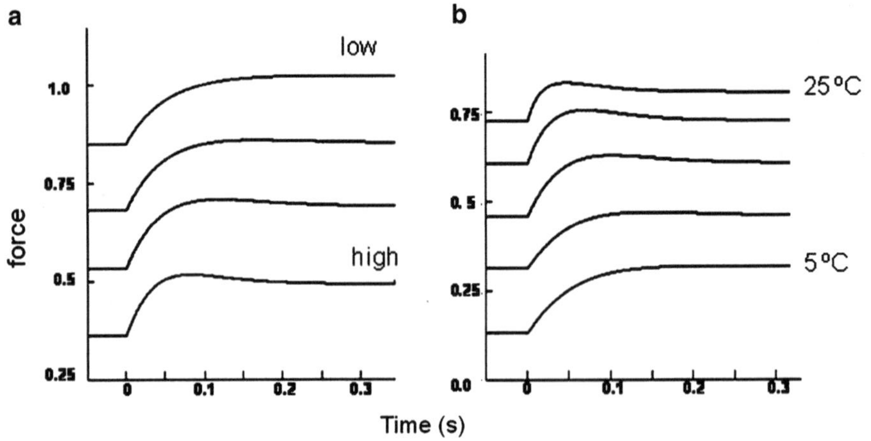

Fig. 9 Simulation – shortening. Simulated tension responses to a T-jump during steady shortening using Scheme 1, where force is normalized to that in isometric case (i.e. P_0 at 10°C). In each case, a shortening was simulated by increasing k+4 and, at steady state, a 5°C T-jump (re-setting k+1) is induced at time zero (see Gutfreund and Ranatunga 1999). (**a**) Simulated tension responses to T-jumps, at ~10°C; initial force is reduced by shortening at different velocities (low to high) and a T-jump induces a faster tension rise at higher velocities. (**b**) T-jump tension responses, when shortening at the same velocity (constant k+4) at different temperatures (5–25°C). The pre- and post-T-jump tension and the rate of tension rise to a T-jump are increased at higher temperatures, as in experiments (adapted from Ranatunga et al. (2007))

5.1 Comparison with Other Studies

As mentioned with results, the notion that force generation step is prior to Pi-release (a transition between two A-M.ADP.P_i states) is well supported from several studies and using different techniques, such as hydrostatic pressure-release (P-jump, Fortune et al. 1991), sinusoidal length oscillation (Kawai and Halvorson 1991) and Pi-jump (Dantzig et al. 1992; Tesi et al. 2000) and Pi-measurement (He et al. 1997). In a detailed examination of mechano-kinetic models, Smith and Sleep (2004) indeed ruled out the alternative possibility that force generation follows Pi-release. Findings from T-jump experiments are consistent with that thesis (Figs. 4 and 6a) and also show that this force generation is endothermic (and entropy driven), as has indeed been emphasized in other studies (Davis 1998; Davis and Epstein 2007; Kawai 2003; Zhao and Kawai 1994).

Following the original experiments and formulations by Huxley and Simmons (1971), quick tension recovery (T_1–T_2 transition) induced by a small length-release step is thought to represent crossbridge power stroke or force generation in muscle; quick tension recovery can be resolved in to two components (Davis and Harrington 1993), labeled here as phase 2a and 2b. The difficulties in correlating T-jump force generation with quick tension recovery after length-release have been raised in previous studies (see Bershitsky et al. 1997). It is found that the T-jump force generation is much slower than tension recovery from length release; however, in the same preparation and at similar temperature, it is faster than phase 2b recovery from stretches (see Ranatunga et al. 2002), indicating that T-jump force rise monitors force generation in crossbridges unstrained by filament sliding (as in length steps); its rate (phase 2b recovery) would be enhanced in length release and decreased following stretch, due to particular strain-dependence of the underlying rate constants (Huxley and Simmons 1971; Ford et al. 1977). Indeed, by extrapolation, the phase 2b recovery rate that corresponds to the isometric point in a length step versus rate of tension recovery plot in experiments on rabbit psoas fibers at ~10°C is ~40–60 s^{-1} (Ranatunga et al. 2002; Coupland et al. 2005) which is comparable to the rate of endothermic force generation in the T-jump experiments; the data also indicated that this extrapolated "isometric phase 2b recovery" from length step experiments is enhanced with P_i and depressed with MgADP, as does T-jump force generation. In general, these observations support the thesis that phase 2 tension recovery after a length step contains a component (phase 2b) that is homologous to the endothermic force generation observed after a rapid T-jump, as originally proposed by Davis and Harrington (1993); see also Davis (1998). This notion gains support from the study of Gilbert and Ford (1988) that showed experimentally that quick tension recovery from length-release is associated with heat absorption. Additionally, the rate of phase 2 tension recovery from length-release is temperature sensitive (Q_{10} of 2–3, Piazzesi et al. 2003), the temperature sensitivity being greater for phase 2b than for phase 2a component of recovery (Davis and Harrington 1993, Davis and Epstein 2007). Thus, like the T-jump force rise, phase 2b tension recovery from length-release represents an endothermic process; question remains as to the kinetic basis of (fast) phase 2a.

Despite the above considerations, however, some fiber mechanics experiments of Bershitsky and Tsaturyan (2002) indicate that different processes underlie tension responses to T-jump and length steps; Ferenczi et al. (2005) considered apparent strain independent, but temperature-dependent, rate-limiting step prior to force generation. Also, whereas T-jump studies above required only one molecular step for force generation, it has been argued from on X-ray diffraction studies using length perturbation that, several molecular steps would be required to complete a working stroke of a crossbridge (Huxley et al. 2006).

5.2 *Mechanism of Endothermic Force Generation*

A conformational change of the acto-myosin crossbridge, resulting in lever arm tilting, is thought to be the molecular-structural mechanism of force muscle generation (see review Geeves and Holmes 1999); whether this is endothermic and how it is associated with increased entropy (disorder) remain unclear. Zhao and Kawai (1994) proposed that altered hydrophobic interaction between actin and myosin-head for the large increase of entropy (see also Kodama (1985)); the transition of non-stereo-specifically attached to stereo-specifically attached crossbridge states would represent such an event, since stereospecific interaction may be hydrophobic. Ferenczi et al. (2005) proposed such a mechanism as a force generating step that is additional to the lever arm tilting step in the crossbridge cycle (see). Davis and Epstein (2007) proposed an alternative temperature-sensitive mechanism, namely, a local unfolding within the crossbridge secondary/tertiary structure and suggested that this might cause the lever arm tilting movement and force generation. From analyses of the rate of tension rise induced by T-jumps at different temperatures as Arrhenius plots, they showed that the observed rate could be resolved into the forward and reverse rate constants for force generation, where the reverse rate constant showed anti-Arrhenius behavior (negative temperature-sensitivity) characteristic for protein folding. Specific experiments aimed at getting direct information are required to elucidate the molecular mechanism.

Studies on myosin-ATPase in solution (Kodama 1985; Millar et al. 1987) have shown that the ATP cleavage step (i.e. in detached crossbridges in fibers) is endothermic and evidence suggests that this is due to a temperature-dependent conformational change(s) in myosin head (Jahn et al. 1999; Malnasi-Csizmadia et al. 2000); such findings from myosin and myosin-ATPase cycle need to be accommodated in the acto-myosin cycle in muscle fibers.

5.3 Some Implications

In principle, the active isometric muscle force is maintained by an "equilibrium" between low- and high-force attached crossbridge states so that a small length-release (a shortening) inducing negative strain and reducing force in attached crossbridges would perturb the equilibrium leading to force generation (Huxley and Simmons 1971; Ford et al. 1977). Conversely, a rapid stretch (a lengthening), inducing positive strain on crossbridges would lead to an inhibition/reversal of the force generation. Implication is that the force generating acto-myosin conformational change is more likely to occur when exposed to negative strain than when exposed to positive strain. Since T-jump perturbs an early molecular step, the T-jump tension response in active muscle can be used as a "signature" of the acto-myosin ATPase cycle. Hence, the enhanced endothermic force generation suggests that the ATPase cycle proceeds more readily during muscle shortening, consistent studies on muscle energetics (see Smith et al. 2005). The steady state force-shortening velocity curve of mammalian muscle was very temperature sensitive such that the maximal mechanical power output increased markedly in warming from 10 to 35°C (Ranatunga 1984, 1998); the finding that the tension in shortening muscle is particularly sensitive to a T-jump suggests that endothermic nature of force generation is a major underlying cause.

T-jump force generation is not seen during lengthening, indicating that the myosin motor fails to undergo the force-generating transition and, hence, the crossbridge/acto-myosin ATPase cycle is short-circuited before phosphate-release and before energy liberation. An implication is that the high force in lengthening muscle arises from the pre-stroke crossbridges attaching, getting strained by stretch and detaching without proceeding through the ATPase cycle, a conclusion consistent with previous studies on muscle fibers (Lombardi and Piazzesi 1990; Getz et al. 1998; Pinniger et al. 2006) and myofibrils (Rassier 2008). The fact that the lengthening muscle tension is insensitive to a T-jump is consistent with similar eccentric forces measured at high and low temperatures in human muscle experiments (De Ruiter and De Haan 2001). The contrasting behaviour of shortening and lengthening muscle to a T-jump suggests a basis for the Fenn effect (Fenn 1924), a cardinal principle of muscle contraction, that energy liberation in muscle is enhanced during shortening and depressed during lengthening. Also, the non-endothermic nature of steady tension during lengthening and a more obvious thermal-expansion effect by a T-jump indicate that stretch of non-crossbridge elements within muscle may contribute to the force and energy storage in lengthening muscle (Edman and Tsuchiya 1996; Pinniger et al. 2006).

Accumulation of products of ATP hydrolysis, particularly Pi, is commonly considered as contributory to in situ muscle fatigue. Temperature studies (e.g. Fig. 3) show that, because of the endothermic nature of force generation and that it is not directly coupled to release of Pi or ADP (products of hydrolysis), the relative effects on force of product accumulation would be less at high physiological temperatures; observations in support of this suggestion have been reported from fatigue-experiments on intact rat fast muscle fibers (see refs in Roots et al. 2009).

Acknowledgements We thank the Wellcome Trust Foundation for financial support of our research, Dr. Gerald Offer (Bristol) for valuable discussions and Blackwell Publishing and Springer for permission to include data we had published in the Journal of Physiology and the Journal of Muscle Research and Cell Motility.

References

Bershitsky SY, Tsaturyan AK (1992) Tension responses to joule temperature jump in skinned rabbit muscle fibres. J Physiol 447:425–448

Bershitsky SY, Tsaturyan AK (2002) The elementary force generation process probed by temperature and length perturbations in muscle fibres from the rabbit. J Physiol 540:971–988

Bershitsky SY, Tsaturyan AK, Bershitskaya ON, Mashanov GI, Brown P, Burns R, Ferenczi MA (1997) Muscle force is generated by myosin heads stereospecifically attached to actin. Nature 388:188–190

Cooke R, Pate, E (1985) The effects of ADP and phosphate on the contraction of muscle fibers. Biophys J 48:789–798

Coupland ME, Ranatunga KW (2003) Force generation induced by rapid temperature jumps in intact mammalian (rat) muscle fibres. J Physiol 548:439–449

Coupland ME, Puchert E, Ranatunga KW (2001) Temperature dependence of active tension in mammalian (rabbit psoas) muscle fibres: effect of inorganic phosphate. J Physiol 536:879–891

Coupland ME, Pinniger GJ, Ranatunga KW (2005) Endothermic force generation, temperature-jump experiments and effects of increased [MgADP] in rabbit psoas muscle fibres. J Physiol 567:471–492

Curtin NA, Davies RE (1973) Chemical and mechanical changes during stretching of activated frog skeletal muscle. Cold Spring Harb Symp Quant Biol 37:619–626

Dantzig JA, Hibberd MG, Trentham DR, Goldman YE (1991) Crossbridge kinetics in the presence of MgADP investigated by photolysis of caged ATP in rabbit psoas muscles. J Physiol 432:639–680

Dantzig JA, Goldman YE, Millar NC, Lacktis J, Homsher E (1992) Reversal of the cross-bridge force-generating transition by photogeneration of phosphate in rabbit psoas muscle fibres. J Physiol 451:247–278

Davis JS (1998) Force generation simplified. Insights from laser temperature-jump experiments on contracting muscle fibers. In Mechanisms of Work Production and Work Absorption in Muscle, ed. Sugi H & Pollack GH, pp. 343–352. Plenum Press, New York

Davis JS, Epstein ND (2007) Mechanism of tension generation in muscle: an analysis of the forward and reverse rate constants. Biophys J 92:2865–2874

Davis JS, Harrington W (1987) Force generation by muscle fibers in rigor: a laser temperature-jump study. Proc Natl Acad Sci U S A 84:975–979

Davis JS, Harrington W (1993) A single order-disorder transition generates tension during the Huxley-Simmons phase 2 in muscle. Biophys J 65:1886–1898

De Ruiter CJ, De Haan A (2001) Similar effects of cooling and fatigue on eccentric and concentric force-velocity relationships in human muscle. J Appl Physiol 6:2109–2116

Edman KA, Tsuchiya T (1996) Strain of passive elements during force enhancement by stretch in frog muscle fibres. J Physiol 490:191–205

Fenn WO (1924) The relationship between the work performed and the energy liberated in muscular contraction. J Physiol 59:373–395

Ferenczi MA, Bershitsky SY, Koubassova N, Siththanandan V, Helsby WI, Pannie P, Roessle M, Narayanan T, Tsaturyan AK (2005) The 'Roll and Lock' mechanism of force generation in muscle. Structure 13:131–141

Ford LE, Huxley AF, Simmons RM (1977) Tension responses to sudden length change in stimulated frog muscle fibres near slack length. J Physiol 269:441–515

Fortune, NS, Geeves MA, Ranatunga, KW (1991) Tension responses to rapid pressure release in glycerinated rabbit muscle fibers. Proc Natl Acad Sci U S A 88:7323–7327

Galler S, Hilber K (1998) Tension/stiffness ratio of skinned rat skeletal muscle fibre types at various temperatures. Acta Physiol Scand 162:119–126

Geeves MA, Holmes KC (1999) Structural mechanism of muscle contraction. Ann Rev Biochem 68:687–728

Getz, EB, Cooke R, Lehman SL (1998) Phase transition in force during ramp stretches of skeletal muscle. Biophys J 75:2971–2983

Gilbert SH, Ford LE (1988) Heat changes during transient tension responses to small releases in active frog muscle. Biophys J 54:611–677

Goldman YE, Hibberd MG, Trentham DR (1984) Relaxation of rabbit psoas muscle fibres from rigor by photochemical generation of adenosine-5′-triphosphate. J Physiol 354:577–604

Goldman YE, McCray JA, Ranatunga KW (1987) Transient tension changes initiated by laser temperature jumps in rabbit psoas muscle fibres. J Physiol 392:71–95

Gutfreund H, Ranatunga KW (1999) Simulation of molecular steps in muscle force generation. Proc Roy Soc B 266:1471–1475

Hadju S. (1951) Behaviour of frog and rat muscle at higher temperatures. Enzymologia 14:187–190

He, Z-H, Chillingworth RK, Brune M, Corrie JET, Trentham DR, Webb MR, Ferenczi MA (1997) ATPase kinetics on activation of rabbit and frog permeabilised isometric muscle fibres: a real time phosphate assay. J Physiol 501:125–148

He, Z-H, Chillingworth RK, Brune M, Corrie JET, Webb MR, Ferenczi MA (1999) The efficiency of contraction in rabbit skeletal muscle fibres, determined from the rate of release of inorganic phosphate. J Physiol 517:839–854

He, Z-H, Bottinelli R, Pellegrino MA, Ferenczi MA, Reggiani C (2000) ATP consumption and efficiency of human single muscle fibers with different myosin isoform composition. Biophys J 79:945–961

Hibberd MG, Dantzig JA, Trentham DR, Goldman YE (1985) Phosphate release and force generation in skeletal muscles fibers. Science 228:1317–1319

Hill AV (1938) The heat of shortening and the dynamic constants of muscle. Proc Roy Soc Lond B 126:136–195

Huxley AF (1957) Muscle structure and theories of contraction. Prog Biophys 7:285–318

Huxley HE (1969) Mechanism of muscle contraction. Science 164:1356–1366

Huxley AF, Simmons RM (1971) Proposed mechanism of force generation in striated muscle. Nature 233:533–538

Huxley HE, Reconditi M, Stewart A, Irving T (2006) X-ray interference studies of crossbridge action in muscle contraction: evidence from muscles during steady shortening. J Mol Biol 363:762–772

Jahn W, Urbanke C, Wray J (1999) Fluorescence temperature-jump studies of myosin S1 structures. Biophys J 76:A146

Katz B (1939) The relations between force and speed in muscular contraction. J Physiol 96:45–64

Kawai M (2003) What do we learn by studying the temperature effect on isometric tension and tension transients in mammalian striated muscle fibres? J Musc Res Cell Motil 24:127–138

Kawai M, Halvorson HR (1991) Two step mechanism of phosphate release and the mechanism of force generation in chemically skinned fibers of rabbit psoas muscle. Biophys J 59:329–342

Kawai M, Kido T, Vogel M, Fink RHA, Ishiwata S (2006) Temperature change does not affect force between regulated actin filaments and heavy meromyosin in single molecule experiments. J Physiol 574:877–878

Kodama T (1985) Thermodynamic analysis of muscle ATPase mechanisms. Physiol Rev 65:467–551

Linari M, Woledge RC, Curtin NA (2003) Energy storage during stretch of active single fibres from frog skeletal muscle. J Physiol 548:461–474

Lombardi V, Piazzesi G (1990) The contractile response during steady lengthening of stimulated frog muscle fibres. J Physiol 431:141–171

Lu Z, Moss RL, Walker JW (1993) Tension transients initiated by photogeneration of MgADP in skinned skeletal muscle fibers. J Gen Physiol 101:867–888

Lu Z, Swartz DR, Metzger JM, Moss RL, Walker JW (2001) Regulation of force development studied by photolysis of caged ADP in rabbit skinned psoas fibers. Biophys J 81:334–344

Lymn RW, Taylor EW (1971) Mechanism of adenosine triphosphate hydrolysis by actomyosin. Biochem 10:4617–4624

Malnasi-Csizmadia A., Woolley RJ, Bagshaw CR (2000) Resolution of conformational states of Dictyostelium myosin II motor domain using tryptophan (W501) mutants: Implications for the open-closed transition identified by crystallography. Biochem 39:16135–16146

Millar NC, Howarth JV, Gutfreund H (1987) A transient kinetic study of enthalpy changes during the reaction of myosin subfragment 1 with ATP. Biochem J 248:683–690

Piazzesi G, Reconditi M, Koubassova N, Decostre V, Linari M, Lucil L, Lombardi V (2003) Temperature dependence of the force-generating process in single from the frog skeletal muscle. J Physiol 549:93–106

Pinniger G J, Ranatunga KW, Offer GW (2006) Crossbridge and non-crossbridge contributions to tension in lengthening rat muscle: force-induced reversal of the power stroke. J Physiol 573:627–643

Ranatunga KW (1984) The force-velocity relation of rat fast- and slow-twitch muscles examined at different temperatures. J Physiol 351:517–529

Ranatunga KW (1994) Thermal stress and Ca-independent contractile activation in mammalian skeletal muscle fibers at high temperatures. Biophys J 66:1531–1541

Ranatunga KW (1996) Endothermic force generation in fast and slow mammalian (rabbit) muscle fibers. Biophys J 71:1905–1913

Ranatunga KW (1998) Temperature dependence of mechanical power output in mammalian (rat) skeletal muscle. Exp Physiol 83:371–376

Ranatunga KW (1999a) Effects of inorganic phosphate on endothermic force generation in muscle. Proc R Soc Lond B 266:1381–1385

Ranatunga KW (1999b) Endothermic force generation in skinned cardiac muscle from rat. J Musc Res Cell Motil 20:489–490

Ranatunga KW, Wylie SR (1983) Temperature-dependent transitions in isometric contractions of rat muscle. J Physiol 339:87–95

Ranatunga KW, Fortune NS, Geeves MA (1990) Hydrostatic compression in glycerinated rabbit muscle fibres. Biophys J 58:1401–1410

Ranatunga KW, Coupland ME, Mutungi G (2002) An asymmetry in the phosphate dependence of tension transients induced by length perturbation in mammalian (rabbit psoas) muscle fibres. J Physiol 542:899–910

Ranatunga KW, Coupland ME, Pinniger GJ, Roots H, Offer GW (2007) Force generation examined by laser temperature-jumps in shortening and lengthening mammalian (rabbit psoas) muscle fibres. J Physiol 585:263–277

Rassier DE (2008) Pre-power stroke cross bridges contribute to force during stretch of skeletal muscle myofibrils. Proc R Soc B 275, 2577–2586

Roots H, Ranatunga KW (2008) An analysis of temperature-dependence of force, during shortening at different velocities, in (mammalian) fast muscle fibres. J Musc Res Cell Motil 29:9–24

Roots H, Ball G, Talbot-Ponsonby J, King M, McBeath K, Ranatunga KW (2009) Muscle fatigue examined at different temperatures in experiments on intact mammalian (rat) muscle fibers. J Appl Physiol 106:378–384

Seow CY, Ford LE (1997) Exchange of ADP on high-force cross-bridges of skinned muscle fibers. Biophys J 72:2719–2735

Siemankowski RF, Wiseman MO, White HD (1985) ADP dissociation from actomyosin sub fragment 1 is sufficiently slow to limit the unloaded shortening velocity in vertebrate muscle. Proc Nat Acad Sci U S A 82:658–662

Smith GA, Sleep J (2004) Mechanokinetics of rapid tension recovery in muscle: the myosin working stroke is followed by a slower release of phosphate. Biophys J 87:442–456

Smith NP, Barclay CJ, Loiselle DS (2005) The efficiency of muscle contraction. Prog Biophys Mol Biol 88:1–58

Tesi C, Colomo F, Nencini S, Piroddi N, Poggesi C (2000) The effect of inorganic phosphate on force generation in single myofibrils from rabbit skeletal muscle. Biophys J 78:3081–3092

Vawda F, Geeves MA, Ranatunga KW (1999) Force generation upon hydrostatic pressure release in tetanized intact frog muscle fibres. J Musc Res Cell Motil 20:477–488

Zhao Y, Kawai M (1994) Kinetic and thermodynamic studies of the cross-bridge cycle in rabbit psoas muscle fibers. Biophys J 67:1655–1668

Efficiency of Cross-Bridges and Mitochondria in Mouse Cardiac Muscle

C.J. Barclay and C. Widén

Abstract The aim of this study was to make cellular-level measurements of the mechanical efficiency of mouse cardiac muscle and to use these measurements to determine (1) the work performed by a cross-bridge in one ATP-splitting cycle and (2) the fraction of the free energy available in metabolic substrates that is transferred by oxidative phosphorylation to free energy in ATP (i.e. mitochondrial thermodynamic efficiency). Experiments were performed using isolated left ventricular mouse papillary muscles (n = 9; studied at 27°C) and the myothermic technique. The production of work and heat was measured during and after 40 contractions at a contraction frequency of 2 Hz. Each contraction consisted of a brief isometric period followed by isovelocity shortening. Work output, heat output and enthalpy output were all independent of shortening velocity. Maximum initial mechanical efficiency (mean ± SEM) was 31.1 ± 1.3% and maximum net mechanical efficiency 16.9 ± 1.5%. It was calculated that the maximum work per cross-bridge cycle was 20 zJ, comparable to values for mouse skeletal muscle, and that mitochondrial thermodynamic efficiency was 72%. Analysis of data in the literature suggests that mitochondrial efficiency of cardiac muscle from other species is also likely to be between 70 and 80%.

Keywords Cardiac muscle • Efficiency • Mitochondria • Cross-bridges • Energy use • ATP hydrolysis

---

C.J. Barclay (✉)
School of Physiotherapy and Exercise Science, Griffith University,
Gold Coast, QLD, Australia
e-mail: c.barclay@griffith.edu.au

D.E. Rassier (ed.), *Muscle Biophysics: From Molecules to Cells*,
Advances in Experimental Medicine and Biology 682,
DOI 10.1007/978-1-4419-6366-6_15,

1 Introduction

Muscles convert the chemical free energy obtained from breaking down metabolic substrates into mechanical work. The free energy conversion takes place in two steps: first, during mitochondrial oxidative phosphorylation free energy is transferred from the metabolic substrate(s) to ATP and, second, the myosin cross-bridges convert free energy from ATP into mechanical work. Efficiency quantifies the fraction of the available free energy that is converted into work. It is relatively straight-forward to determine the efficiency with which cross-bridges perform work but how to determine the efficiency of mitochondrial free energy transfer is less obvious. For processes operating in series (e.g. the production of work by cross-bridges at the expense of ATP followed by the oxidative regeneration of ATP), the overall efficiency is the product of the efficiencies of each process. It is possible to measure the overall efficiency of contraction and the efficiency of work production by the cross-bridges so this relationship can be used to calculate the efficiency of oxidative phosphorylation. The main purpose of this study is to determine the efficiency of oxidative phosphorylation in mouse cardiac muscle, which has not been precisely specified (Gibbs and Barclay 1995).

In general, efficiency is the ratio of work produced to the energy expended to produce that work but precise definitions depend on the index of energy expenditure (for a review, see Smith et al. 2005). When trying to understand fundamental aspects of contraction, as outlined in the preceding paragraph, it is the work produced relative to the free energy change (ΔG) associated with that work that is of relevance because it is only free energy that can be converted into work. Efficiency calculated using ΔG is called thermodynamic efficiency. The practical difficulty with thermodynamic efficiency is that ΔG cannot be measured. However, it is possible to measure a related quantity, the enthalpy change (ΔH), which is the sum of the heat and work produced. When efficiency is calculated using ΔH as the index of energy use it is called mechanical efficiency. The relevance of ΔH is that the amount of enthalpy produced by a contracting muscle is proportional to the extent of the biochemical reactions that underlie contraction and that mechanical efficiency can be used to estimate thermodynamic efficiency.

In the current study, we used the myothermic or heat measurement technique with isolated mouse papillary muscles. This technique allows enthalpy output from a contracting muscle to be separated into the component attributable to ATP-consuming processes (the initial enthalpy output) and that due to oxidative ATP regeneration (recovery enthalpy output) (Mast et al. 1990; Barclay et al. 2003; Barclay and Weber 2004). The measurements reported in this study indicate that the efficiency of the mitochondria in cardiac muscle is 72% and an analysis of the literature indicates that this value is most likely independent of animal species.

2 Methods

2.1 *Papillary Muscle Preparation*

The preparation, apparatus and experimental protocol have been described in detail previously (Widén and Barclay 2006) and only a brief description is provided here. Six to twelve week old male Swiss mice were rendered unconscious by inhalation of 80% CO_2–20% O_2 gas mixture and killed by cervical dislocation. All animal-handling procedures were approved by the Griffith University Animal Ethics Committee. Papillary muscles were dissected from the left ventricle with the heart immersed in oxygenated (95% O_2–5% CO_2) Krebs-Henseleit solution of the following composition (mM): 118 NaCl, 4.75 KCl, 1.18 KH_2PO_4, 1.18 $MgSO_4$, 24.8 $NaHCO_3$, 2.5 $CaCl_2$, 10 glucose, 30 2,3-butanedione monoxime (BDM). BDM was included in the solution only during dissection of the muscles.

2.2 *Experimental Apparatus*

Muscles were connected via fine stainless steel wires to a force transducer (AE801, SensorOne, CA, USA) at one end and a servo-controlled motor (322B, Aurora Scientific Inc., Ontario, Canada) at the other end. The muscle lay along the active thermocouples of a thin-film, antimony-bismuth thermopile (Mulieri et al. 1977; Barclay et al. 1995), which was used to measure changes in muscle temperature. The recording region of the thermopile was 6 mm long, contained 24 thermocouples and produced 1.25 $mV°C^{-1}$. The muscle and thermopile were enclosed in an isothermal chamber maintained at 27°C by water circulated through the chamber walls from a thermostatically-controlled reservoir.

Muscle heat production was calculated from the records of the change in muscle temperature and external work output was calculated as the integral of force with respect to change in muscle length.

Records of energy output, as heat and work, from papillary muscles were made during and after a contraction protocol designed to mimic the cyclic strain patterns of papillary muscles in vivo (Semafuko and Bowie 1975). At the start of experiments, muscle length was adjusted so that sarcomere length was 2.1 μm (Widén and Barclay 2006); this length was designated L_0. In each contraction cycle, muscle length remained at L_0 for 60 ms after the stimulus pulse was delivered. Then an isovelocity shortening (amplitude, 15% L_0) was applied and continued until force relaxation was complete. At that time, muscle length was returned to L_0, also at a constant velocity. A range of shortening velocities between 0.4 and 1.7 $L_0 s^{-1}$ was used and 40 contractions were performed at a contraction frequency of 2 Hz.

2.3 Efficiency Calculations

Mechanical efficiency is the ratio of mechanical work output (W) to ΔH. ΔH is the sum of the heat (Q) and work (W) outputs; i.e. ΔH = W + Q. Net mechanical efficiency (ε_{net}) was defined the ratio of the total work produced in the 40 contractions to the total, supra-basal enthalpy produced during and after the contractions series (ΔH_{Net}).

$$\varepsilon_{Net} = \frac{W}{\Delta H_{Net}} \tag{1}$$

The initial mechanical efficiency (ε_I) was defined as the ratio of work produced in the first three twitches of a series to the enthalpy output produced in the same time (ΔH_I).

$$\varepsilon_I = \frac{W}{\Delta H_I} \tag{2}$$

ΔH_I is the enthalpy change that arises from the breakdown of phosphate-containing compounds. At the end of the third contraction cycle, (i.e. 1.5 s after the first stimulus pulse), oxidative recovery metabolism accounted for <4% of the enthalpy produced (Widén and Barclay 2006).

The cost of recovery metabolism relative to initial metabolism was quantified by the ratio of the enthalpy output due to complete oxidative recovery, ΔH_{Net}, to the initial enthalpy output, ΔH_I, – the recovery to initial ratio (R/I). R/I can also be calculated from e_I and ε_{Net} (Widén and Barclay 2006).

$$\frac{R}{I} = \frac{\varepsilon_I}{\varepsilon_{Net}} - 1 \tag{3}$$

2.4 Data Normalization and Statistical Analysis

At the end of experiments the platinum clips were cut off the preparation and the muscle was lightly blotted and weighed. Rates of energy output were normalized by blotted muscle mass. Data are presented as mean ± standard error. The statistical significance of variations in initial and net work, heat, enthalpy and efficiency with shortening velocity was assessed using one-way analysis of variance. Decisions concerning statistical significance were made at the 95 % level of confidence.

3 Results

Experiments were performed with nine papillary muscles (mean mass, 1.3 ± 0.2 mg; mean length, 3.4 ± 0.2 mm). The work and heat outputs were measured during and after a contraction protocol taking 20 s and which was repeated

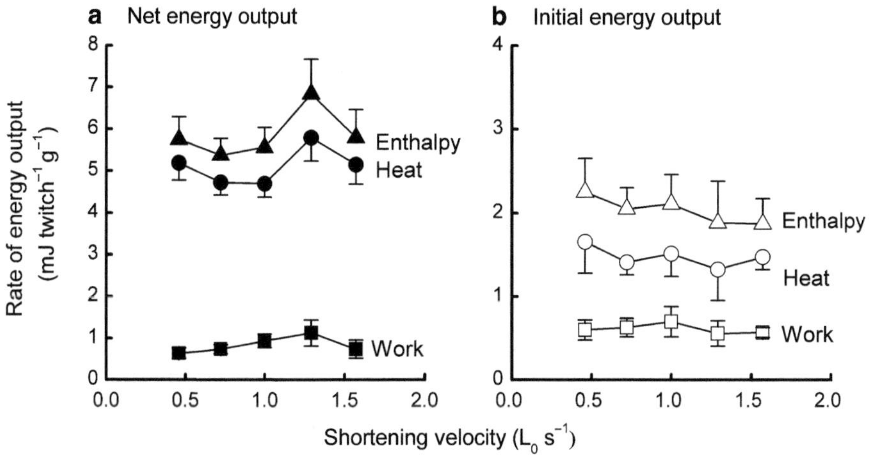

Fig. 1 Effects of shortening velocity on energy output. (**a**) Net energy output was the cumulative work, heat and enthalpy produced during and, for heat and enthalpy, after the 40 contractions until recovery was complete (~60s). Values are expressed as the average amounts per twitch. There was no significant effect of shortening velocity on any of these variables. Net enthalpy output is the energetic equivalent of the supra-basal O_2 consumption and thus reflects the mitochondrial response to the contraction protocol. (**b**) The amounts of work, heat and enthalpy produced in the first three contractions, expressed as average quantities per twitch. Over this time, negligible oxidative recovery occurred so the enthalpy arises from net breakdown of creatine phosphate. Shortening velocity had no significant effect on any of the measured initial variables

for each muscle for different velocities of shortening. In Fig. 1a, the variations in net energy output (work, heat and enthalpy) with shortening velocity are shown. All data are the total amounts produced in the 40 contractions divided by the number of contractions, giving an average energy output per twitch. The rate of enthalpy output is the sum of the rates of work and heat production. The total enthalpy output was all the enthalpy, in excess of that due to the basal metabolism, produced during and after the series of contractions. Figure 1b shows the corresponding data for initial energy output. Shortening velocity had no significant effect on any of the variables, net or initial, measured. That is, the average work per twitch, heat per twitch and enthalpy per twitch were all independent of shortening velocity.

The variations of initial and net mechanical efficiency with shortening velocity are shown in Fig. 2. There was no significant effect of shortening velocity on initial mechanical efficiency ($F=1.55$, $p=0.22$). Peak initial efficiency was 31.1 ± 1.3 % (upper line, Fig. 2). Shortening velocity had a small but significant effect on net mechanical efficiency ($F=3.52$, $p=0.021$); the value at 1 $L_0 s^{-1}$ was significantly greater than that at the lowest velocity tested (lower line, Fig. 2). The peak net mechanical efficiency was 16.9 ± 1.5 %. R/I was also independent of shortening velocity and its value, averaged across all velocities tested, was 1.20 ± 0.09 (30 observations from 9 preparations).

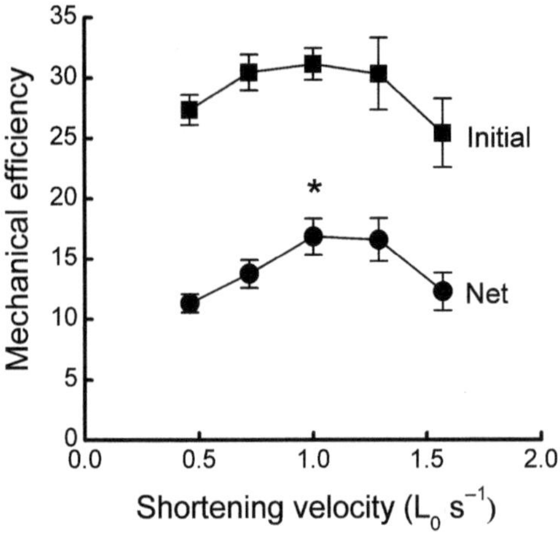

Fig. 2 Effects of shortening velocity on mechanical efficiency. Initial mechanical efficiency (*filled square*) is the ratio of the work and enthalpy produced in the first three contractions, when oxidative recovery contributes little to the enthalpy output. Initial mechanical efficiency was independent of shortening velocity. Net mechanical efficiency (*filled circle*) is the ratio of the cumulative work and cumulative enthalpy output over the complete protocol. Shortening velocity had a small but significant effect on net mechanical efficiency: ε_{Net} at a velocity of 1 $L_0 s^{-1}$ (indicated by *) was significantly higher than that at the lowest velocity tested

4 Discussion

In this study we measured the initial and net mechanical efficiencies of mouse cardiac muscle, which were calculated using rates of enthalpy output as the index of energy cost. It is notable that the all the energetic variables measured were largely unaffected by shortening velocity. In particular, neither initial nor net mechanical efficiency varied much across the range of shortening velocities tested. Ultimately the generation of work by cross-bridges involves conversion into work of part of the free energy produced when ATP is hydrolyzed. To gain insight into the molecular processes underlying the energy conversion process, the initial mechanical efficiency can be used to calculate the work performed by a cross-bridge for each ATP consumed. The fraction of the free energy of ATP hydrolysis (ΔG_{ATP}) that is converted into work can also be estimated; this is the thermodynamic efficiency of cross-bridge work generation.

It is also possible to use the mechanical efficiencies and the R/I ratio to calculate the efficiency of mitochondrial oxidative phosphorylation (η_R). This efficiency quantifies how much of the free energy produced by substrate oxidation is transferred to free energy in ATP.

4.1 *Efficiency of Work Production by Cross-Bridges*

The mechanical work produced in a contraction cycle arises solely from cross-bridge cycling but the enthalpy produced arises from both cross-bridges and from ion pumps.

In mouse cardiac papillary muscles, ~80% of the enthalpy produced in an isometric twitch is attributable to cross-bridge cycling (Widén and Barclay 2006). Enthalpy output per twitch is the same in isometric and shortening contractions so the measured maximum ε_I of 32% is equivalent to a cross-bridge mechanical efficiency of 32/0.8 = 40%. Initial enthalpy output in this study, averaged across all velocities, was 2.1 ± 0.2 mJ g^{-1} $twitch^{-1}$ and the cross-bridge component of this is 0.8 × 2.1 = 1.7 mJ g^{-1} $twitch^{-1}$. Initial enthalpy arises from PCr splitting so ΔH_I can be converted to the corresponding rate of ATP splitting knowing the molar enthalpy of PCr splitting (ΔH_{PCr}). Under the conditions in muscle cells at 27°C, ΔH_{PCr} is 37 kJ mol^{-1}. If it is assumed that one PCr was hydrolysed for each ATP consumed, then the rate of ATP splitting by cross-bridges is 1.7 mJ g^{-1} s^{-1}/37 kJ mol^{-1} = 4.6×10^{-8} mol g^{-1} $twitch^{-1}$. Average work output was 0.59 ± 0.06 mW g^{-1} $twitch^{-1}$ so the amount of work performed for each ATP split = 0.59×10^{-3} J g^{-1} s^{-1}/($4.6 \times 10^{-8} \times 6.022 \times 10^{23}$) = 22 ± 3 zJ ATP^{-1} (where z = zepto, the SI prefix for 10^{-21}). The standard error was calculated by combining relative errors of the work, initial enthalpy and non-cross-bridge heat and ignoring uncertainty in ΔH_{PCr}. That is, if one ATP is split in each cross-bridge cycle, then the work performed by one cross-bridge in one cycle is 22 zJ. The work per cross-bridge cycle can also be expressed relative to ΔG_{ATP}. In cardiac muscle under aerobic conditions and moderate workloads, ΔG_{ATP} is ~60 kJ mol^{-1} (see following section) which is 100 zJ $molecule^{-1}$. Then the fraction of the free energy available that is turned into mechanical work, that is, the cross-bridge thermodynamic efficiency (η_{CB}), is 22/100 = 22%.

How does the calculated work per cross-bridge cycle for mouse cardiac muscle compare to that for mouse skeletal muscle? A similar contraction protocol to that used in this study was previously used to measure the initial mechanical efficiency of mouse fast- and slow-twitch muscles (Barclay and Weber 2004), providing data that can be used for a comparison. Note that this protocol is unlikely to produce the maximum efficiency because it included an isometric phase at the start during which enthalpy was produced but no work performed. However, the strain pattern was very similar to that used in the current study and in both cases the protocols were designed to mimic in vivo strain patterns. The work per cross-bridge cycle and thermodynamic efficiencies of the skeletal muscles were analyzed in the same manner as presented above for cardiac muscle and taking account of non-cross-bridge energy use (35% of the isometric energy output, Barclay 1996; Barclay et al. 2008) and the higher rate of enthalpy output from shortening skeletal muscle compared to isometrically contracting muscle (Barclay et al. 1993). The results of the analysis are shown in Table 1. The work performed per cross-bridge cycle in the skeletal muscles was slightly lower than that in cardiac muscle and was lower in fast-twitch muscle than slow-twitch muscle. Interestingly, if the calculation for skeletal muscle is performed using data from isovelocity shortening (work and enthalpy measured only during shortening) at the velocity giving maximum efficiency (Barclay 1996), then the work per ATP consumed is ~20 and 30zJ for fast and slow skeletal muscles of the mouse, respectively.

The similarities in work performed per ATP split for mouse cardiac and skeletal muscle lend weight to the view that the properties of cross-bridges in cardiac muscle are the same as those in skeletal muscle. Other evidence consistent with this

Table 1 Comparison of efficiencies of cardiac and skeletal muscles of the mouse

	Cardiac	Fast skeletal[a]	Slow skeletal[a]
Work/ATP (zJ)	22	13	17
η_{CB} (%)	22	12	17
η_R (%)[b]	75±4	86±3	72±4

[a]For fully-activated mouse soleus and EDL muscles during isovelocity shortening, values for work/ATP are 21 and 31 zJ, respectively. Calculated using data from Barclay (1996)
[b]ΔG_{ATP}: cardiac muscle, 60 kJ mol^{-1}. Fast-twitch skeletal muscle, 66 kJ mol^{-1}. Slow-twitch skeletal muscle, 60 kJ mol^{-1}. Values for mouse skeletal muscle calculated from intracellular concentrations of ATP, ADP and P_i (Kushmerick et al. 1992) and assuming the standard free energy for ATP hydrolysis is 31 kJ mol^{-1} (Rosing and Slater 1972)

idea includes the observations that myofibrillar forces are the same in skeletal and cardiac muscle (Colomo et al. 1997), that the T_2 curve, determined from the rapid, partial force recovery following a fast, small change in muscle length and interpreted as representing the cross-bridge force-extension relationship (Huxley and Simmons 1971), is same in frog cardiac and skeletal muscle (Colomo et al. 1994) and that the same forces are produced by cross-bridges purified from skeletal and cardiac muscle (VanBuren et al. 1995).

4.2 *Efficiency of Mitochondrial Oxidative Phosphorylation*

The thermodynamic efficiency of mitochondrial energy conversion, η_R, expressed as a fraction, can be calculated as follows (Barclay and Weber 2004).

$$\eta_R = \frac{\varepsilon_{Net}}{\varepsilon_I} \cdot \frac{\Delta G_{ATP}}{\Delta H_{PCr}} \tag{4}$$

Underlying this calculation is the assumption that the molar values of ΔG and ΔH for substrate oxidation are the same in which case ε_{Net} is equal to η_{Net}, the net thermodynamic efficiency; in fact, for glucose oxidation ΔH and ΔG differ by only 2% (2,810 and 2,870 kJ mol^{-1}, respectively) so this assumption is realistic. Substituting (3) into (4) and re-arranging gives an expression relating ΔG_{ATP} to η_R, ΔH_{PCr} and R/I.

$$\Delta G_{ATP} = \eta_R \cdot \Delta H_{PCr} \cdot (R/I + 1) \tag{5}$$

ΔH_{PCr} has been measured in conditions comparable to those in muscle cells and at 27°C is 37 kJ mol^{-1} (Woledge and Reilly 1988). η_R must, by definition, be ≤1 so from (5) an upper limit for the value of ΔG_{ATP} is $37 \times (1.2+1) = 81$ kJ mol^{-1}. In the remaining text and in Fig. 3, η_R has been expressed as a percentage. Equation (5) was used to construct a graph showing the relationship between ΔG_{ATP} and ΔH_{PCr} for mouse cardiac muscle (i.e. assuming R/I = 1.2) for η_R values between 60 and 90% (Fig. 3a). The dashed lines indicate the likely values of ΔG_{ATP} and ΔH_{PCr} in

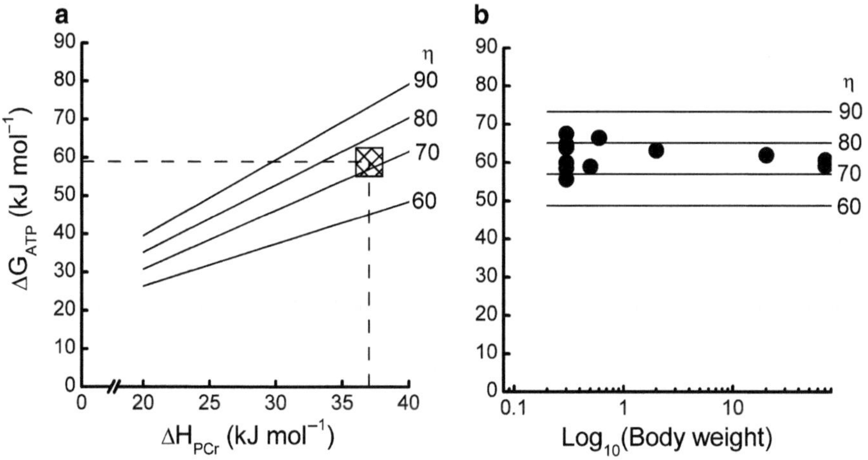

Fig. 3 Free energy of ATP hydrolysis and mitochondrial efficiency. (**a**) The relationship between ΔG_{ATP}, ΔH_{PCr} and η_R. The sloping lines correspond to different η_R values, as indicated. The lines were calculated from (5) assuming R/I was 1.2, as measured in the current study. The vertical dashed line indicates the value of ΔH_{PCr} corresponding to 27°C and intracellular pH of 7.2. The horizontal dashed line indicates the mean value of ΔG_{ATP} from 9 reported values for mouse heart (see text for details); the mean value is 58.9 kJ mol^{-1}. For these values η_R is 72%.The hatched region indicates likely ranges of ΔH_{PCr} and ΔG_{ATP} values. The ranges of ΔH_{PCr} indicated are those for intracellular pH of 7.1 and 7.3 (Woledge and Reilly 1988). The range of ΔG_{ATP} values indicated is bounded by the limits of the 95% confidence interval of the reported values for mouse cardiac muscle, 55.1–62.6 kJ mol^{-1}. (**b**) Literature values of ΔG_{ATP} (*filled circle*) for cardiac muscle plotted as a function of the base 10 logarithm of body weight. Data are shown for heart muscle from mouse, rat, ferret, rabbit, dog and human (Kammermeier et al. 1982; Allen et al. 1985; Headrick et al. 1994; Dobson and Headrick 1995; Balschi et al. 1997; Chacko et al. 2000; Dobson and Himmelreich 2002; Javadpour et al. 2003). The horizontal lines were calculated using (5) for η_R of 0.6, 0.7, 0.8 and 0.9 and assuming ΔH_{PCr} was 37 kJ mol^{-1} and R/I was 1.2

mouse cardiac muscle under the conditions used in this study. The intersection of the lines indicates that η_R is 72%. The shaded region in Fig. 3a is bounded vertically by the 95% confidence intervals of reported ΔG_{ATP} values for mouse cardiac muscle[1] (mean, 58.9 kJ mol^{-1}; n, 9; 95% confidence interval, 55.1–62.6 kJ mol^{-1}) and horizontally by the ΔH_{PCr} values for pH of 7.1 and 7.3, a range that encompasses reported values for cardiac muscle. This analysis indicates that extreme range of likely η_R values is 70–80%. This is comparable to estimates for rat cardiac muscle (Barclay et al. 2003) and also similar to values for mouse skeletal muscle (Table 1).

It is of interest to consider how the efficiency of free energy transfer by the mitochondria calculated here for mouse muscle might compare to that for other species. There is no evidence of differences in the intracellular environment that would alter ΔH_{PCr} among species and ΔH_{PCr} is practically independent of temperature

[1] Values for ΔG_{ATP} in mouse heart muscle from Saupe et al. (1998, 2000), Spindler et al. (1998, 2002), Chacko et al. (2000), Dobson and Himmelreich (2002), Weiss et al. (2002), Javadpour et al. (2003), and Day et al. (2006).

at pH > 7 (Woledge and Reilly 1988). The R/I ratio measured here is similar both to the theoretical value for oxidative recovery (12) and to values measured using isolated papillary muscles from other species (e.g. rabbit, Mast et al. 1990; rat, Barclay et al. 2003). Finally, ΔG_{ATP} for cardiac muscle does not appear to vary systematically amongst species. In Fig. 3b, a comparison is shown of reported ΔG_{ATP} values for cardiac muscle for species ranging in size from mouse to human. Body mass has no systematic effect on ΔG_{ATP} and the average across all species is ~61 kJ mol^{-1}. The horizontal lines in Fig. 3b show the values of ΔG_{ATP} that correspond to values of η_R of 60, 70, 80 and 90% which were calculated assuming ΔH_{PCr} of 37 kJ mol^{-1} and R/I of 1.2. The location of the majority of the data between the lines corresponding to η_R of 70 and 80% indicate that it is likely that the efficiency of oxidative recovery is between 70 and 80% for cardiac muscle from all species.

5 Conclusion

The analysis presented here is an example of molecular level inferences based on cellular level experimental measurements. Mouse cardiac muscle displayed stable energetic properties across a broad range of shortening velocities. The calculated average amount of work performed by cross-bridges per cycle is very similar to that for mouse skeletal muscle, consistent with the idea that the fundamental cross-bridge properties are the same in both cardiac and skeletal muscle. The analysis also indicates that mitochondria in cardiac muscle, regardless of species, transfer a large fraction (about three-quarters) of substrate free energy to ATP.

References

Allen DG, Morris PG, Orchard CH, Pirolo JS (1985) A nuclear magnetic resonance study of metabolism in the ferret heart during hypoxia and inhibition of glycolysis. J Physiol 361:185–204

Balschi JA, Shen H, Madden MC, Hai JO, Bradley EL, Jr., Wolkowicz PE (1997) Model systems for modulating the free energy of ATP hydrolysis in normoxically perfused rat hearts. J Mol Cell Cardiol 29:3123–3133

Barclay CJ (1996) Mechanical efficiency and fatigue of fast and slow muscles of the mouse. J Physiol 497:781–794

Barclay CJ, Weber CL (2004) Slow skeletal muscles of the mouse have greater initial efficiency than fast muscles but the same net efficiency. J Physiol 559:519–533

Barclay CJ, Constable JK, Gibbs CL (1993) Energetics of fast- and slow-twitch muscles of the mouse. J Physiol 472:61–80

Barclay CJ, Arnold PD, Gibbs CL (1995) Fatigue and heat production in repeated contractions of mouse skeletal muscle. J Physiol 488:741–752

Barclay CJ, Widén C, Mellors LJ (2003) Initial mechanical efficiency of isolated cardiac muscle. J Exp Biol 206:2725–2732

Barclay CJ, Lichtwark GA, Curtin NA (2008) The energetic cost of activation in mouse fast-twitch muscle is the same whether measured using reduced filament overlap or N-benzyl-p-toluenesulphonamide. Acta Physiol 193:381–391

Chacko VP, Aresta F, Chacko SM, Weiss RG (2000) MRI/MRS assessment of in vivo murine cardiac metabolism, morphology, and function at physiological heart rates. Am J Physiol Heart Circ Physiol 279:H2218–H2224

Colomo F, Poggesi C, Tesi C (1994) Force responses to rapid length changes in single intact cells from frog heart. J Physiol 475:347–350

Colomo F, Piroddi N, Poggesi C, Te Kronnie G, Tesi C (1997) Active and passive forces of isolated myofibrils from cardiac and fast skeletal muscle of the frog. J Physiol 500:535–548

Day SM, Westfall MV, Fomicheva EV, Hoyer K, Yasuda S, Cross NC, D'Alecy LG, Ingwall JS, Metzger JM (2006) Histidine button engineered into cardiac troponin I protects the ischemic and failing heart. Nat Med 12:181–189

Dobson GP, Headrick JP (1995) Bioenergetic scaling: metabolic design and body-size constraints in mammals. Proc Natl Acad Sci USA 92:7317–7321

Dobson GP, Himmelreich U (2002) Heart design: free ADP scales with absolute mitochondrial and myofibrillar volumes from mouse to human. Bioch Biophys Acta 1553:261–267

Gibbs CL, Barclay CJ (1995) Cardiac efficiency. Cardiov Res 30:627–634

Headrick JP, Dobson GP, Williams JP, McKirdy JC, Jordan L, Willis RJ (1994) Bioenergetics and control of oxygen consumption in the in situ rat heart. Am J Physiol Heart Circ Physiol 267:H1074–H1084

Huxley AF, Simmons RM (1971) Proposed mechanism of force generation in striated muscle. Nature 233:533–538

Javadpour MM, Tardiff JC, Pinz I, Ingwall JS (2003) Decreased energetics in murine hearts bearing the R92Q mutation in cardiac troponin T. J Clin Invest 112:768–775

Kammermeier H, Schmidt P, Jungling E (1982) Free energy change of ATP-hydrolysis: a causal factor of early hypoxic failure of the myocardium? J Mol Cell Cardiol 14:267–277

Kushmerick MJ, Moerland TS, Wiseman RW (1992) Mammalian skeletal muscle fibers distinguished by contents of phosphocreatine, ATP, and Pi. Proc Natl Acad Sci U S A 89:7521–7525

Mast F, Woledge RC, Elzinga G (1990) Analysis of thermopile records from contracting isolated cardiac muscle. Am J Physiol Heart Circ Physiol 259:H1601–H1605

Mulieri LA, Luhr G, Trefry J, Alpert NR (1977) Metal-film thermopiles for use with rabbit right ventricular papillary muscles. Am J Physiol Cell Physiol 233:C146–C156

Rosing J, Slater EC (1972) The value of ΔG^{o} for the hydrolysis of ATP. Bioch Biophys Acta 267:275–290

Saupe KW, Spindler M, Tian R, Ingwall JS (1998) Impaired cardiac energetics in mice lacking muscle-specific isoenzymes of creatine kinase. Circ Res 82:898–907

Saupe KW, Spindler M, Hopkins JC, Shen W, Ingwall JS (2000) Kinetic, thermodynamic, and developmental consequences of deleting creatine kinase isoenzymes from the heart. Reaction kinetics of the creatine kinase isoenzymes in the intact heart. J Biol Chem 275:19742–19746

Semafuko WE, Bowie WC (1975) Papillary muscle dynamics: in situ function and responses of the papillary muscle. Am J Physiol 228:1800–1807

Smith NP, Barclay CJ, Loiselle DS (2005) The efficiency of muscle contraction. Prog Biophys Mol Biol 88:1–58

Spindler M, Saupe KW, Christe ME, Sweeney HL, Seidman CE, Seidman JG, Ingwall JS (1998) Diastolic dysfunction and altered energetics in the alphaMHC403/+ mouse model of familial hypertrophic cardiomyopathy. J Clin Invest 101:1775–1783

Spindler M, Niebler R, Remkes H, Horn M, Lanz T, Neubauer S (2002) Mitochondrial creatine kinase is critically necessary for normal myocardial high-energy phosphate metabolism. Am J Physiol Heart Circ Physiol 283:H680–687

VanBuren P, Harris DE, Alpert NR, Warshaw DM (1995) Cardiac V1 and V3 myosins differ in their hydrolytic and mechanical activities in vitro. Circ Res 77:439–444

Weiss RG, Chatham JC, Georgakopolous D, Charron MJ, Wallimann T, Kay L, Walzel B, Wang Y, Kass DA, Gerstenblith G, Chacko VP (2002) An increase in the myocardial PCr/ATP ratio in GLUT4 null mice. FASEB 16:613–615

Widén C, Barclay CJ (2006) ATP splitting by half the cross-bridges can explain the twitch energetics of mouse papillary muscle. J Physiol 573:5–15

Woledge RC, Reilly PJ (1988) Molar enthalpy change for hydrolysis of phosphorylcreatine under conditions in muscle cells. Biophys J 54:97–104

Mechanisms of Skeletal Muscle Weakness

Håkan Westerblad, Nicolas Place, and Takashi Yamada

Abstract Skeletal muscle weakness is an important feature of numerous pathological conditions and it may also be a component in normal ageing. Decreased muscular strength can be due to decreased muscle mass and/or intrinsic defects in the muscle cells. In this chapter we will discuss decreased force production due to mechanisms intrinsic to skeletal muscle cells. We will mainly use data from mouse disease models to exemplify defects at various sites in the cellular activation-contraction pathway. We will show that depending on the underlying problem, muscle weakness can be due decreased Ca^{2+} release from the sarcoplasmic reticulum, reduced myofibrillar Ca^{2+} sensitivity and/or decreased ability of the cross-bridges to generate force.

Keywords Skeletal muscle • Force • Weakness • Ca^{2+} • Mitochondria • Reactive oxygen species • Sarcoplasmic reticulum • Exercise

1 Introduction

Skeletal muscle weakness is an important feature of numerous pathological conditions. These include primary muscle diseases (e.g. inherited myopathies and muscle dystrophies) and more generalized disorders (e.g. general inflammatory diseases, cancer and in association with intensive care treatment). Muscle weakness can also be a consequence of a life style with too little physical activities and it is a part of the normal ageing process. Muscle weakness can limit the independency and quality of life because daily activities requires much more effort with weak muscles (e.g. to climb a few stairs can be a serious challenge). Furthermore, it has been

H. Westerblad (✉)
Department of Physiology and Pharmacology, Karolinska Institutet, 171 77 Stockholm, Sweden
e-mail: Hakan.Westerblad@ki.se

D.E. Rassier (ed.), *Muscle Biophysics: From Molecules to Cells*,
Advances in Experimental Medicine and Biology 682,
DOI 10.1007/978-1-4419-6366-6_16,

shown that elderly people with weak muscles are more susceptible to fall-related injuries than age-matched controls with stronger muscles (Gehlsen and Whaley 1990; Whipple et al. 1987).

Decreased muscular strength can simply be due to decreased muscle mass, that is, muscle atrophy. However, in many conditions muscle weakness is observed despite normal muscle mass or the decrease in muscle mass cannot explain the decrease in muscle performance. In other words, muscle weakness can be caused by intrinsic defects in the muscle cells. In this chapter we will discuss decreased force production due to mechanisms intrinsic to skeletal muscle cells. The starting point of our discussion will be the chain of events in muscle cells that starts with an action potential and ends with cross-bridge formation and force production. We will mainly use data from mouse disease models to exemplify defects at various sites in the cellular activation-contraction pathway.

2 The Usage of Intact Single Muscle Fibers as an Experimental Tool

Ideally experiments on muscle contractile dysfunction should be performed on intact muscles activated by their α-motor neurons and with their blood circulation intact. However, with this approach it is difficult to explore cellular mechanisms. For instance, mammalian skeletal muscles generally contain a mixture of fiber types with markedly different functional properties, which then can be differently affected in conditions with muscle weakness. Moreover, intracellular Ca^{2+} handling has a central role in the assessment of muscle contractile dysfunctions and techniques available at present do not allow detailed measurements of the free myoplasmic Ca^{2+} concentration ($[Ca^{2+}]_i$) in fibers of fully intact muscles. Another experimental approach is to use isolated whole muscles, stimulated and perfused in a muscle bath. An important problem with isolated whole muscles is the lack the normal circulation, which means that diffusion gradients will develop with deeper parts facing limited access to O_2, changes in ion composition and accumulation of metabolites (Barclay 2005; Zhang et al. 2006).

Many of the problems associated with studying mechanisms underlying weakness in whole muscles in vivo or in vitro are overcome by using intact single fibers dissected from whole muscles. For example, problems with O_2 supply and extracellular gradients of K^+ and other products do not occur. Furthermore, the force measured is coming from only one cell and can be directly correlated with e.g. $[Ca^{2+}]_i$, which can be readily measured with fluorescent indicators. However, there are also disadvantages. For instance, the techniques for dissection of single fibers may be difficult to learn. To simplify the dissection procedure, our single muscle fiber experiments are mostly performed on a superficial foot muscle, flexor digitorum brevis (FDB), in which the fibers are short (<1 mm) and fast-twitch type IIA or IIX. In some instances it is questionable if the cells of this muscle are representative and therefore it is important to compare the results obtained with this preparation with results from fibers of other muscles (e.g. Sect. 5 below).

Another experimental approach to study muscle weakness is to use skinned muscle fibers; that is, muscle fibers with their surface membrane removed either chemically or mechanically. The solution surrounding the myofibrils is directly controlled in skinned fibers. This means that it is possible to directly study the effect on contraction of specified changes in the "intracellular" milieu. A disadvantage in this context is that soluble compounds in the intracellular fluid will be diluted in the bath solution and this may affect contractile function. Moreover, protein modifications due to, for instance, phosphorylation, oxidation and nitrosylation may be different in the skinned as compared to the intact cell and this can modify contractile function. Traditionally skinned fibers were activated by means of increasing the bath [Ca^{2+}], but recently methods have been developed by which skinned fibers can be activated by electrical current pulses triggering the normal action potential-mediated pathway (Posterino et al. 2000).

Immature muscle cells generated from satellite cells or cultured muscle cell lines are generally not suitable for studies of muscle weakness because even if these cells can be stimulated to generate force, this force cannot be accurately measured because the cells lack tendons. Furthermore, cellular Ca^{2+} handling is an essential component in the control of muscle contraction and the Ca^{2+} handling in immature muscle-like cells show little resemblance to the rapid and strictly controlled system in adult skeletal muscle fibers (cf. Kubis et al. (2003) and Baylor and Hollingworth (2003)).

To sum up, there are advantages and disadvantages with all models frequently used to study muscle contractile performance. In the present chapter we discuss cellular mechanisms underlying muscle weakness. Key experiments have mostly been performed on intact single muscle fibers, because we believe that this preparation is better suited for simultaneous detection of changes in individual steps in the cellular activation-contraction pathway than currently available alternatives.

3 The Intracellular Activation-Contraction Pathway in Skeletal Muscle Cells

Skeletal muscle cells are activated by acetyl choline (Ach) released from the end terminals of α-motor neurones. ACh binding to its receptor in the sarcolemma results in a localized depolarization, which in turn activates nearby voltage-dependent Na^+ channels and an action potential is generated. This neuromuscular transmission has a large safety margin and is not limiting under any normal conditions. However, the process is defective in myasthenia gravis, a disease with antibodies against the ACh receptors resulting in early fatigue development and muscle weakness (Meriggioli 2009).

The action potential is conducted along the surface membrane of the muscle cells and also into the transverse tubular (t-tubular) system. Action potentials activate the t-tubular voltage sensor, the dihydropyridine receptor (DHPR), which is located in close connection to the Ca^{2+} channel of the sarcoplasmic reticulum (SR), the ryanodine receptor 1 (RyR1) The activated DHPR:s mechanically interact with RyR1:s, which open and Ca^{2+} is released from the SR into the myoplasm (Dulhunty 2006). Ca^{2+} then binds to troponin, which results in a conformational change that is transmitted to tropomyosin. The myosin heads (cross-bridges) can then bind to actin, cross-bridge cycling starts and the muscle cell contracts. Ca^{2+} is continuously pumped back into the SR by the SR Ca^{2+} pumps (SERCA:s). When α-motor neurone activity ceases and SR Ca^{2+} release is stopped, the free Ca^{2+} concentration in the myoplasm ($[Ca^{2+}]_i$) decreases and the muscle cell relaxes.

Muscle weakness can result from impaired function of any of the events described above. In this chapter we will exemplify how defective function of specific processes can cause muscle weakness.

4 Skeletal-Muscle-Specific Tfam Deficient Mice: A Mitochondrial Myopathy Model that Displays Muscle Weakness Due to Decreased SR Ca^{2+} Stores and Impaired Cross-Bridge Force Production

Genetic mutations influencing the function of the mitochondrial respiratory chain can cause diseases with variable appearance regarding severity of symptoms, organs affected and age of onset (Larsson and Oldfors 2001; Thorburn et al. 2004; Thorburn 2004; Schapira 2006), although tissues with high-energy demand, such as skeletal muscle, are often affected (Rossignol et al. 2000). Muscle weakness and exercise intolerance are prominent symptoms in patients with mitochondrial myopathies (Tarnopolsky and Raha 2005).

Mice with skeletal-muscle-specific disruption of the nuclear gene for mitochondrial transcription factor A (Tfam KO mice) have been used as a model of mitochondrial myopathy. Skeletal muscles of these mice show important signs of mitochondrial myopathy, such as accumulation of abnormally appearing mitochondria, deteriorating respiratory chain function and reduced phosphocreatine concentration (Wredenberg et al. 2002, 2006).

An initial study on skeletal muscle of the Tfam KO mice shows decreased force production (Wredenberg et al. 2002). In a subsequent study, changes in the cellular activation-contraction pathway that may underlie the muscle weakness were studied (Aydin et al. 2009). In principle, the weakness can be due to (1) decreased SR Ca^{2+} release, (2) decreased myofibrillar Ca^{2+} sensitivity, (3) and/or decreased ability of cross-bridges to produce force (Westerblad and Allen 1996). To distinguish between these alternatives, force and $[Ca^{2+}]_i$ were simultaneously measured in intact single FDB fibers. $[Ca^{2+}]_i$ was measured with the fluorescent Ca^{2+} indicator indo-1, which was microinjected into the isolated cell (Andrade et al. 1998). The results show a lower tetanic $[Ca^{2+}]_i$ and force in Tfam KO than in control fibers (Fig. 1). The force-$[Ca^{2+}]_i$ relationship was obtained from tetanic contractions at different frequencies and showed no difference in myofibrillar Ca^{2+} sensitivity

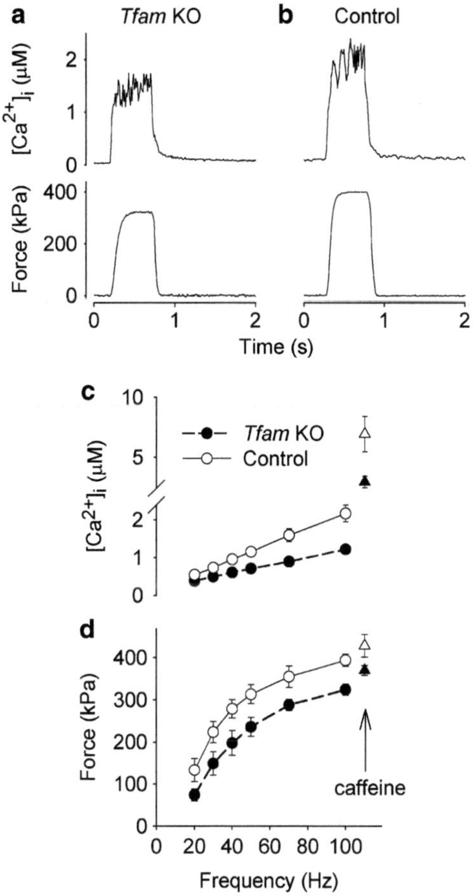

Fig. 1 Tetanic $[Ca^{2+}]_i$ and force are reduced in FDB fibers of mitochondrial myopathy mice. Original $[Ca^{2+}]_i$ (**a**) and force (**b**) records from 100 Hz tetanic contractions in Tfam KO and control single muscle fibers. Mean data (±SEM) of $[Ca^{2+}]_i$ (**c**) and force (**d**) in tetani of different frequencies in Tfam KO (*filled symbols*, *dashed lines*) and control (*open symbols*, *full lines*) fibers; 100 Hz tetani were also produced in the presence of 5 mM caffeine. Figure adapted from Aydin et al. (2009)

between Tfam KO and control cells, whereas the force at saturating $[Ca^{2+}]_i$ was about 15% lower in Tfam KO cells. Thus, the lower force in Tfam KO muscle fibers was due to reduced SR Ca^{2+} release and a decreased ability of cross-bridges to generate force. The latter feature might be explained by a higher inorganic phosphate ion (Pi) concentration in Tfam KO than in control cells (Wredenberg et al. 2006), because increased [Pi] is known to decrease cross-bridge force production (Millar and Homsher 1990; Pate and Cooke 1989). In addition, a relatively larger fraction of the muscle fiber volume is occupied by mitochondria in Tfam KO than in control cells (Wredenberg et al. 2002) and this leads to an equivalent decrease in myofibrillar content.

The next question to address was why tetanic $[Ca^{2+}]_i$ was decreased in Tfam KO muscle fibers. Tetanic stimulation in the presence of caffeine can be used to assess the amount of Ca^{2+} stored in the SR (Allen and Westerblad 1995). $[Ca^{2+}]_i$ was lower in Tfam KO than in control cells during 100 Hz stimulation in the presence of 5 mM caffeine, which indicates a decreased SR Ca^{2+} storage in the former. Analyses of SR Ca^{2+} handling proteins showed a specific down-regulation of the skeletal muscle isoform of calsequestrin, CASQ1, in Tfam KO muscle both at the mRNA and protein level (Aydin et al. 2009). The classical role of CASQ1 is as a high capacity, low affinity Ca^{2+} buffer that keeps the free $[Ca^{2+}]$ in the SR lumen close to 1 mM during contraction, despite the large fluxes where Ca^{2+} is first released from and subsequently pumped back into the SR (Beard et al. 2004). In addition, it appears that CASQ1 and SR Ca^{2+} interact with the RyR1 and in this way affect Ca^{2+} release during contraction (Beard et al. 2004). Fast-twitch muscle cells of mice lacking CASQ1 show decreased $[Ca^{2+}]_i$ with electrical stimulation and caffeine application (Paolini et al. 2007). Thus, the decreased tetanic $[Ca^{2+}]_i$ in Tfam KO muscle can be explained by decreased CASQ1 expression leading to reduced SR Ca^{2+} storage and possibly direct effects on the RyR1.

Mitochondrial $[Ca^{2+}]$ was measured with the fluorescent indicator Rhod-2 during fatigue in single Tfam KO and control FDB fibers (Aydin et al. 2009). Tfam KO fiber showed a marked increase in mitochondrial $[Ca^{2+}]$ (three- to fourfold increase in Rhod-2 fluorescence) during fatigue and it was still increased 10 min after the end of stimulation. Conversely, control fibers displayed no change in mitochondrial $[Ca^{2+}]$ during fatigue or recovery. Mitochondrial Ca^{2+} uptake has been shown to buffer $[Ca^{2+}]_i$ transients in neurons (David et al. 1998, David 1999) and in several non-excitable cell types (Tinel et al. 1999; Montero et al. 2002; Malli et al. 2003). The increased mitochondrial Ca^{2+} uptake in Tfam KO cells might then contribute to the decrease tetanic $[Ca^{2+}]_i$ seen in these cells. However, mitochondria do not contain any high capacity Ca^{2+} binding proteins and the Ca^{2+} content of mitochondria in muscle cells is only a small fraction (<2%) of that in the SR (Allen et al. 2008). Thus, it is unlikely that mitochondrial Ca^{2+} uptake contributed to the decreased tetanic $[Ca^{2+}]_i$ in Tfam KO fibers (Aydin et al. 2009).

Modest and transient increases in mitochondrial $[Ca^{2+}]$ can stimulate mitochondrial ATP production (McCormack and Denton 1993; Jouaville et al. 1999; Rutter et al. 1996), whereas a prolonged increase can trigger cellular damage and death, possibly involving opening of the mitochondrial permeability transition pore (mtPTP) (Brookes et al. 2004; Duchen 2000; Madsen et al. 1996). Cyclosporin A (CSA) is a drug that binds to the mitochondrial protein cyclophilin D, which is considered to be part of the mtPTP. CSA treatment has a protective effect in collagen VI myopathies (Ulrich congenital muscular dystrophy and Bethlem myopathy) where dysfunctional mitochondria have a central role in the disease process (Irwin et al. 2003; Angelin et al. 2007; Merlini et al. 2008). CSA is considered to inhibit opening of mtPTP in stressed and Ca^{2+}-overloaded mitochondria and in this way block mitochondrial Ca^{2+} release and depolarization leading to cell damage and death (Duchen 2000; Brookes et al. 2004). The results from Tfam KO fibers show an alternative mechanism for CSA action; CSA actually inhibited the contraction-induced increase in mitochondrial [Ca^{2+}] in Tfam KO fibers (Aydin et al. 2009). This contentious finding can be explained by an electrochemical gradient in the direction of Ca^{2+} flowing into the mitochondria when [Ca^{2+}]$_i$ increases during contraction (Bruton et al. 2003; Lännergren et al. 2001). Mitochondria of Tfam KO cells displayed no signs of Ca^{2+} overload in the rested state or mitochondrial depolarization during fatiguing stimulation. Therefore the electrochemical gradient would drive Ca^{2+} into the mitochondria of Tfam KO fibers during fatiguing stimulation resulting in increased mitochondrial [Ca^{2+}]. The fact that mitochondrial [Ca^{2+}] did not increase during fatigue in control fibers indicate that the permeability of the mitochondrial membrane for Ca^{2+} is very limited in these cells. Thus, Ca^{2+} passes the mitochondrial membrane in Tfam KO fibers and this involves a CSA/cyclophilin D dependent pathway. This means that the positive effects of CSA treatment in myopathies with mitochondrial engagement might involve inhibition of mitochondrial Ca^{2+} accumulation (Angelin et al. 2007; Irwin et al. 2003; Merlini et al. 2008; Millay et al. 2008), which reduces the deleterious effects of a prolonged increase of mitochondrial [Ca^{2+}] (Brookes et al. 2004; Duchen 2000; Madsen et al. 1996).

Exercise intolerance is a key symptom in human mitochondrial myopathies (Tarnopolsky and Raha 2005) and an obvious mechanism behind this would be a decreased mitochondrial energy production. Fatigue was induced in Tfam KO and control fibers with a series of 50 repeated tetanic stimulations (Aydin et al. 2009). Unexpectedly, the results showed no difference between the Tfam KO and control single fibers regarding the rate of force decrease during fatiguing stimulation, although the absolute force was lower in Tfam KO fibers throughout fatiguing stimulation because of their lower initial force (Fig. 2). Recovery after fatigue was also similar in Tfam KO and control fibers. The lack of an effect on fatigue development in Tfam KO muscle can be explained by the respiratory chain defects being compensated by an increased mitochondrial content (Wredenberg et al. 2002) and possibly a stimulation-induced increase in mitochondrial [Ca^{2+}] (Aydin et al. 2009), which would stimulate mitochondrial respiration (McCormack and Denton 1993; Jouaville et al. 1999; Rutter et al. 1996). Moreover, the lower tetanic [Ca^{2+}]$_i$ and force in Tfam muscle fibers would result in reduced energy consumption during fatiguing stimulation. It should be noted that during human exercise, task failure may occur when the upper limit of discomfort or sensation of excessive effort is reached (Marcora et al. 2009). Thus, during exercise in vivo, the muscle weakness observed in Tfam KO muscle would lead to early fatigue development not because the muscles fibers actually fatigue more rapidly but because muscles always have to work at a higher fraction of their maximal capacity (Allen et al. 2008).

It is often proposed that increased production reactive oxygen species (ROS) and oxidative stress have a central role in the pathology of primary mitochondrial diseases (Esposito et al. 1999; Tarnopolsky and Raha 2005). However, recent studies on mitochondrial disease models did not show any signs of increased ROS production or

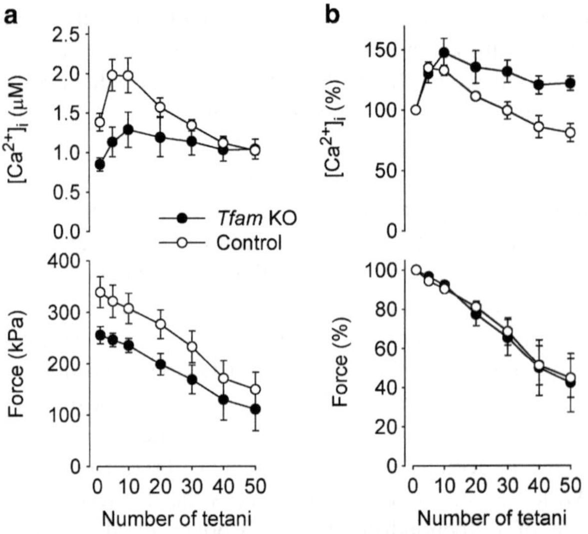

Fig. 2 Weakness is present throughout fatiguing stimulation but fatigue properties per se are not altered in muscle fibers from mitochondrial myopathy mice. (**a**) Mean (±SEM) tetanic $[Ca^{2+}]_i$ and force during fatigue induced by 50 tetanic contractions in single FDB fibers of Tfam KO (*filled circle*) and control (*open circle*) mice. (**b**) Same data as in (**a**) expressed relative to initial values, which were set to 100%. Figure adapted from Aydin et al. (2009)

ROS induced cell damage (Trifunovic et al. 2005; Kujoth et al. 2005). In accordance with these results, Tfam KO fibers showed no signs of increased ROS production (Aydin et al. 2009). Thus, increased ROS and oxidative stress are not obligate findings in mitochondrial diseases. In this context it is worth emphasizing that the "general" belief that ROS are deleterious has to be modified. For instance, ROS appears to have important signaling functions in skeletal muscle and impaired muscle function is seen both with decreased and increased ROS (Andrade et al. 1998).

To conclude, Tfam KO skeletal muscle fibers adapt to impaired mitochondrial respiration (1) by increasing mitochondrial mass and Ca^{2+} uptake, which increases ATP production, and (2) by decreasing SR Ca^{2+} release, which reduces the energy expenditure. The downside of these adaptations is that increased mitochondrial $[Ca^{2+}]$ may cause cellular damage and reduced SR Ca^{2+} release causes muscle weakness and exercise intolerance.

5 Skeletal Muscle of Cold-Acclimated UCP1 Deficient Mice Displays Muscle Weakness Due to Defective Function of the SR Ca^{2+} Release Channels

Mammals respond to an acute exposure to a cold environment by repeatedly activating some of their skeletal muscles, i.e. they shiver (Hemingway 1963). This results in heat generation and hence the body core temperature can be maintained. Another heat generating system, adaptive nonshivering thermogenesis, is activated during prolonged cold exposure. Heat is then generated by increased metabolism in brown adipose tissue due to activation of uncoupling protein-1 (UCP1) (Harper and Himms-Hagen 2001; Cannon and Nedergaard 2004). Mice deficient in UCP1 (UCP1 KO) cannot activate the UCP1-dependent adaptive nonshivering thermogenesis and remain dependent on shivering thermogenesis during prolonged cold exposure (Enerbäck et al. 1997; Golozoubova et al. 2001).

The activity of the sympathetic nervous system is increased during cold exposure (Motelica 1969; LeBlanc et al. 1967; Roth et al. 1988). A prolonged increase in β-adrenergic activity has been associated with impaired SR Ca^{2+} release due to protein kinase A (PKA) -induced hyperphosphorylation of RyR1 and dissociation of the channel-stabilizing subunit calstabin1 (also known as FKBP12) (Reiken et al. 2003; Ward et al. 2003; Bellinger et al. 2008a, b). The effect of cold exposure on muscle function was recently studied in WT and UCP1 KO mice (Aydin et al. 2008). Mice were either kept at room temperature (24°C; control) or in the cold (4°C, cold-acclimated). Measurements were performed in two muscles; the deep, postural soleus muscles, which participate in the shivering response, and the distally and superficially located FDB muscles, which would not shiver (Hemingway 1963). Soleus muscles of cold-acclimated mice showed lower tetanic $[Ca^{2+}]_i$ and force in UCP1 KO as compared to WT (Fig. 3a, b). Moreover, soleus muscles of cold-acclimated UCP1 KO mice displayed marked PKA hyperphosphorylation of RyR1 and calstabin1 depletion from the RyR1 channel complex, whereas cold exposure only caused a minor decrease in RyR1 calstabin1 binding in WT mice (Fig. 3c, d). On the other hand, there was no difference in tetanic $[Ca^{2+}]_i$ or force in FDB muscles of cold-acclimated WT and UCP1 KO mice (Fig. 4a, b). Furthermore, in FDB muscles cold exposure induced only a slight increase in RyR1 PKA phosphorylation and decrease in RyR1 calstabin1 binding (Fig. 4c, d).

In conclusion, only minor changes in the RyR1 channel complex and no contractile dysfunction was induced in FDB muscles and in WT soleus muscles of cold-acclimated mice. These muscles would not participate in continued shivering during cold exposure and hence they reflect the response to an increased β-adrenergic stimulation on its own. Conversely, soleus muscles of cold-acclimated UCP1 KO mice showed RyR1 PKA hyperphosphorylation and marked calstabin1 depletion from the RyR1 channel complex accompanied by decreased SR Ca^{2+} release and muscle weakness. These cold-acclimated UCP1 KO soleus muscles were exposed to both an increased β-adrenergic drive and the stress induced by shivering. Thus, it appears that the combination of β-adrenergic stress and continuous contractile activity during cold exposure is required to drive muscle cells into a defective state.

6 Muscle Weakness After Fatiguing Stimulation Can Be Due to Either Decreased SR Ca^{2+} Release or Reduced Myofibrillar Ca^{2+} Sensitivity

Fatigue develops during intense or prolonged activation of skeletal muscle (Fitts 1994; Allen et al. 1995, 2008). The recovery of contractile function after fatiguing activities can be very slow, sometimes requiring days for full recovery. The delayed force

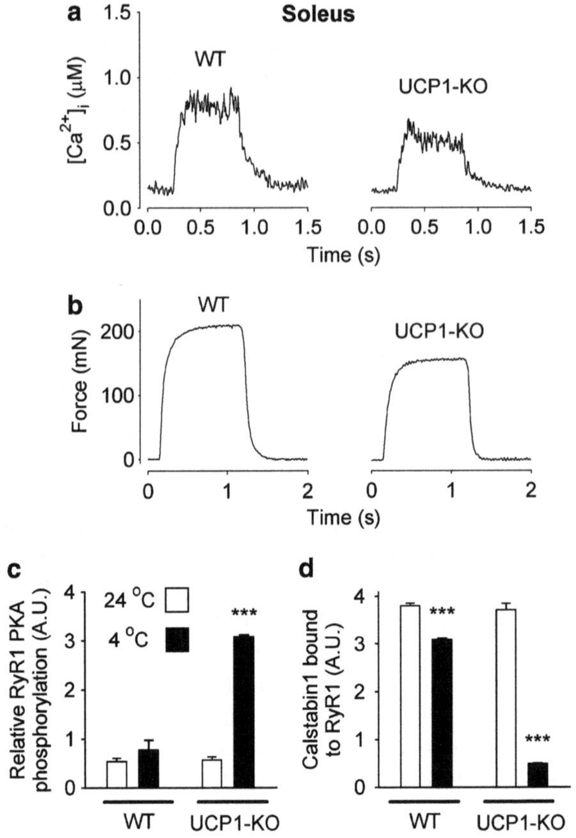

Fig. 3 Defective SR Ca^{2+} release causes muscle weakness in soleus muscle of cold-acclimated UCP1 KO mice. Representative $[Ca^{2+}]_i$ (**a**) and force (**b**) records from 100-Hz tetanic contractions produced in a WT (*left*) and a UCP1 KO (*right*) soleus muscle. Mean data (±SEM) of relative RyR1 PKA phosphorylation (**c**) and binding of calstabin1 (**d**). *White bars*, soleus muscles from control mice kept at 24°C; *black bars*, soleus muscles from cold-acclimated mice kept at 4°C. ***P<0.001 for cold-acclimated mice vs. their respective controls. Figure adapted from Aydin et al. (2008)

recovery was first described in human muscles and it was shown to be most marked at low stimulation frequencies (Edwards et al. 1977), which resemble the firing frequencies of human motor units during movements requiring low to moderate forces (Marsden et al. 1971). This delayed recovery has been observed in numerous human muscles and also in various animal muscle preparations (Westerblad et al. 2000; Jones 1996; Keeton and Binder-Macleod 2006; Martin et al. 2004). It was originally named 'low-frequency fatigue' but this term is now frequently used to describe other fatigue-related features, e.g. the force decrease induced by low frequency stimulation. It has therefore been suggested that this delayed recovery should be named 'prolonged low-frequency force depression (PLFFD)' to better describe the phenomenon and to avoid confusion with other

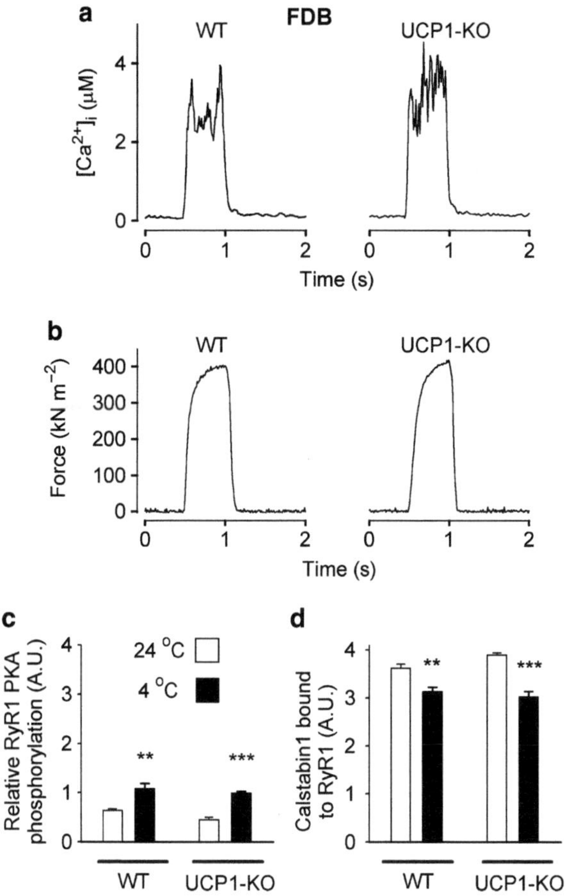

Fig. 4 SR Ca^{2+} release is not impaired in FDB fibers of cold-acclimated UCP1 KO mice. Representative $[Ca^{2+}]_i$ (**a**) and force (**b**) records from 100-Hz tetanic contractions produced in a WT (*left*) and a UCP1 KO (*right*) FDB fiber. Mean data (±SEM) of relative RyR1 PKA phosphorylation (**c**) and binding of calstabin1 (**d**). *White bars*, FDB muscles from control mice kept at 24°C; *black bars*, FDB muscles from cold-acclimated mice kept at 4°C. $^{**}P<0.01$, $^{***}P<0.001$ for cold-acclimated mice vs. their respective controls. Figure adapted from Aydin et al. (2008)

aspects of fatigue (Fitts 1994; Allen et al. 1995, 2008). PLFFD results in an increased sense of effort that may limit the physical performance and it is implicated in various neuromuscular disorders (Keeton and Binder-Macleod 2006; Polkey and Moxham 2001; Smith and Reid 2006; Supinski and Callahan 2005; Supinski et al. 1999).

Based on the sigmoidal shape of the force-$[Ca^{2+}]_i$ relationship in muscle, PLFFD can be due to decreased myofibrillar Ca^{2+} sensitivity and/or reduced SR Ca^{2+} release, whereas decreased cross-bridge force generation will lead to an equal reduction of force production at all stimulation frequencies (Westerblad et al. 2000;

Jones 1996; Allen et al. 2008). The cellular mechanisms underlying PLFFD was recently studied in mouse and rat FDB fibers with focus on the effects of ROS (Bruton et al. 2008). Since the myofibrillar Ca^{2+} sensitivity is very sensitive to hydrogen peroxide (Andrade et al. 2001), it was hypothesized that the cause of PLFFD (i.e. decreased SR Ca^{2+} release or reduced myofibrillar Ca^{2+} sensitivity) depends on differences in ROS metabolism and the activity of superoxide dismutase (SOD), which converts superoxide to hydrogen peroxide.

Rat FDB fibers fatigued by repeated tetanic stimulation showed a marked PLFFD without any significant decrease in tetanic $[Ca^{2+}]_i$, which means that the force decrease was due to decreased myofibrillar Ca^{2+} sensitivity. Mouse FDB fibers fatigued by repeated tetani, on the other hand, display PLFFD that can be explained by decreased tetanic $[Ca^{2+}]_i$ (Westerblad et al. 1993). Measurements of SOD activities showed about twofold higher activities of SOD1 (cytoplasmic copper-zinc-dependent), SOD2 (mitochondrial manganese-dependent), and total SOD in rat as compared to mouse FDB muscles (Bruton et al. 2008). Moreover, myoplasmic ROS increased during fatiguing stimulation in rat but not in mouse fibers. Thus, all these results fit with a larger increase in H_2O_2 during fatigue in rat FDB fibers, which results in decreased myofibrillar Ca^{2+} sensitivity. Conversely, in mouse FDB fibers there would be a larger increase in superoxide, which then decreases SR Ca^{2+} release. Interestingly, application of the reducing agent dithiothreitol partially reversed the fatigue-induced prolonged decrease in myofibrillar Ca^{2+} sensitivity in rat fibers, but had no significant effect on the decreased SR Ca^{2+} release in mouse fibers.

To further address this issue, experiments were performed on FDB fibers from a transgenic mouse that overexpresses SOD2 (Silva et al. 2005). Muscles of these SOD2 overexpressing mice display increased SOD activity and decreased superoxide levels (Silva et al. 2005). Under control conditions, there was no difference between control and SOD2 overexpressing FDB fibers regarding tetanic force and $[Ca^{2+}]_i$ and myofibrillar Ca^{2+} sensitivity was also similar (Bruton et al. 2008). SOD2 overexpressing fibers were more fatigue resistant than control fibers. Both groups showed a marked PLFFD after fatiguing stimulation (Fig. 5). This was accompanied by decreased tetanic $[Ca^{2+}]_i$ in control fibers, which agrees with previous results from mouse FDB fibers (Westerblad et al. 1993). On the other, PLFFD occurred without any marked changes in tetanic $[Ca^{2+}]_i$ in SOD2 overexpressing fibers, which is similar to the situation in rat FDB fibers.

The increased mitochondrial respiration during repeated contractions has been associated with increased production of superoxide (Smith and Reid 2006). In mouse FDB fibers, mitochondria are mainly located close to the Z-disc, which is close to the sites of SR Ca^{2+} release but at some distance from the myosin filaments (Lännergren et al. 1999). Superoxide can also be produced by NADPH oxidase and this enzyme can be activated by the increase in $[Ca^{2+}]_i$ during repeated contractions. NADPH oxidase is preferentially localized to the transverse tubular membrane, where it might affect the SR Ca^{2+} release channels (Espinosa et al. 2006; Hidalgo et al. 2006). The superoxide anion is unlikely to have widespread effects because it cannot easily diffuse through membranes. In addition to being metabolized to hydrogen peroxide, superoxide can be converted to hydroxyl radicals or interact with nitric oxide to form peroxynitrite. Both these highly reactive radicals can react with almost

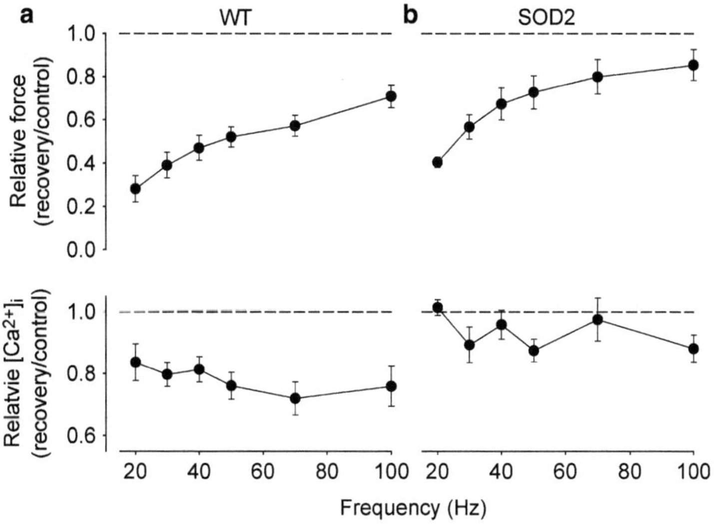

Fig. 5 Both WT and SOD2 overexpressing FDB fibers display muscle weakness especially at low stimulation frequencies (PLFFD), but the underlying mechanism differs. Mean data (±SEM) of the relative change in tetanic force (*top*) and $[Ca^{2+}]_i$ (*bottom*) in WT (**a**) and SOD2 overexpressing (**b**) fibers. Relative changes were calculated as ratio 30 min after (recovery) to before (control) fatigue induced by repeated tetanic stimulation. Figure from Bruton et al. (2008)

any molecule at a diffusion limited rate (Murrant and Reid 2001). Thus, prolonged exposure to increased superoxide might directly or indirectly impair SR Ca^{2+} release and this is a likely mechanism behind PLFFD in control mouse FDB fibers (Bruton et al. 2008). FDB fibers of rats and SOD2 overexpressing mice display a higher SOD activity than control mouse fibers, which limits the superoxide accumulation because of an effective conversion to hydrogen peroxide. Hydrogen peroxide can readily pass through membranes and appears to preferentially affect myofibrillar Ca^{2+} sensitivity (Andrade et al. 1998; Andrade et al. 2001). Thus, PLFFD in FDB fibers of rats and SOD2 overexpressing mice was mainly due to decreased myofibrillar Ca^{2+} sensitivity, which is likely to be related to a high capacity to convert superoxide to hydrogen peroxide during fatiguing stimulation (Bruton et al. 2008).

To conclude, fatigue-induced muscle weakness at low stimulation frequencies, PLFFD, can be caused by two ROS-dependent mechanisms: reduced SR Ca^{2+} release associated with low SOD activity and decreased myofibrillar Ca^{2+} sensitivity associated with high SOD activity.

7 Concluding Remarks

In this chapter we have exemplified various cellular mechanisms that can induce muscle weakness. While identifying basic mechanisms has a clear scientific relevance, it is legitimate to ask whether it can also have some practical consequences.

Our opinion is that this is the case. For instance, defective SR Ca^{2+} release due to RyR1 dysfunction and calstabin1 depletion can be counteracted by compounds that stabilizes the RyR1 channel complex (Bellinger et al. 2008a, b), even in dystrophic muscle where muscle weakness is a key problem (Bellinger et al. 2009). Furthermore, ROS-mediated decreases in myofibrillar Ca^{2+} sensitivity might be prevented or limited by antioxidants and reversed by reducing agents (Moopanar and Allen 2005, 2006, Andrade et al. 1998, 2001, Mishima et al. 2005, Bruton et al. 2008). Exercise intolerance can be caused by muscle weakness rather than by decreased fatigue resistance of the muscle fibers per se (Allen et al. 2008). When designing treatment to counteract early fatigue development it can then be important to distinguish between these two possibilities; strength training rather than endurance exercise would be preferred if muscle weakness is the major factor behind the increased fatigability.

References

Allen DG, Westerblad H (1995) The effects of caffeine on intracellular calcium, force and the rate of relaxation of mouse skeletal muscle. J Physiol 487:331–342

Allen DG, Lännergren J, Westerblad H (1995) Muscle cell function during prolonged activity: cellular mechanisms of fatigue. Exp Physiol 80:497–527

Allen DG, Lamb GD, Westerblad H (2008) Skeletal muscle fatigue: cellular mechanisms. Physiol Rev 88:287–332

Andrade FH, Reid MB, Allen DG, Westerblad H (1998) Effect of hydrogen peroxide and dithiothreitol on contractile function of single skeletal muscle fibres from the mouse. J Physiol 509:565–575

Andrade FH, Reid MB, Westerblad H (2001) Contractile response of skeletal muscle to low peroxide concentrations: myofibrillar calcium sensitivity as a likely target for redox-modulation. FASEB J 15:309–311

Angelin A, Tiepolo T, Sabatelli P, Grumati P, Bergamin N, Golfieri C, Mattioli E, Gualandi F, Ferlini A, Merlini L, Maraldi NM, Bonaldo P, Bernardi P (2007) Mitochondrial dysfunction in the pathogenesis of Ullrich congenital muscular dystrophy and prospective therapy with cyclosporins. Proc Natl Acad Sci U S A 104:991–996

Aydin J, Shabalina IG, Place N, Reiken S, Zhang SJ, Bellinger AM, Nedergaard J, Cannon B, Marks AR, Bruton JD, Westerblad H (2008) Nonshivering thermogenesis protects against defective calcium handling in muscle. FASEB J 22:3919–3924

Aydin J, Andersson DC, Hänninen SL, Wredenberg A, Tavi P, Park CB, Larsson NG, Bruton JD, Westerblad H (2009) Increased mitochondrial Ca^{2+} and decreased sarcoplasmic reticulum Ca^{2+} in mitochondrial myopathy. Hum Mol Genet 18:278–288

Barclay CJ (2005) Modelling diffusive O_2 supply of isolated preparations of mammalian skeletal and cardiac muscle. J Muscle Res Cell Motil 26:225–235

Baylor SM, Hollingworth S (2003) Sarcoplasmic reticulum calcium release compared in slow-twitch and fast-twitch fibres of mouse muscle. J Physiol 551:125–138

Beard NA, Laver DR, Dulhunty AF (2004) Calsequestrin and the calcium release channel of skeletal and cardiac muscle. Prog Biophys Mol Biol 85:33–69

Bellinger AM, Mongillo M, Marks AR (2008a) Stressed out: the skeletal muscle ryanodine receptor as a target of stress. J Clin Invest 118:445–453

Bellinger AM, Reiken S, Dura M, Murphy PW, Deng SX, Landry DW, Nieman D, Lehnart SE, Samaru M, LaCampagne A, Marks AR (2008b) Remodeling of ryanodine receptor complex causes "leaky" channels: a molecular mechanism for decreased exercise capacity. Proc Natl Acad Sci U S A 105:2198–2202

Bellinger AM, Reiken S, Carlson C, Mongillo M, Liu X, Rothman L, Matecki S, Lacampagne A, Marks AR (2009) Hypernitrosylated ryanodine receptor calcium release channels are leaky in dystrophic muscle. Nat Med 15:325–330

Brookes PS, Yoon Y, Robotham JL, Anders MW, Sheu SS (2004) Calcium, ATP, and ROS: a mitochondrial love-hate triangle. Am J Physiol Cell Physiol 287:C817–C833

Bruton J, Tavi P, Aydin J, Westerblad H, Lännergren J (2003) Mitochondrial and myoplasmic Ca^{2+} in single fibres from mouse limb muscles during repeated tetanic contractions. J Physiol 551:179–190

Bruton JD, Place N, Yamada T, Silva JP, Andrade FH, Dahlstedt AJ, Zhang SJ, Katz A, Larsson NG, Westerblad H (2008) Reactive oxygen species and fatigue-induced prolonged low-frequency force depression in skeletal muscle fibres of rats, mice and SOD2 overexpressing mice. J Physiol 586:175–184

Cannon B, Nedergaard J (2004) Brown adipose tissue: function and physiological significance. Physiol Rev 84:277–359

David G (1999) Mitochondrial clearance of cytosolic Ca^{2+} in stimulated lizard motor nerve terminals proceeds without progressive elevation of mitochondrial matrix [Ca^{2+}]. J Neurosci 19:7495–7506

David G, Barrett JN, Barrett EF (1998) Evidence that mitochondria buffer physiological Ca^{2+} loads in lizard motor nerve terminals. J Physiol 509:59–65

Duchen MR (2000) Mitochondria and Ca^{2+} in cell physiology and pathophysiology. Cell Calcium 28:339–348

Dulhunty AF (2006) Excitation-contraction coupling from the 1950s into the new millennium. Clin Exp Pharmacol Physiol 33:763–772

Edwards RH, Hill DK, Jones DA, Merton PA (1977) Fatigue of long duration in human skeletal muscle after exercise. J Physiol 272:769–778

Enerbäck S, Jacobsson A, Simpson EM, Guerra C, Yamashita H, Harper ME, Kozak LP (1997) Mice lacking mitochondrial uncoupling protein are cold-sensitive but not obese. Nature 387:90–94

Espinosa A, Leiva A, Peña M, Müller M, Debandi A, Hidalgo C, Carrasco MA, Jaimovich E (2006) Myotube depolarization generates reactive oxygen species through NAD(P)H oxidase; ROS-elicited Ca^{2+} stimulates ERK, CREB, early genes. J Cell Physiol 209:379–388

Esposito LA, Melov S, Panov A, Cottrell BA, Wallace DC (1999) Mitochondrial disease in mouse results in increased oxidative stress. Proc Natl Acad Sci U S A 96:4820–4825

Fitts RH (1994) Cellular mechanisms of muscle fatigue. Physiol Rev 74:49–94

Gehlsen GM, Whaley MH (1990) Falls in the elderly: Part II, Balance, strength, and flexibility. Arch Phys Med Rehabil 71:739–741

Golozoubova V, Hohtola E, Matthias A, Jacobsson A, Cannon B, Nedergaard J (2001) Only UCP1 can mediate adaptive nonshivering thermogenesis in the cold. FASEB J 15:2048–2050

Harper ME, Himms-Hagen J (2001) Mitochondrial efficiency: lessons learned from transgenic mice. Biochim Biophys Acta 1504:159–172

Hemingway A (1963) Shivering. Physiol Rev 43:397–422

Hidalgo C, Sánchez G, Barrientos G, Aracena-Parks P (2006) A transverse tubule NADPH oxidase activity stimulates calcium release from isolated triads via ryanodine receptor type 1 S-glutathionylation. J Biol Chem 281:26473–26482

Irwin WA, Bergamin N, Sabatelli P, Reggiani C, Megighian A, Merlini L, Braghetta P, Columbaro M, Volpin D, Bressan GM, Bernardi P, Bonaldo P (2003) Mitochondrial dysfunction and apoptosis in myopathic mice with collagen VI deficiency. Nat Genet 35:367–371

Jones DA (1996) High-and low-frequency fatigue revisited. Acta Physiol Scand 156:265–270

Jouaville LS, Pinton P, Bastianutto C, Rutter GA, Rizzuto R (1999) Regulation of mitochondrial ATP synthesis by calcium: evidence for a long-term metabolic priming. Proc Natl Acad Sci U S A 96:13807–13812

Keeton RB, Binder-Macleod SA (2006) Low-frequency fatigue. Phys Ther 86:1146–1150

Kubis HP, Hanke N, Scheibe RJ, Meissner JD, Gros G (2003) Ca^{2+} transients activate calcineurin/NFATc1 and initiate fast-to-slow transformation in a primary skeletal muscle culture. Am J Physiol Cell Physiol 285:C56–C63

Kujoth GC, Hiona A, Pugh TD, Someya S, Panzer K, Wohlgemuth SE, Hofer T, Seo AY, Sullivan R, Jobling WA, Morrow JD, Van Remmen H, Sedivy JM, Yamasoba T, Tanokura M, Weindruch R, Leeuwenburgh C, Prolla TA (2005) Mitochondrial DNA mutations, oxidative stress, and apoptosis in mammalian aging. Science 309:481–484

Lännergren J, Bruton JD, Westerblad H (1999) Vacuole formation in fatigued single muscle fibres from frog and mouse. J Muscle Res Cell Motil 20:19–32

Lännergren J, Westerblad H, Bruton JD (2001) Changes in mitochondrial Ca^{2+} detected with Rhod-2 in single frog and mouse skeletal muscle fibres during and after repeated tetanic contractions. J Muscle Res Cell Motil 22:265–275

Larsson NG, Oldfors A (2001) Mitochondrial myopathies. Acta Physiol Scand 171:385–393

LeBlanc J, Robinson D, Sharman DF, Tousignant P (1967) Catecholamines and short-term adaptation to cold in mice. Am J Physiol 213:1419–1422

Madsen K, Ertbjerg P, Djurhuus MS, Pedersen PK (1996) Calcium content and respiratory control index of skeletal muscle mitochondria during exercise and recovery. Am J Physiol Endo Met 271:E1044–E1050

Malli R, Frieden M, Osibow K, Graier WF (2003) Mitochondria efficiently buffer subplasmalemmal Ca^{2+} elevation during agonist stimulation. J Biol Chem 278:10807–10815

Marcora SM, Staiano W, Manning V (2009) Mental fatigue impairs physical performance in humans. J Appl Physiol 106:857–864

Marsden CD, Meadows JC, Merton PA (1971) Isolated single motor units in human muscle and their rate of discharge during maximal voluntary effort. J Physiol 217:12P–13P

Martin V, Millet GY, Martin A, Deley G, Lattier G (2004) Assessment of low-frequency fatigue with two methods of electrical stimulation. J Appl Physiol 97:1923–1929

McCormack JG, Denton RM (1993) Mitochondrial Ca^{2+} transport and the role of intramitochondrial Ca^{2+} in the regulation of energy metabolism. Dev Neurosci 15:165–173

Meriggioli MN (2009) Myasthenia gravis with anti-acetylcholine receptor antibodies. Front Neurol Neurosci 26:94–108

Merlini L, Angelin A, Tiepolo T, Braghetta P, Sabatelli P, Zamparelli A, Ferlini A, Maraldi NM, Bonaldo P, Bernardi P (2008) Cyclosporin A corrects mitochondrial dysfunction and muscle apoptosis in patients with collagen VI myopathies. Proc Natl Acad Sci U S A 105:5225–5229

Millar NC, Homsher E (1990) The effect of phosphate and calcium on force generation in glycerinated rabbit skeletal muscle fibers. A steady-state and transient kinetic study. J Biol Chem 265:20234–20240

Millay DP, Sargent MA, Osinska H, Baines CP, Barton ER, Vuagniaux G, Sweeney HL, Robbins J, Molkentin JD (2008) Genetic and pharmacologic inhibition of mitochondrial-dependent necrosis attenuates muscular dystrophy. Nat Med 14:442–447

Mishima T, Yamada T, Matsunaga S, Wada M (2005) N-acetylcysteine fails to modulate the in vitro function of sarcoplasmic reticulum of diaphragm in the final phase of fatigue. Acta Physiol Scand 184:195–202

Montero M, Alonso MT, Albillos A, Cuchillo-Ibáñez I, Olivares R, Villalobos C, Alvarez J (2002) Effect of inositol 1,4,5-trisphosphate receptor stimulation on mitochondrial [Ca^{2+}] and secretion in chromaffin cells. Biochem J 365:451–459

Moopanar TR, Allen DG (2005) Reactive oxygen species reduce myofibrillar Ca^{2+} sensitivity in fatiguing mouse skeletal muscle at 37°C. J Physiol 564:189–199

Moopanar TR, Allen DG (2006) The activity-induced reduction of myofibrillar Ca^{2+} sensitivity in mouse skeletal muscle is reversed by dithiothreitol. J Physiol 571:191–200

Motelica I (1969) Urinary excretion of catecholamines and vanilmandelic acid in rats exposed to cold. Acta Physiol Scand 76:393–395

Murrant CL, Reid MB (2001) Detection of reactive oxygen and reactive nitrogen species in skeletal muscle. Microsc Res Tech 55:236–248

Paolini C, Quarta M, Nori A, Boncompagni S, Canato M, Volpe P, Allen PD, Reggiani C, Protasi F (2007) Reorganized stores and impaired calcium handling in skeletal muscle of mice lacking calsequestrin-1. J Physiol 583:767–784

Pate E, Cooke R (1989) Addition of phosphate to active muscle fibers probes actomyosin states within the powerstroke. Pflügers Arch 414:73–81

Polkey MI, Moxham J (2001) Clinical aspects of respiratory muscle dysfunction in the critically ill. Chest 119:926–939

Posterino GS, Lamb GD, Stephenson DG (2000) Twitch and tetanic force responses and longitudinal propagation of action potentials in skinned skeletal muscle fibres of the rat. J Physiol 527:131–137

Reiken S, Lacampagne A, Zhou H, Kherani A, Lehnart SE, Ward C, Huang F, Gaburjakova M, Gaburjakova J, Rosemblit N, Warren MS, He KL, Yi GH, Wang J, Burkhoff D, Vassort G, Marks AR (2003) PKA phosphorylation activates the calcium release channel (ryanodine receptor) in skeletal muscle: defective regulation in heart failure. J Cell Biol 160:919–928

Rossignol R, Letellier T, Malgat M, Rocher C, Mazat JP (2000) Tissue variation in the control of oxidative phosphorylation: implication for mitochondrial diseases. Biochem J 347:45–53

Roth J, Zeisberger E, Schwandt HJ (1988) Influence of increased catecholamine levels in blood plasma during cold-adaptation and intramuscular infusion on thresholds of thermoregulatory reactions in guinea-pigs. J Comp Physiol [B] 157:855–863

Rutter GA, Burnett P, Rizzuto R, Brini M, Murgia M, Pozzan T, Tavaré JM, Denton RM (1996) Subcellular imaging of intramitochondrial Ca^{2+} with recombinant targeted aequorin: significance for the regulation of pyruvate dehydrogenase activity. Proc Natl Acad Sci U S A 93:5489–5494

Schapira AH (2006) Mitochondrial disease. Lancet 368:70–82

Silva JP, Shabalina IG, Dufour E, Petrovic N, Backlund EC, Hultenby K, Wibom R, Nedergaard J, Cannon B, Larsson NG (2005) SOD2 overexpression: enhanced mitochondrial tolerance but absence of effect on UCP activity. EMBO J 24:4061–4070

Smith MA, Reid MB (2006) Redox modulation of contractile function in respiratory and limb skeletal muscle. Respir Physiol Neurobiol 151:229–241

Supinski GS, Callahan LA (2005) Diaphragmatic free radical generation increases in an animal model of heart failure. J Appl Physiol 99:1078–1084

Supinski G, Stofan D, Callahan LA, Nethery D, Nosek TM, DiMarco A (1999) Peroxynitrite induces contractile dysfunction and lipid peroxidation in the diaphragm. J Appl Physiol 87:783–791

Tarnopolsky MA, Raha S (2005) Mitochondrial myopathies: diagnosis, exercise intolerance, and treatment options. Med Sci Sports Exerc 37:2086–2093

Thorburn DR (2004) Mitochondrial disorders: prevalence, myths and advances. J Inherit Metab Dis 27:349–362

Thorburn DR, Sugiana C, Salemi R, Kirby DM, Worgan L, Ohtake A, Ryan MT (2004) Biochemical and molecular diagnosis of mitochondrial respiratory chain disorders. Biochim Biophys Acta 1659:121–128

Tinel H, Cancela JM, Mogami H, Gerasimenko JV, Gerasimenko OV, Tepikin AV, Petersen OH (1999) Active mitochondria surrounding the pancreatic acinar granule region prevent spreading of inositol trisphosphate-evoked local cytosolic Ca^{2+} signals. EMBO J 18:4999–5008

Trifunovic A, Hansson A, Wredenberg A, Rovio AT, Dufour E, Khvorostov I, Spelbrink JN, Wibom R, Jacobs HT, Larsson NG (2005) Somatic mtDNA mutations cause aging phenotypes without affecting reactive oxygen species production. Proc Natl Acad Sci U S A 102:17993–17998

Ward CW, Reiken S, Marks AR, Marty I, Vassort G, Lacampagne A (2003) Defects in ryanodine receptor calcium release in skeletal muscle from post-myocardial infarct rats. FASEB J 17:1517–1519

Westerblad H, Allen DG (1996) Mechanisms underlying changes of tetanic $[Ca^{2+}]_i$ and force in skeletal muscle. Acta Physiol Scand 156:407–416

Westerblad H, Duty S, Allen DG (1993) Intracellular calcium concentration during low-frequency fatigue in isolated single fibers of mouse skeletal muscle. J Appl Physiol 75:382–388

Westerblad H, Bruton JD, Allen DG, Lännergren J (2000) Functional significance of Ca^{2+} in long-lasting fatigue of skeletal muscle. Eur J Appl Physiol 83:166–174

Whipple RH, Wolfson LI, Amerman PM (1987) The relationship of knee and ankle weakness to falls in nursing home residents: an isokinetic study. J Am Geriatr Soc 35:13–20

Wredenberg A, Wibom R, Wilhelmsson H, Graff C, Wiener HH, Burden SJ, Oldfors A, Westerblad H, Larsson NG (2002) Increased mitochondrial mass in mitochondrial myopathy mice. Proc Natl Acad Sci U S A 99:15066–15071

Wredenberg A, Freyer C, Sandström ME, Katz A, Wibom R, Westerblad H, Larsson NG (2006) Respiratory chain dysfunction in skeletal muscle does not cause insulin resistance. Biochem Biophys Res Commun 350:202–207

Zhang SJ, Bruton JD, Katz A, Westerblad H (2006) Limited oxygen diffusion accelerates fatigue development in mouse skeletal muscle. J Physiol 572:551–559

Stretch-Induced Membrane Damage in Muscle: Comparison of Wild-Type and *mdx* Mice

David G. Allen, Bao-ting Zhang, and Nicholas P. Whitehead

Abstract One component of stretch-induced muscle damage is an increase in the permeability of the cell membrane. As a result soluble myoplasmic proteins leak out of the muscle into the plasma, extracellular proteins can enter the muscle, and extracellular ions, including calcium, are driven down their electrochemical gradient into the myoplasm. In Duchenne muscular dystrophy, caused by the absence of the cytoskeletal protein dystrophin, stretch-induced membrane damage is much more severe. The most popular theory to explain the occurrence of stretch-induced membrane damage is that stretched-contractions cause transient mechanically-induced defects in the membrane (tears or rips). Dystrophin, which is part of a mechanical link between the contractile machinery and the extracellular matrix, is thought to contribute to membrane strength so that in its absence mechanically-induced defects are worse. In our view the evidence that stretch-induced muscle damage causes increased membrane permeability is overwhelming but the evidence that the increased permeability is caused by mechanically-induced defects is weak. Instead we review the substantial evidence that the membrane permeability is a secondary consequence of the mechanical events in which elevated intracellular calcium and reactive oxygen species are important intermediaries.

Keywords Membrane damage • Strech-induced muscle damage • Skeletal muscle • Dystrophin • Duchenne muscular dystrophy • Intracellular calcium

1 Stretch-Induced Muscle Damage

Any type of muscle contraction, if repeated many times, can result in muscle damage but contractions in which the muscle is stretched (eccentric contractions) are particularly prone to causing damage (Newham et al. 1983b). Stretch-induced

D.G. Allen (✉)
Bosch Institute and School of Medical Sciences, University of Sydney, F13, Sydney, NSW 2006, Australia
e-mail: david.allen@sydney.edu.au

D.E. Rassier (ed.), *Muscle Biophysics: From Molecules to Cells*, Advances in Experimental Medicine and Biology 682,
DOI 10.1007/978-1-4419-6366-6_17,

muscle damage is characterized by an immediate weakness and a slowly developing soreness, stiffness and swelling (Hough 1902; for review see Clarkson and Hubal 2002). The soreness, most apparent on the day after the exercise, is a striking feature and has led to the name delayed-onset muscle soreness (DOMS). For example, walking down a mountain, in which the quadriceps group is stretched on each step by the body weight, commonly leads to soreness and stiffness on the following day and these symptoms and the initial weakness take several days to recover. Fridén et al. (1981) established that, following stretch-induced muscle damage, the sarcomeres were disorganized, with a mixture of overstretched and understretched sarcomeres and Z lines out-of-register across the fiber. These and other observation led to the 'popping sarcomere' theory (Morgan 1990) which proposed that, during rapid extension of sarcomeres on the descending limb of the tension-length curves, the weakest sarcomeres are unstable and can suddenly stretch (pop) to a length where they are stabilized by elastic elements such as titin. This theory is capable of explaining many of the mechanical properties of muscles following stretched contractions (Morgan and Allen 1999; Proske and Morgan 2001). Another feature of stretch-induced muscle damage is a rise in the concentration of soluble muscle proteins in the blood, such as creatine kinase (Newham et al. 1983a), which typically peaks 1–4 days after the damage. McNeil and Khakee (1992) proposed, on the basis of uptake of albumin, that stretched muscles had mechanically-induced membrane defects. Further evidence of increased membrane permeability came from measurements of depolarization following stretched contractions (McBride et al. 2000). These observations together provide evidence of damage to the surface membrane which is the focus of the present article. By damage to the surface membrane we mean increased permeability of the membrane to large (and small) molecules without any implication as to the mechanism. One purpose of the present chapter is to review possible mechanisms which cause this membrane damage.

2 Duchenne Muscular Dystrophy and the *mdx* Mouse

Duchenne muscular dystrophy (DMD) is a severe degenerative disease of muscle which affects boys with a mutation in the dystrophin gene on the Y chromosome. A range of mutations are recognized but the net effect is the absence of dystrophin in muscle. Dystrophin is a cytoskeletal protein which links intracellular γ-actin of the cytoskeleton to a group of proteins in the cell membrane, the dystrophin-associated protein complex (DAPC). The DAPC are further linked to the extracellular matrix through laminins (Fig. 1a). In DMD not only is dystrophin absent but the DAPC is greatly reduced (Ervasti and Campbell 1991) while several other proteins normally associated with the DAPC have increased expression (Gervasio et al. 2008) (Fig. 1b). While the primary cause of the disease is the absence of dystrophin, the complex pathways which link the absence of dystrophin to profound muscle damage and wasting are unclear.

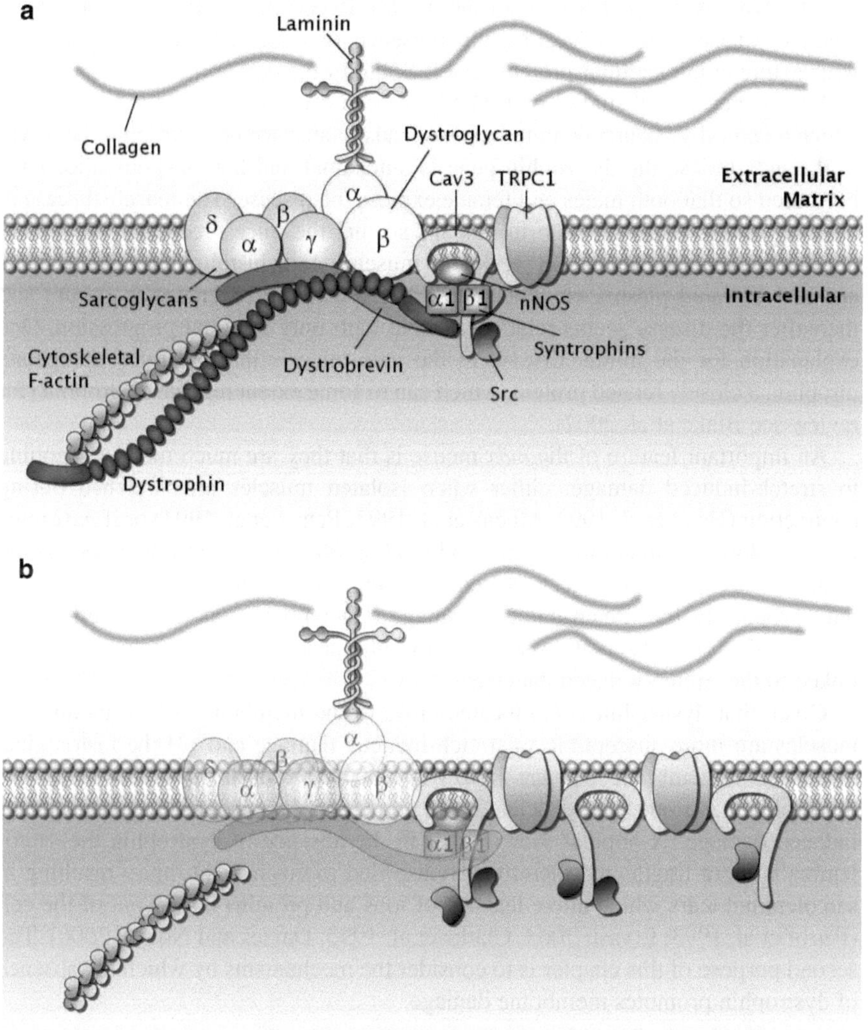

Fig. 1 Schematic diagram illustrating the relationship between dystrophin, the dystrophin-associated protein complex and the extracellular matrix. *Panel* (**a**) shows the connections in a wild-type muscle. *Panel* (**b**) indicates some of the protein changes observed in dystrophic muscle. From Allen et al. (2010)

Early studies on DMD described discontinuities of the surface membrane visible in the EM and noted uptake of extracellular proteins into the myoplasm in these areas (Mokri and Engel 1975). This led to the hypothesis that extracellular calcium might enter the fibers at these discontinuities and subsequent study confirmed that the total muscle calcium was greatly increased in damaged fibers (Bodensteiner and

Engel 1978). A cardinal feature of the disease is very large elevations of plasma creatine kinase again suggesting the presence of damaged membrane through which soluble intracellular proteins can leak out of the cell.

Most experimental studies on DMD have been carried out in the *mdx* mouse, in which a naturally occurring mutation also leads to absence of dystrophin. However in the *mdx* mouse the dystrophin gene is autosomal and homozygous mice have been bred so that both males and female express the disease. The muscle disease in the *mdx* mouse is milder than in humans so that the mice breed and live a near normal life span. At about 2–3 weeks the muscles show histological signs of damage and elevated plasma creatine kinase but a wave of regeneration occurs and thereafter the disease seems relatively stable with only minimal progression. One explanation for the milder disease in the *mdx* mice is that the mice upregulate utrophin, a closely related protein, which can to some extent replace dystrophin (for review see Blake et al. 2002).

An important feature of the *mdx* mouse is that they are much more susceptible to stretch-induced damage, either when isolated muscles are stretched during contraction (Head et al. 1992; Moens et al. 1993; Petrof et al. 1993) or if *mdx* mice run on a downhill treadmill (Whitehead et al. 2006). Of special importance is the finding that expression of a shortened dystrophin in *mdx* muscles greatly reduces their propensity for stretch-induced damage (Deconinck et al. 1996; DelloRusso et al. 2002). This observation suggests that the absence of dystrophin is closely linked to the stretch-induced damage.

Given that dystrophin is (1) located close to the membrane, (2) in its absence muscles are more susceptible to stretch-induced damage and (3) the histological evidence of membrane damage in DMD and *mdx* muscle, it seems clear that dystrophin is in some way associated with protecting the membrane against stretch-induced damage. A popular view is that in the absence of dystrophin the sarcolemma is more fragile and therefore predisposed to mechanical injury resulting in sarcolemmal tears which allow leakage of ions and proteins in and out of the cell (Petrof et al. 1993; Ervasti 2003; Clarke et al. 1993; Davies and Nowak 2006). The second purpose of this chapter is to consider the mechanisms by which the absence of dystrophin promotes membrane damage.

3 The Mechanical Properties of the Surface Membrane

Muscle has to be capable of stretching and shortening over a considerable range. Cell volume change is minimal during contraction, so the surface area of the muscle must increase during stretch and decrease during shortening. The situation is complicated by the presence of the T-tubule network which must remain in register with the contractile proteins and the sarcolemma despite changes in muscle length and diameter. The lateral mobility of lipid molecules within the membrane is thought to be high (Sengupta et al. 2007) so that the sarcolemma is capable of rearrangement of shape but has very little capacity to stretch once shape rearrangement is no longer possible. This has been demonstrated with red cells which can be sucked into a small bore pipette with a very low pressure representing the force require to change the shape (shear modulus) (Evans et al. 1976) (e.g. Fig. 2a). The shear modulus is highly correlated with the spectrin content and is thought to depend on the cytoskeletal network (Waugh and Agre 1988). Once the remaining cell outside the pipette becomes spherical a much higher pressure is required to extend the surface area (area expansion modulus) (Evans et al. 1976) (e.g. Fig. 2b). Notably the area expansion modulus is 103–104 times higher than the shear modulus. Finally at some pressure the cell bursts (e.g. Fig. 2c) defining the tensile strength of the membrane. Hutter et al. (1991) have performed comparable experiments on vesicles obtained from skeletal muscle treated with collagenase and placed in a 140 mM KCl solution (which exerts no long term osmotic pressure and therefore causes swelling). Under these conditions, the fibers swell, the membrane lifts off the myofibrils forming blebs and, subsequently, the blebs form sealed vesicles which are released from the muscle. As Hutter (1992) points out, during bleb formation the T-tubules must disconnect from the surface membrane but, since Tryphan blue is excluded, the surface membrane must seal rapidly. If the loose vesicles are sucked into large bore

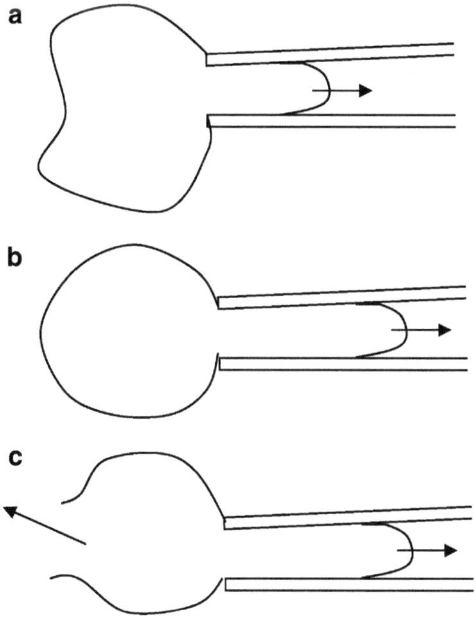

Fig. 2 Diagram illustrating measurements of membrane strength in red cells or membrane vesicles. (**a**) Free vesicle is sucked into the lumen of a large bore glass micropipette. A very low pressure is required to change the shape of the vesicle represented by the shear modulus. (**b**) Once the external membrane becomes spherical a relatively large pressure is required to stretch the membrane area, representing the area expansion modulus. (**c**) After the membrane has increased its area by a few percent, failure of the membrane occurs defined by the bursting pressure representing the tensile strength of the membrane. Modified from video images of Nichol and Hutter (1996b)

pipette, a U shaped membrane forms in the pipette which lengthens until the outside membrane is spherical (Fig. 2b). Further increments of pressure lead to area expansion followed rapidly by failure (Fig. 2c). Notably the increase in area was only 2% before failure. The bursting pressure or tensile strength was 60 μdynes/cm in vesicles from normal muscle which is similar to the value obtained by suction into a pipette (Hutter et al. 1991) and with red cells (Evans et al. 1976).

The implications of this work for intact muscle are that when a muscle shortens there will be 'loose' membrane which can be seen as 'festoons' attached at the costameres (Shear and Bloch 1985). Conversely when a muscle is stretched, 'spare' membrane is required to make up the additional area during stretch and circumferential folds of membrane have been visualized which fulfill this role (Dulhunty and Franzini-Armstrong 1975). Once folds are utilized, which occurs at a sarcomere length of 3.0 μm, caveolae provide an additional supply of additional membrane up to sarcomere lengths of 8.0 μm at which length membrane failure occurs.

Given these properties of the membrane, it seems unlikely that the membrane alone contributes much to longitudinal fiber stiffness and in fact measurements of stiffness are not notably affected by mechanical skinning at least at sarcomere lengths of less than 3.2 μm (Natori 1954; Podolsky 1964). Most of the resting stiffness is caused by titin which is a very long and elastic protein which connects the Z disc to the thick filaments (for review see Granzier and Labeit 2006).

Skinning a single muscle fiber has little effect on the maximal Ca^{2+}-activated tension, presumably because the forces are transmitted longitudinally along the muscle fiber (Hellam and Podolsky 1969). Thus a single fiber, as in Fig. 4a, gives similar force when the intact fiber is maximally tetanised or when the skinned fiber is maximally Ca^{2+}-activated. This result implies that dystrophin, which remains attached to the sarcolemma after skinning (Zubrzycka-Gaarn et al. 1991), does not have a role in longitudinal force transmission at least within one fiber. Dystrophin probably plays a role in longitudinal force transmission at the tendinous insertion, where dystrophin expression is increased (Samitt and Bonilla 1990). This is apparent in zebra fish in which a mutation which leads to absence of dystrophin, causes disconnection of fibers from the tendons (Bassett et al. 2003).

What contribution, if any, does dystrophin make to the mechanical properties of the membrane and/or muscle? To understand this issue we need to know its distribution. A strong body of opinion suggests that dystrophin has a costameric, subsarcolemmal distribution. The costameres are protein structures in the surface membrane which run circumferentially around the fiber forming parallel lines with a sarcomeric spacing (Pardo et al. 1983). Dystrophin is connected at its N-terminal to γ-actin in its filamentous (F-actin) form. The C-terminal domain binds to β-dystroglycan and to α-dystrobrevin which are part of the DAPC resident in the membrane (Davies and Nowak 2006). Dystrophin, and its binding partners, β-dystroglycan and γ-actin have all been found to colocalize with each other and with other costameric proteins (spectrin, vinculin, talin, a-actinin) (Porter et al. 1992; Rybakova et al. 2000; Rahkila et al. 2001). However others have identified γ-actin only in the Z-discs. (Papponen et al. 2009) and show little co-localization with β-dystroglycan. On the former interpretation, dystrophin has a sub-sarcolemmal distribution and might, like spectrin, contribute to shear stiffness. On either interpretation, the DAPC forms the mechanical link through laminins to the extracellular collagen and matrix and dystrophin is therefore part of the system which provides lateral connections between neighboring fibers.

It is possible that dystrophin makes a minor contribution to the longitudinal stiffness of muscle through its longitudinal connections. In support of this possibility Kumar et al. (2004) found the longitudinal stiffness of *mdx* diaphragm from 3 week old mice was substantially reduced compared to wild-type muscle. Given that longitudinal fiber stiffness below 3.2 μm probably unrelated to the membrane (Natori 1954; Podolsky 1964), this result may not be a direct response to dystrophin absence. For instance it could reflect elevated $[Ca^{2+}]i$ causing activation of calpain and proteolytic damage to titin (Goll et al. 2003; Zhang et al. 2008). In contrast, Stedman et al. (1991) observed a large increase in *mdx* diaphragm stiffness in old *mdx* mice (1.5 year) and suggested that this was related to the large increase in extracellular fibrosis present at that age.

A more likely suggestion, given the predominately lateral nature of the dystrophin/DAPC/laminin connections, is that this complex contributes to the lateral stiffness of whole muscles. Street (1983) clearly demonstrated that lateral connections between fibers could provide a longitudinal force pathway under some circumstances. She dissected a single fiber which was partly cleaned and partly surrounded by a 'splint' of damaged fibers and connective tissue (Fig. 4b). When the middle end of the splint was pinned to the chamber and force measured only at the tendon of the splinted end (Fig. 4c), stimulation of the single fiber, whose clean tendon was disconnected, was still able to record 85% of the force produced by the single fiber with both ends fixed (Fig. 4b). This demonstrated that lateral connections between the fiber and the splint could transfer most of the longitudinal force of the fiber to the splint. Several proteins contribute to these lateral connections. The dystrophin complex contributes to the extracellular component whereas desmin connects the Z discs of adjacent myofibrils and connects the Z-disc to the costameres in the sarcolemma (Ervasti 2003). As one might expect from this distribution the desmin KO diaphragm has essentially normal longitudinal resting stiffness but stiffness measured laterally across many fibers is substantially reduced (Boriek et al. 2001). Such lateral, as opposed to longitudinal connections, could have at least two important roles. One is maintaining the appropriate localization of the T-tubules at the region of overlap of the thick and thin filaments. Figure 5a illustrates schematically the spatial relationship of the T-tubules and the myofilaments and shows the positioning of the Z-lines and their connections by cytoskeletal proteins, such as desmin, to the costamere. The right hand sarcomere in Fig. 5a is shown as overstretched with the thick filament misplaced to the left side, a frequent finding in EMs of muscles damaged by stretched contractions (Brown and Hill 1991). In this situation if the costameric connections hold, the sarcolemma will be locally stretched; alternatively, as illustrated, if the costameric connections fail then the location of the T-tubule near the junction of the thick and thin filaments will lead to substantial distortion of the T-tubule. The latter situation may be more common in muscular dystrophy or after several stretched contraction when desmin appears to be damaged (Lieber et al. 1996; Zhang et al. 2008). A second role for the extracellular lateral connections is illustrated in Fig. 5b. In large muscles the fibers do not run the full length and are connected by intramuscular tendons. If in a non-maximal contraction the colored fibers are activated but the clear fibers are not, then, without lateral connections, the muscle will develop little tension as each active fiber will shorten down as it stretches the passive fiber in series. In practice extensive lateral connections maintain all fibers in register and longitudinal force is effectively transmitted by the combination of longitudinal and lateral connections.

4 What Happens to the Mechanical Properties of the Membrane When Dystrophin Is Missing?

A number of studies have tested mechanical properties of the sarcolemma from *mdx* compared to wild-type mice. Menke and Jockusch (1991) showed that the ability of isolated muscle fibers to survive a hypo-osmotic shock was reduced in *mdx* fibers. Blebs of membrane were raised by the osmotic challenge in both normal and *mdx* fibers but *mdx* fibers were more likely to proceed to cell death judged by Trypan blue uptake or pyruvate kinase release. The movement of large molecules in and out of the cells shows that membrane damage had occurred but its mechanism is unclear. Several studies have measured the pressure required for membrane rupture either in intact cell using a patch electrode (Fig. 3a) or in vacuoles prepared

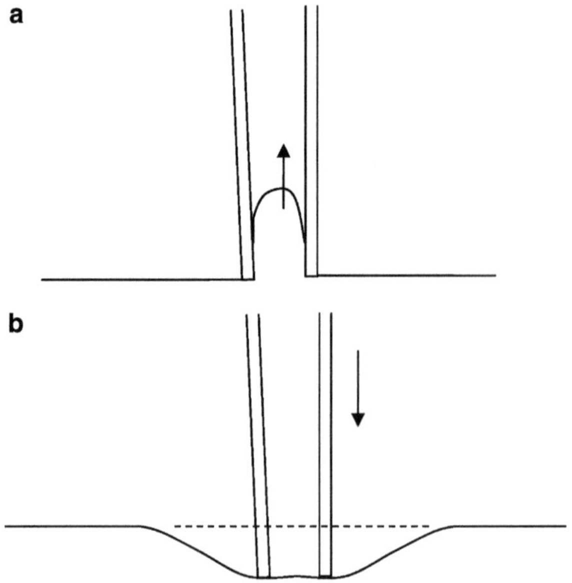

Fig. 3 Measurements of membrane strength in intact muscle cells. (**a**) Sucking membrane into a patch pipette. Membrane strength can be estimated by the pressure at which the membrane fails. (**b**) Indentation pressure. The force required to indent the membrane a given distance is the indentation stiffness of the membrane

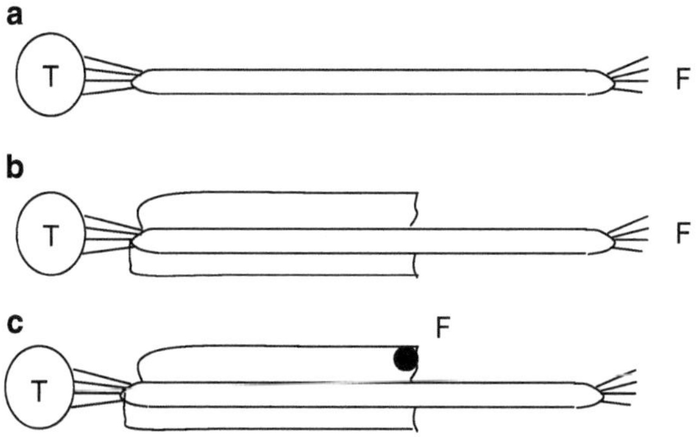

Fig. 4 Demonstration of the importance of lateral connections in the development of force in a multicellular muscle preparation. (**a**) Conventional single fiber with force measured by a tension transducer (T) connected to the tendon at one end while the tendon is fixed (F) at the other end. (**b**) The same force is recorded if a single fiber is dissected but with a substantial cuff of inactive muscle at one end. (**c**) If the muscle is now fixed at F (by a pin through the inactive muscle) approximately 85% of the original force can be recorded when the fiber is activated. Lateral connections between the intact fiber and the inactive muscle allow the transfer of force between the inactive muscle to the intact fiber. Illustration of the work of Street (1983)

by hypo-osmotic shock (Fig. 2) and found little or no difference between *mdx* and wild type (Franco and Lansman 1990; Hutter et al. 1991). Hutter et al. found a small but significant difference between wildtype and *mdx* muscle (bursting pressure 60 vs. 52 μN/cm in wild type vs. *mdx*) but felt that these small differences were unlikely to be the basis for the much greater susceptibility of *mdx* muscle to stretch-induced damage. This suggests that some other protein or pathway may be abnormal in *mdx* and responsible for the osmotic instability. One possibility is that Ca^{2+} is more elevated in *mdx* muscle following osmotic shock (Wang et al. 2005; Jung et al. 2008) and that the elevated $[Ca^{2+}]i$ makes the membrane more liable to rupture. The susceptibility of isolated membranes to elevated $[Ca^{2+}]i$ has been demonstrated by Nichol and Hutter (1996a). They used Ca^{2+} ionophores to equilibrate extracellular and intracellular Ca^{2+} and found that at little as 1–2 μM intracellular $[Ca^{2+}]$ led to a rapid and irreversible increase in the pressure-induced failure of muscle membranes. Extracellular Ca^{2+} had no obvious effect (Nichol and Hutter 1996b).

Another kind of stiffness, resistance to indentation, can be determined from the force required to indent the surface of a single fiber (Fig. 4b). Pasternak et al. (1995) showed that this type of indentation stiffness was reduced fourfold in *mdx* fibers. This suggests that the network of connections provided by the dystrophin/DAPC/laminin complex and the extracellular matrix provide resistance to lateral indentation. These results have recently been extended using an atomic force microscope to measure the force required for indentation (Puttini et al. 2009). In wild-type mice, the indentation stiffness was reduced about fourfold by cytochalasin-D, which disrupts the actin cytoskeleton, and the lower value was similar to that observed in *mdx* mice. Furthermore, either use of an exon-skipping

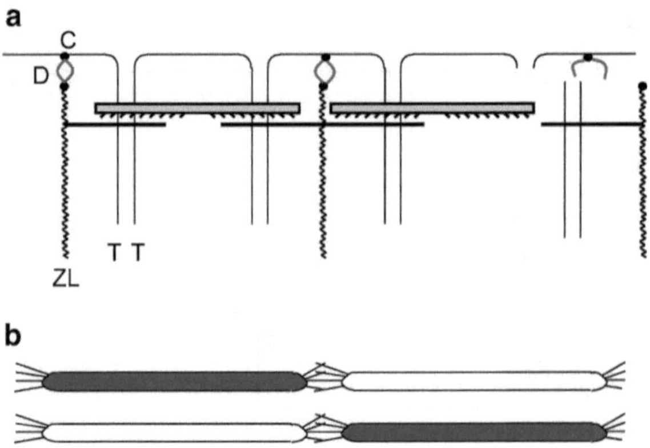

Fig. 5 Possible roles for lateral connections in muscle. (**a**) Illustrates contractile proteins, thick and thin filaments, with thin filaments connected to the Z line (ZL). Z line is connected to the costamere (C), an array of proteins attached to the membrane and forming a ring of proteins encircling the fiber at each Z line. Connections between the costamere and the Z line are provided by desmin (D) and dystrophin through its connections to the cytoskeletal γ-actin filaments. The left hand sarcomere has a normal structure with equal overlap of thick and thin filaments in both half-sarcomeres. The right hand sarcomere shows the type of disruption often observed after stretched contractions. The sarcomere is stretched and the thick filament is displaced to the left so that it lies asymmetrically in the sarcomere. If the costamere remains connected to the Z line, the local sarcolemma and T-tubule (TT) will be substantially stretched; alternatively, as shown, the connection between the Z line and costamere may fail. (**b**) Illustration of a long muscle in which the muscle fibres are connected by intramuscular tendons. In a submaximal contraction not all fibers are activated and the purple fibers are assumed activated. If there were no lateral connections, very little force would be produced at the ends of the muscle because each active fiber would stretch inactive fibers in series with it. In practice the lateral connection maintain all fibers in register even in submaximal contractions and force is transmitted along the fiber by both longitudinal and lateral connections

approach, which leads to the expression of a shortened dystrophin, or transfection with utrophin both partly restored the normal indentation stiffness. These results suggest that both the actin cytoskeleton and dystrophin are required for normal indentation stiffness but whether this physical property is related to the propensity to stretch-induced membrane damage is uncertain.

5 What Causes Membrane Damage After Stretched Contractions?

The mechanical measurements described above lead to a model of the membrane as a structure with little intrinsic strength and little resistance to stretch but spare membrane is available which is utilized during stretch. The costamere of the membrane is attached to the Z-discs by an array of proteins, prominent of which are dystrophin and desmin and these provide the lateral resistance to stretch and maintain fibers in register across the muscle even when only a small fraction of muscles are activated. On a model like this it is far from clear why contraction, with or without stretching, causes membrane damage. Perhaps the answer involves the T-tubular network which forms a series of interconnecting tubules crossing the fiber in the region of overlap of thick and thin filaments (Fig. 5a). If the T-tubules do not remain in register with the contractile proteins then the connections between the T-tubules and the membrane will be stretched and it is easy to imagine that the narrow T-tubules could be sheared from their connections to the surface membrane (Fig. 5a). In support of this possibility is the finding that the T-tubular network is severely disrupted following stretched contractions (Takekura et al. 2001).

If the cause of membrane damage is mechanically-induced (tears or rips), then one might predict that they would occur at the time of the mechanical event. While the tear remains, any molecules which have an electrochemical gradient across the membrane will tend to equilibrate through the tear. With this in mind we have imaged Ca^{2+} and Na^{+} inside fibers during and immediately after eccentric contractions expecting to see highly localized regions of high Ca^{2+} or Na^{+} just inside the membrane as Ca^{2+} and/or Na^{+} enter down their electrochemical gradients (Balnave et al. 1997, Yeung et al. 2003). We have never successfully observed such events which could mean (1) they do not exist, (2) they are infrequent, (3) they are below the resolution of our methods. Studies by Bansal et al. (2003), in which artificial holes in the membrane were generated by intense laser pulses, showed that holes of this sort were repaired in about a minute, judged by rate of increase of fluorescence from FM 1–43 which stains lipid membranes exposed by the mechanical defect. These authors identified a dystrophic muscle disease, dysferlinopathy, characterized by mutations to dysferlin, in which repair was greatly slowed; interestingly in *mdx* mice repair was unchanged from wild type.

The discussion above suggests that if tears in the membrane are the cause of the membrane damage, the increased permeability should start immediately after the stretched contraction and probably terminating within a minute as the tears are repaired. In fact measurements of the increase in Na^{+} and Ca^{2+} after the stretched contractions do not show an immediate increase, rather there is a slow rise reaching a maximum after 10–20 min (Balnave and Allen 1995; Yeung et al. 2003). Furthermore this increase in ion levels is blocked by drugs which block stretch-activated channels (Yeung et al. 2003; Sonobe et al. 2008) suggesting that the cause is channel activation rather than mechanical defects to the membrane. A similar conclusion was reached by (McBride et al. 2000) who studied membrane depolarization in muscles subjected to stretched contractions. A depolarization of ~20 mV was apparent 1–2 h following the stretched contraction and still present at 1 day but recovered after 2 days. This depolarization was prevented by addition of streptomycin to the drinking water; alternatively addition of either streptomycin or Gd^{3+} to the muscle bathing solution prevented the depolarization. These results establish that opening of stretch activated channels can explain increases in $[Na^{+}]i$, $[Ca^{2+}]i$ and depolarization.

What then is the cause of the membrane permeability to large molecules which is so clearly evident if it is not tears in the membrane? There is substantial literature establishing that elevated intracellular calcium ($[Ca^{2+}]i$) contributes to membrane damage. Early studies established that increasing $[Ca^{2+}]i$ with ionophore or Ca^{2+} channel opening drugs led to loss of enzymes from muscle and that protection was provided by phospholipase A2 inhibitors and by ROS scavengers (Duncan and Jackson 1987; Howl and Publicover 1990). As noted above, studies on isolated membrane fragments have confirmed that the tensile strength is greatly decreased by elevation of $[Ca^{2+}]i$ (Nichol and Hutter 1996a). Thus it appears that elevated $[Ca^{2+}]i$ weakens the membrane either by damage to phospholipids caused by activity of phospholipase A2 or by excessive production of ROS leading to lipid peroxidation. It is also known that eccentric contractions lead to early damage to desmin, titin and dystrophin (Lieber et al. 1996) and recent evidence confirms that this is caused by activation of the calpain by entry of extracellular Ca^{2+} (Zhang et al. 2008). It now seems likely that the calpain involved is μ-calpain as a calpain-3 knockout mouse still shows signs of titin disruption after elevation of $[Ca^{2+}]i$ (Verburg et al. 2009). Thus an alternative explanation for the increased membrane permeability is that it is a consequence of elevated $[Ca^{2+}]i$ caused by opening of stretch-activated channels. The elevated $[Ca^{2+}]i$ causes increased membrane permeability by damage to phospholipids caused by either phospholipase A2 or by lipid peroxidation. A feature of this pathway is that the increased permeability will be delayed at least until a sufficient rise in the Ca^{2+} has occurred which on current evidence would seem likely to be some minutes. This time delay potentially allows this mechanism to be distinguished from mechanical tears, which would be expected to cause increased permeability starting at the time of the stretched contractions. Viewed in this light, the study of McNeil and Khakee (1992) is probably incapable of distinguishing the two theories as the exercise period was 1 h, so that even if the first protein uptake measurements were taken immediately after the cessation of exercise, there would have been a delay of between zero and 1 h, between the contractions and the first measurement.

6 Repair of Membrane Damage

There is abundant evidence that stretched contractions can cause increased membrane permeability but many studies show that muscles eventually make a full recovery. Studies of the repair process generally induce membrane damage often by scrape injury or, in a recent approach, by intense laser stimulation of a localized region of the membrane (Bansal et al. 2003). Studies of membrane repair in many tissues have shown that repair involves entry of extracellular Ca^{2+} through the defect. Ca^{2+} causes accumulation of vesicles in the cytoplasm near the defect and the these vesicles subsequently fuse and form a patch of membrane which seals the defect in the sarcolemma (for review see McNeil and Steinhardt 2003). It also seems that induction of Ca^{2+} entry channels near the site of the defect (McCarter and Steinhardt 2000) and activation of calpain (Mellgren et al. 2007) are elements in the repair process. It has been shown that dysferlin has a role in the fusion of the vesicles

(Bansal et al. 2003) and that mutations of dysferlin cause a myopathy. Very recently it has been shown that another protein (TRIM72 or MG53) is also involved and that the membrane defect leads to oxidative stress which cause oligomerization of TRIM72 which is then involved in vesicle recruitment to the damage site (Cai et al. 2009a, b). Mice which lack functional TRIM72 also develop a myopathy. It is clear that lack of the machinery for membrane repair can cause diseases with features in common with muscular dystrophy but there is no evidence for this type of mechanism in DMD or the *mdx* mouse.

7 Why Is Membrane Permeability Greater in *mdx* Muscle?

A characteristic of DMD is the high plasma levels of creatine kinase indicating membrane damage in muscle. Furthermore large molecules, such as fibronectin, albumin and Evans blue (which binds to albumin), are excluded from normal fibers but can be found in many fibers of the *mdx* mouse, particularly after intense or eccentric exercise (Clarke et al. 1993; Hamer et al. 2002; Whitehead et al. 2006). Thus there is abundant evidence that muscle damage is greater and more easily induced in *mdx* compared to wild-type muscle.

It is widely assumed that this results from the absence of dystrophin, which results in reduced mechanical stability of the membrane and leads to more frequent mechanical tears in the membrane (Petrof et al. 1993; Ervasti and Sonnemann 2008; Davies and Nowak 2006). However, as reviewed above, the only measured differences in membrane properties are the indentation stiffness and the susceptibility to hypo-osmotic damage. Perhaps the absence of dystrophin makes T-tubular detachment from the surface membrane more frequent but measurements of conduction of the AP through the T-system found no difference between *mdx* and wild-type muscle suggesting that T-tubular continuity is normal (Woods et al. 2005). Furthermore Hutter (1992) pointed out that when T-tubules are disrupted from the sarcolemma, repair of the sarcolemma appears rapid. Arguing against the mechanical tears hypothesis, we found that the increased ionic influx after a series of stretch contractions in *mdx* muscle is very slow and blocked by drugs which block stretch-activated channels (Yeung et al. 2005). In a recent study we examined the entry of dyes into wild-type and *mdx* muscle fibers following contractions with an imposed stretch (Whitehead et al. 2006). As expected membrane damage, assessed by dye entry, was much greater in the *mdx* muscles. We found that the time course of dye entry was slow (still increasing after 60 min) which is not compatible with tears that occur immediately and seal over within a few minutes. Furthermore, the membrane damage was reduced by stretch-activated channel blockers which would have prevented the rise in $[Ca^{2+}]i$ (Yeung et al. 2005). This result suggests that dye entry is a secondary consequence of stretch-activated channel opening and the resulting elevation of $[Ca^{2+}]i$. In a subsequent study we found that application of N-acetyl cysteine, a reactive oxygen species scavenger, also reduced the membrane damage judged by dye entry (Whitehead et al. 2008). This suggests that

ROS may contribute to the membrane damage by lipid peroxidation (Nguyen and Tidball 2003) or that ROS may be involved in turning on the Ca^{2+} entry pathway via stretch-activated channels (Gervasio et al. 2008).

In summary, while it is clear that membranes become permeable after stretched contractions and that this permeability change is greater in *mdx* than wild-type muscle, there is little evidence that mechanically-induced membrane damage is the cause of the permeability increase. Instead there is good evidence that stretched contractions cause a rise of $[Ca^{2+}]i$ by activation of stretch-activated channels. The increase in $[Ca^{2+}]i$ causes a loss of membrane strength and/or damage to the lipids which are the most likely cause of the increased membrane permeability.

Acknowledgements We are grateful to the National Health and Medical Research Council of Australia who provided financial support for this work.

References

Allen DG, Gervasio OL, Yeung EW, Whitehead NP (2010) Calcium and the damage pathways in muscular dystrophy. Canadian J Physiol Pharmacol 88: 83–91

Balnave CD, Allen DG (1995) Intracellular calcium and force in single mouse muscle fibres following repeated contractions with stretch. J Physiol 488:25–36

Balnave CD, Davey DF, Allen DG (1997) Distribution of sarcomere length and $[Ca^{2+}]i$ in single fibres from mouse skeletal muscle following stretch-induced injury. J Physiol 502:649–659

Bansal D, Miyake K, Vogel SS, Groh S, Chen CC, Williamson R, McNeil PL, Campbell KP (2003) Defective membrane repair in dysferlin-deficient muscular dystrophy. Nature 423:168–172

Bassett DI, Bryson-Richardson RJ, Daggett DF, Gautier P, Keenan DG, Currie PD (2003) Dystrophin is required for the formation of stable muscle attachments in the zebrafish embryo. Development 130:5851–5860

Blake DJ, Weir A, Newey SE, Davies KE (2002) Function and genetics of dystrophin and dystrophin-related proteins in muscle. Physiol Rev 82:291–329

Bodensteiner JB, Engel AG (1978) Intracellular calcium accumulation in Duchenne dystrophy and other myopathies: a study of 567,000 muscle fibers in 114 biopsies. Neurology 28:439–446

Boriek AM, Capetanaki Y, Hwang W, Officer T, Badshah M, Rodarte J, Tidball JG (2001) Desmin integrates the three-dimensional mechanical properties of muscles. Am J Physiol Cell Physiol 280:C46–C52

Brown LM, Hill L (1991) Some observations on variations in filament overlap in tetanized muscle fibres and fibres stretched during a tetanus, detected in the electron microscope after rapid fixation. J Mus Res Cell Mot 12:171–182

Cai C, Masumiya H, Weisleder N, Matsuda N, Nishi M, Hwang M, Ko JK, Lin P, Thornton A, Zhao X, Pan Z, Komazaki S, Brotto M, Takeshima H, Ma J (2009a) MG53 nucleates assembly of cell membrane repair machinery. Nat Cell Biol 11:56–64

Cai C, Masumiya H, Weisleder N, Pan Z, Nishi M, Komazaki S, Takeshima H, Ma J (2009b) MG53 regulates membrane budding and exocytosis in muscle cells. J Biol Chem 284:3314–3322

Clarke MS, Khakee R, McNeil PL (1993) Loss of cytoplasmic basic fibroblast growth factor from physiologically wounded myofibers of normal and dystrophic muscle. J Cell Sci 106:121–133

Clarkson PM, Hubal MJ (2002) Exercise-induced muscle damage in humans. Am J Phys Med Rehabil 81:S52–S69

Davies KE, Nowak KJ (2006) Molecular mechanisms of muscular dystrophies: old and new players. Nat Rev Mol Cell Biol 7:762–773

Deconinck N, Ragot T, Marechal G, Perricaudet M, Gillis JM (1996) Functional protection of dystrophic mouse (mdx) muscles after adenovirus-mediated transfer of a dystrophin minigene. Proc Natl Acad Sci U S A 93:3570–3574

DelloRusso C, Scott JM, Hartigan-O'Connor D, Salvatori G, Barjot C, Robinson AS, Crawford RW, Brooks SV, Chamberlain JS (2002) Functional correction of adult mdx mouse muscle using gutted adenoviral vectors expressing full-length dystrophin. Proc Natl Acad Sci U S A 99:12979–12984

Dulhunty AF, Franzini-Armstrong C (1975) The relative contributions of the folds and caveolae to the surface membrane of frog skeletal muscle fibres at different sarcomere lengths. J Physiol 250:513–539

Duncan CJ, Jackson MJ (1987) Different mechanisms mediate structural changes and intracellular enzyme efflux following damage to skeletal muscle. J Cell Sci 87:183–188

Ervasti JM (2003) Costameres: the Achilles' heel of Herculean muscle. J Biol Chem 278:13591–13594

Ervasti JM, Campbell KP (1991) Membrane Organization of the Dystrophin-Glycoprotein complex. Cell 66:1121–1131

Ervasti JM, Sonnemann KJ (2008) Biology of the striated muscle dystrophin-glycoprotein complex. Int Rev Cytol 265:191–225

Evans EA, Waugh R, Melnik L (1976) Elastic area compressibility modulus of red cell membrane. Biophys J 16:585–595

Franco A Jr, Lansman JB (1990) Calcium entry through stretch-inactivated ion channels in mdx myotubes. Nature 344:670–673

Fridén J, Sjöström M, Ekblom B (1981) A morphological study of delayed muscle soreness. Experientia 37:506–507

Gervasio OL, Whitehead NP, Yeung EW, Phillips WD, Allen DG (2008) TRPC1 binds to caveolin-3 and is regulated by Src kinase: role in Duchenne muscular dystrophy. J Cell Sci 121:2246–2255

Goll DE, Thompson VF, Li H, Wei W, Cong J (2003) The calpain system. Physiol Rev 83:731–801

Granzier HL, Labeit S (2006) The giant muscle protein titin is an adjustable molecular spring. Exerc Sport Sci Rev 34:50–53

Hamer PW, McGeachie JM, Davies MJ, Grounds MD (2002) Evans Blue Dye as an in vivo marker of myofibre damage: optimising parameters for detecting initial myofibre membrane permeability. J Anat 200:69–79

Head SI, Williams DA, Stephenson DG (1992) Abnormalities in structure and function of limb skeletal muscle fibres of dystrophic mdx mice. Proc R Soc Lond B Biol Sci 248:163–169

Hellam DC, Podolsky RJ (1969) Force measurements in skinned muscle fibres. J Physiol 200:807–819

Hough T (1902) Ergographic studies in muscular soreness. Am J Physiol 7:76–92

Howl JD, Publicover SJ (1990) Permeabilisation of the sarcolemma in mouse diaphragm exposed to Bay K 8644 in vitro: time course, dependence on Ca^{2+} and effects of enzyme inhibitors. Acta Neuropathol 79:438–443

Hutter OF (1992) The membrane hypothesis of Duchenne muscular dystrophy: quest for functional evidence. J Inherit Metab Dis 15:565–577

Hutter OF, Burton FL, Bovell DL (1991) Mechanical properties of normal and mdx mouse sarcolemma: bearing on function of dystrophin. J Muscle Res Cell Motil 12:585–589

Jung C, Martins AS, Niggli E, Shirokova N (2008) Dystrophic cardiomyopathy: amplification of cellular damage by Ca^{2+} signaling and reactive oxygen species-generating pathways. Cardiovasc Res 77(4):766–773

Kumar A, Khandelwal N, Malya R, Reid MB, Boriek AM (2004) Loss of dystrophin causes aberrant mechanotransduction in skeletal muscle fibers. FASEB J 18:102–113
Lieber RL, Thornell LE, Friden J (1996) Muscle cytoskeletal disruption occurs within the first 15 min of cyclic eccentric contraction. J Appl Physiol 80:278–284
McBride TA, Stockert BW, Gorin FA, Carlsen RC (2000) Stretch-activated ion channels contribute to membrane depolarization after eccentric contractions. J Appl Physiol 88:91–101
McCarter GC, Steinhardt RA (2000) Increased activity of calcium leak channels caused by proteolysis near sarcolemmal ruptures. J Membr Biol 176:169–174
McNeil PL, Khakee R (1992) Disruptions of muscle fiber plasma membranes. Role in exercise-induced damage. Am J Pathol 140:1097–1109
McNeil PL, Steinhardt RA (2003) Plasma membrane disruption: repair, prevention, adaptation. Annu Rev Cell Dev Biol 19:697–731
Mellgren RL, Zhang W, Miyake K, McNeil PL (2007) Calpain is required for the rapid, calcium-dependent repair of wounded plasma membrane. J Biol Chem 282:2567–2575
Menke A, Jockusch H (1991) Decreased osmotic stability of dystrophin-less muscle cells from the mdx mouse. Nature 349:69–71
Moens P, Baatsen PH, Marechal G (1993) Increased susceptibility of EDL muscles from mdx mice to damage induced by contractions with stretch. J Muscle Res Cell Motil 14:446–451
Mokri B, Engel AG (1975) Duchenne dystrophy: electron microscopic findings pointing to a basic or early abnormality in the plasma membrane of the muscle fiber. Neurology 25:1111–1120
Morgan DL (1990) New insights into the behavior of muscle during active lengthening. Biophys J 57:209–221
Morgan DL, Allen DG (1999) Early events in stretch-induced muscle damage. J Appl Physiol 87:2007–2015
Natori R (1954) The property and contraction process of isolated myofibrils. Jikeikai Med J 1:119–126
Newham DJ, Jones DA, Edwards RH (1983a) Large delayed plasma creatine kinase changes after stepping exercise. Muscle Nerve 6:380–385
Newham DJ, Mills KR, Quigley BM, Edwards RH (1983b) Pain and fatigue after concentric and eccentric muscle contractions. Clin Sci 64:55–62
Nguyen HX, Tidball JG (2003) Null mutation of gp91phox reduces muscle membrane lysis during muscle inflammation in mice. J Physiol 553:833–841
Nichol JA, Hutter OF (1996a) Ca^{2+} loading reduces the tensile strength of sarcolemmal vesicles shed from rabbit muscle. J Physiol 493:199–209
Nichol JA, Hutter OF (1996b) Tensile strength and dilatational elasticity of giant sarcolemmal vesicles shed from rabbit muscle. J Physiol 493:187–198
Papponen H, Kaisto T, Leinonen S, Kaakinen M, Metsikko K (2009) Evidence for gamma-actin as a Z disc component in skeletal myofibers. Exp Cell Res 315:218–225
Pardo JV, Siliciano JD, Craig SW (1983) A vinculin-containing cortical lattice in skeletal muscle: transverse lattice elements ("costameres") mark sites of attachment between myofibrils and sarcolemma. Proc Natl Acad Sci U S A 80:1008–1012
Pasternak C, Wong S, Elson EL (1995) Mechanical function of dystrophin in muscle cells. J Cell Biol 128:355–361
Petrof BJ, Shrager JB, Stedman HH, Kelly AM, Sweeney HL (1993) Dystrophin protects the sarcolemma from stresses developed during muscle contraction. Proc Natl Acad Sci U S A 90:3710–3714
Podolsky RJ (1964) The maximum sarcomere length for contraction of isolated myofibrils. J Physiol 170:110–123
Porter GA, Dmytrenko GM, Winkelmann JC, Bloch RJ (1992) Dystrophin colocalizes with beta-spectrin in distinct subsarcolemmal domains in mammalian skeletal muscle. J Cell Biol 117:997–1005
Proske U, Morgan DL (2001) Muscle damage from eccentric exercise: mechanism, mechanical signs, adaptation and clinical applications. J Physiol 537:333–345

Puttini S, Lekka M, Dorchies OM, Saugy D, Incitti T, Ruegg UT, Bozzoni I, Kulik AJ, Mermod N (2009) Gene-mediated restoration of normal myofiber elasticity in dystrophic muscles. Mol Ther 17:19–25

Rahkila P, Takala TE, Parton RG, Metsikko K (2001) Protein targeting to the plasma membrane of adult skeletal muscle fiber: an organized mosaic of functional domains. Exp Cell Res 267:61–72

Rybakova IN, Patel JR, Ervasti JM (2000) The dystrophin complex forms a mechanically strong link between the sarcolemma and costameric actin. J Cell Biol 150:1209–1214

Samitt CE, Bonilla E (1990) Immunocytochemical study of dystrophin at the myotendinous junction. Muscle Nerve 13:493–500

Sengupta P, Baird B, Holowka D (2007) Lipid rafts, fluid/fluid phase separation, and their relevance to plasma membrane structure and function. Semin Cell Dev Biol 18:583–590

Shear CR, Bloch RJ (1985) Vinculin in subsarcolemmal densities in chicken skeletal muscle: localization and relationship to intracellular and extracellular structures. J Cell Biol 101:240–256

Sonobe T, Inagaki T, Poole DC, Kano Y (2008) Intracellular calcium accumulation following eccentric contractions in rat skeletal muscle in vivo: role of stretch-activated channels. Am J Physiol Regul Integr Comp Physiol 294:R1329–R1337

Stedman HH, Sweeney HL, Shrager JB, Maguire HC, Panettieri RA, Petrof B, Narusawa M, Leferovich JM, Sladky JT, Kelly AM (1991) The mdx mouse diaphragm reproduces the degenerative changes of Duchenne muscular dystrophy. Nature 352:536–539

Street SF (1983) Lateral transmission of tension in frog myofibers: a myofibrillar network and transverse cytoskeletal connections are possible transmitters. J Cell Physiol 114:346–364

Takekura H, Fujinami N, Nishizawa T, Ogasawara H, Kasuga N (2001) Eccentric exercise-induced morphological changes in the membrane systems involved in excitation-contraction coupling in rat skeletal muscle. J Physiol 533:571–583

Verburg E, Murphy RM, Richard I, Lamb GD (2009) Involvement of calpains in Ca^{2+}-induced disruption of excitation-contraction coupling in mammalian skeletal muscle fibers. Am J Physiol Cell Physiol 296:C1115–C1122

Wang X, Weisleder N, Collet C, Zhou J, Chu Y, Hirata Y, Zhao X, Pan Z, Brotto M, Cheng H, Ma J (2005) Uncontrolled calcium sparks act as a dystrophic signal for mammalian skeletal muscle. Nat Cell Biol 7:525–530

Waugh RE, Agre P (1988) Reductions of erythrocyte membrane viscoelastic coefficients reflect spectrin deficiencies in hereditary spherocytosis. J Clin Invest 81:133–141

Whitehead NP, Streamer M, Lusambili LI, Sachs F, Allen DG (2006) Streptomycin reduces stretch-induced membrane permeability in muscles from mdx mice. Neuromuscular Disorders 16:845–854

Whitehead NP, Pham C, Gervasio OL, Allen DG (2008) N-acetylcysteine ameliorates skeletal muscle pathophysiology in mdx mice. J Physiol 586:2003–2014

Woods CE, Novo D, DiFranco M, Capote J, Vergara JL (2005) Propagation in the transverse tubular system and voltage dependence of calcium release in normal and mdx mouse muscle fibres. J Physiol 568:867–880

Yeung EW, Ballard HJ, Bourreau JP, Allen DG (2003) Intracellular sodium in mammalian muscle fibers after eccentric contractions. J Appl Physiol 94:2475–2482

Yeung EW, Whitehead NP, Suchyna TM, Gottlieb PA, Sachs F, Allen DG (2005) Effects of stretch-activated channel blockers on $[Ca^{2+}]i$ and muscle damage in the mdx mouse. J Physiol 562:367–380

Zhang BT, Yeung SS, Allen DG, Qin L, Yeung EW (2008) Role of the calcium-calpain pathway in cytoskeletal damage after eccentric contractions. J Appl Physiol 105:352–357

Zubrzycka-Gaarn EE, Hutter OF, Karpati G, Klamut HJ, Bulman DE, Hodges RS, Worton RG, Ray PN (1991) Dystrophin is tightly associated with the sarcolemma of mammalian skeletal muscle fibers. Exp Cell Res 192:278–288

Cellular and Whole Muscle Studies of Activity Dependent Potentiation

Brian R. MacIntosh

Abstract With a single activation, a skeletal muscle fiber, motor unit or whole muscle will yield a twitch contraction. The twitch is not an "all-or-none" response, but a submaximal response that can vary from one time to another. Prior activation causes myosin regulatory light chain (RLC) phosphorylation, by an enzyme called myosin light chain kinase. This phosphorylation dissipates slowly over the next several minutes due to a slow activity of light chain phosphatase. Phosphorylation of the RLC increases the mobility of the S1 head of myosin, bringing the S1 head in closer proximity to the myosin binding sites on actin. This increased mobility increases the rate of engagement of cross-bridges and increases the rate of force development and contraction magnitude on subsequent twitch or other submaximal contractions. We call this increased contractile response "activity dependent potentiation". With sequential twitches or incompletely fused tetanic contractions, the term staircase is used to describe the progressive increase in amplitude of contraction. If a twitch is elicited after a tetanic contraction, we call the enhanced response posttetanic potentiation. Stretching a muscle fiber to a longer length will also bring the actin filaments close to the myosin heads. This increases the Ca^{2+} sensitivity, independent of RLC phosphorylation. At long sarcomere lengths, the impact of RLC phosphorylation is diminished, because Ca^{2+} sensitivity is already increased. Similarly, lowering the temperature at which the muscle is tested increases Ca^{2+} sensitivity. At low temperatures, staircase and posttetanic potentiation are diminished, but RLC phosphorylation still occurs. Activity dependent potentiation is an important functional modulator of contractile response.

Keywords Calcium sensitivity • Staircase • Posttetanic potentiation • Myosin regulatory light chains • Twitch • Acidosis • Muscle length

---

B.R. MacIntosh (✉)
Faculty of Kinesiology, University of Calgary, 2500 University Dr NW,
Calgary, AB, Canada T2N1N4
e-mail: brian@kin.ucalgary.ca

D.E. Rassier (ed.), *Muscle Biophysics: From Molecules to Cells*,
Advances in Experimental Medicine and Biology 682,
DOI 10.1007/978-1-4419-6366-6_18, © Springer Science+Business Media, LLC 2010

1 Introduction

Everyone has experienced the negative consequences of repeated activation that we call muscle fatigue. However, there is also a positive outcome of prior activation that has been less extensively studied, and certainly less pervasively experienced. Potentiation or enhancement of contractile performance is a short-term consequence of muscle activation. This chapter will focus on the mechanical responses and features of staircase and posttetanic potentiation, the two forms of activity dependent potentiation resulting from stimulation of muscle. This chapter presents a description of the types of activity dependent potentiation, a brief historical perspective, and a description of the process of excitation–contraction coupling. Following this, the known mechanism underlying activity dependent potentiation will be presented, and this will be followed by a description of the mechanical features of activity dependent potentiation and interactions with other factors that alter Ca^{2+} sensitivity. Finally, the interactions of fatigue and staircase will be considered and the consequences of aging and disuse on activity dependent potentiation will be presented.

2 Preparations

In contrast with most of this book, this chapter will rely on in situ muscle contractile properties to relate the mechanical features of activity dependent potentiation. This is out of necessity; most of the work on the mechanics of activity dependent potentiation has been done with live animals and human subjects, rather than with single fiber experiments. There does not appear to be any isolated molecule mechanical work specifically dealing with activity dependent potentiation. This may be due to the fact that repetitive activation is difficult to control in a laser trap. There is only one study that uses the in vitro motility assay to study the mechanism of activity dependent potentiation. The main reason for this paucity of experimental evidence at the molecular level is probably related to the fact that the assumed mechanism of activity dependent potentiation relies on an intact myofilament lattice. This criterion can only be achieved with myofibrils or higher levels of organization. Each of these preparations along with their advantages and disadvantages is briefly described below.

2.1 *In Vivo*

In vivo studies of muscle mechanics involve measurements of joint moment in response to voluntary or stimulated contractions. Stimulation can be electrical over the muscle or nerve innervating the muscle of interest, or magnetic, over the nerve or appropriate motor cortex region of the brain. Usually, in vivo studies involve an entire muscle group with an intact circulation, though in some cases the circulation can be omitted by an appropriately placed pneumatic cuff. Individual muscles can be studied with neural stimulation. The advantage of the in vivo preparation is that it represents how muscles behave in reality. The disadvantage is that the responses come from a multitude of muscle fibers, often of mixed fiber-type and certainly of various and (often) undetermined size. The specific mechanism(s) influencing a given contractile response are difficult to elucidate with in vivo study of muscle. However, recent advances in imaging technology have improved our ability to visualize the changes that occur beneath the skin.

2.2 *In Situ*

An in situ muscle preparation is one for which the muscle is intact, but the tendon is disconnected from the bone, it is usually connected directly to a force transducer. This is the most common preparation for the study of activity dependent potentiation, and the preparation that was used for much of the mechanical material presented in this chapter. Advantages of this preparation are: normal circulation is maintained; muscle fibers are activated at their neuromuscular junction, the preparation is usually maintained at physiological temperature, the extracellular space is realistic and direct measurement of force is relatively simple. Disadvantages include: the response is still typically from a collection of muscle fibers, often of mixed fiber-type; biochemical measurements are not easily made sequentially over time without using a large number of animals and subcellular imaging is difficult if not impossible.

2.3 *In Vitro*

When tissue, cells or subcellular parts are studied “ex vivo”, the preparation is considered to be in vitro, (literally translates to, “in glass”). Studies related to activity dependent potentiation have used in vitro conditions with whole muscles, fiber bundles, intact single fibers, skinned fibers, myofibrils and isolated filaments. This approach offers the advantage that the environment is totally under the control of the investigator. This is the intracellular environment in the case of skinned fibers, myofibrils and filaments, but is otherwise the extracellular environment. In addition to oxygen, pH and ion composition, drugs or chemical inhibitors can be tested without consideration of the potential effects on parts of the animal other than the muscle under study. This latter point would be a major concern for in situ or in vivo study using inhibitors of specific enzymes of channels. At the single fiber level, the consequences of any intervention are easier to interpret, because typically there is only one fiber-type present. However, hybrid fibers can contain more than one myosin isoform (Billeter et al. 1981; Staron and Pette 1993). On the other hand, this preparation removes the muscle from its natural environment, and imposes a virtually infinite extracellular space that permits greater exchange across the plasmalemma, and minimizes the impact of any change of the extracellular concentration of the substances exchanged. A complication of the in vitro approach is that such studies are often conducted at room temperature or colder. This is done for a variety of reasons. The low temperature and the extremely large extra-cellular space make this preparation somewhat unphysiological. Caution must be exercised when interpreting these experiments.

3 Definitions of Terms

There are several terms that are used to describe the positive mechanical consequences of prior activation. These terms include staircase, posttetanic potentiation and postactivation potentiation. Before definitions of these terms are presented, twitch and tetanic contractions will be briefly defined. These definitions provide a context for understanding activity dependent potentiation.

3.1 Twitch

A twitch contraction is a transient rise and fall of force following a single activation (see Fig. 1). A twitch can be elicited in mechanically skinned fibers, intact single fibers, muscle fiber bundles, single or multiple motor units, whole muscles and groups of muscles operating over a common joint. The twitch is an elegant response that represents the smallest unit of functional elicitation of contractile response from a muscle. It involves all steps of excitation–contraction coupling, which are described below. Although the twitch has been referred to as an "all-or-none" response, the amplitude of a twitch can be modulated by a variety of factors that influence the rate and duration of the rising and/or falling phase of the contraction and therefore its magnitude. Figure 1 illustrates these theoretical ways that the twitch contraction magnitude can be altered. With the possible exception of amphibian muscle studied close to 0°C, a twitch contraction is a submaximal contraction. In mammalian muscle at 37°C, the twitch: tetanus ratio is typically 0.1–0.2.

3.2 Tetanic Contraction

When sequential activations of muscle occur at an interval that is short enough to prevent complete relaxation between mechanical responses, the resulting contraction is considered to be a tetanic contraction. When mechanical oscillations

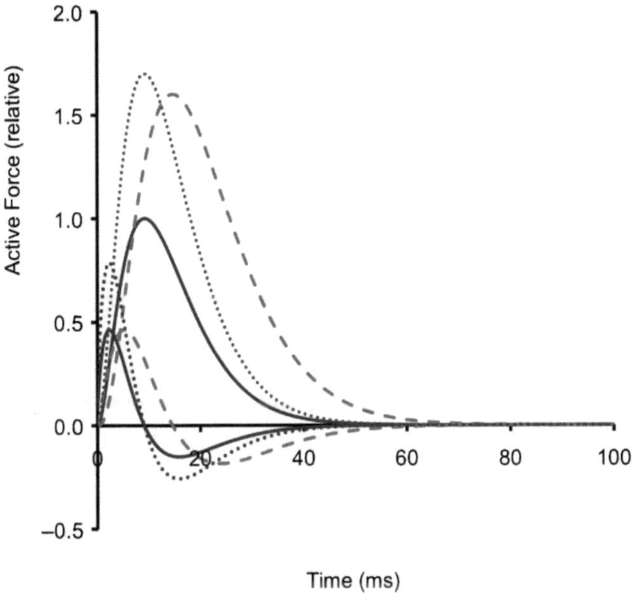

Fig. 1 Control and enhanced twitch contractions. Active force and rate of force change (arbitrary units) for a control twitch (*solid line*), and theoretical potentiated twitches illustrating the possible enhancement primarily by increasing the intensity of activation (*dotted line*) or the duration of activation (*dashed line*). Note that the time to peak (marked by the crossover from positive to negative rate of force change) is not altered by change in intensity of activation, but is prolonged with increased duration of activation. Simulated twitch contractions, representing rat fast-twitch skeletal muscle at 37°C

between activations are apparent, the contraction is called an incompletely fused tetanic contraction. When there is no apparent relaxation between sequential activations, the contraction is called a completely fused tetanic contraction. A tetanic contraction can be very brief (as few as two stimulating pulses) or very long, lasting several seconds. Long tetanic contractions have both potentiating and inhibiting (fatigue) influences that affect the magnitude of response.

3.3 Activity Dependent Potentiation

When an otherwise submaximal contraction is enhanced in amplitude by prior activation, this can be referred to as activity dependent potentiation. There are three subcategories of activity dependent potentiation: staircase, posttetanic potentiation and postactivation potentiation. Each of these forms of enhanced response relies on artificially evoked contractile response for detection and quantification. The evoked response is typically a twitch contraction, but can also be any submaximal tetanic contraction; an incompletely fused tetanic contraction can be enhanced by prior contractile activity if the contraction is still very brief (still "submaximal"). This means the force–frequency relationship is altered by prior muscle activity.

3.4 *Staircase*

Staircase has traditionally been considered to be the progressive enhancement of twitch contractile response during repeated stimulation at low frequencies (Close and Hoh 1968; Krarup 1981a, b). An example of staircase is shown in Fig. 2. The actual frequency that elicits staircase depends on the species of animal under study, and the temperature at which the muscle was studied. We have studied twitch staircase with the rat medial gastrocnemius muscle with 5 and 10 Hz stimulation (MacIntosh 1991; MacIntosh and Rassier 2002). We have also used repeated brief submaximal tetanic contractions to study staircase. In this case the train frequency (rate of occurrence of contractions) is typically 2 Hz or less (MacIntosh and Willis 2000). The frequency of stimulation within each train can vary from that level that yields minimal fusion to a completely fused tetanic contraction, though in the latter case, very brief contractions are needed to obtain enhancement

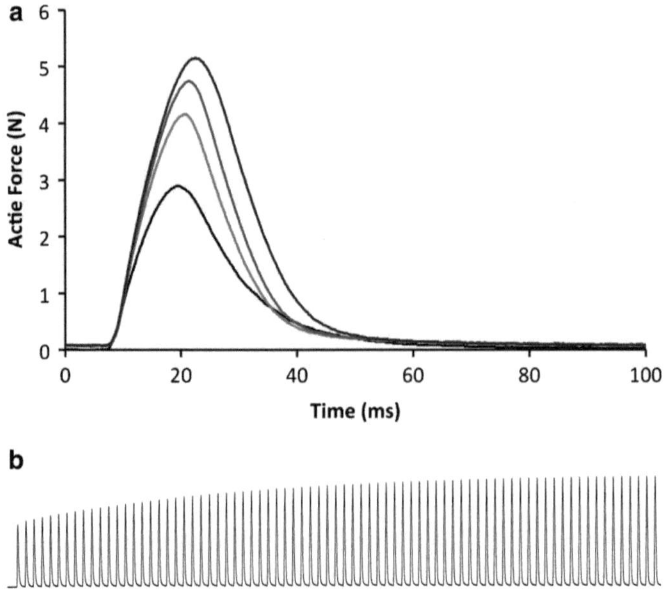

Fig. 2 Twitch contractions showing staircase at 10 Hz. In (**a**), superimposed twitch contractions are shown for contractions at 0, 2, 4, and 10 s. In (**b**), sequential contractions are shown for the first 8 s of 10 Hz stimulation. Each contraction is enhanced by the prior contractile activity, but the rate of change of magnitude of enhancement decreases with time

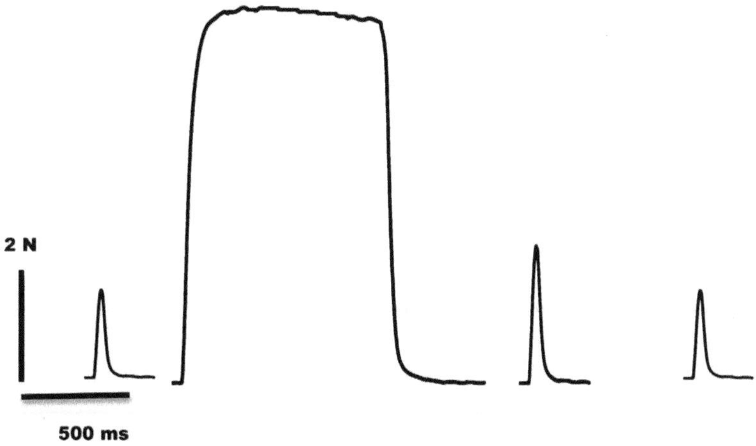

Fig. 3 Posttetanic potentiation in the rat gastrocnemius muscle. An initial twitch contraction is shown, followed by a 1 s tetanic contraction. The first twitch contraction after the tetanic contraction is enhanced, and the enhancement dissipates over the next few minutes, as shown by a twitch obtained after 5 min or inactivity

(MacIntosh et al. 2008). The key feature of staircase is that sequential contractions with the same pattern of stimulation are evaluated, and a progressive enhancement of contractile magnitude occurs.

3.5 Posttetanic Potentiation

The magnitude of a twitch contraction is typically enhanced following an electrically elicited tetanic contraction. This form of activity dependent potentiation is known as posttetanic potentiation. Figure 3 provides an example of posttetanic potentiation, and illustrates the time-course of decay of this enhancement. Note that these contractions represent synchronous activation of all motor units in the muscle. A similar pattern of change in active force could be obtained with a single motor unit or a single fiber. Enhancement of any submaximal contraction after a tetanic contraction could be considered posttetanic potentiation.

3.6 Postactivation Potentiation

When the contraction that elicits subsequent potentiation is originated by voluntary activation, then the enhancement of a subsequent submaximal contraction is considered to be postactivation potentiation. Evaluation of such enhancement requires an evoked response to be sure the enhancement occurs for the same stimulation, as that used prior to the voluntary contraction. There are several papers in the literature that use this term to represent improved physical performance of a task after a warm-up, without confirmation that an evoked response has been enhanced by the warm-up. This use of the term is inappropriate.

4 Historical Perspective

The fact that prior stimulation had a positive influence on subsequent contractions was first noted by Guttman et al. (1936) and was studied further by Rosenblueth and Morison (1937). These authors confirmed with electromyography that the potentiation resulted from changes within the muscle cell, rather than at the neuromuscular junction or muscle membrane. It was not until the 1960s that a potential mechanism was proposed. Desmedt and Hainaut (1968) demonstrated that the duration of contraction was not prolonged and proposed that the intensity of activation appeared to be enhanced during staircase. This pattern of enhancement would be represented by the dotted line in Fig. 1. Consistent with this mechanism, Györke (1993) observed an increase in peak [Ca^{2+}] in sequential contractions. Allen et al. (1989) have observed increments in sequential submaximal tetanic contractions, and Decostre et al. (2000) have reported increased peak [Ca^{2+}] during posttetanic twitch contractions. These observations, that Ca^{2+} release is enhanced during activity dependent potentiation has been largely ignored as a potential mechanism of activity dependent potentiation since the discovery that potentiation correlates strongly with phosphorylation of the regulatory light chains (RLC) of myosin (Klug et al. 1982; Manning and Stull 1982; Moore and Stull 1984).

It was first reported by Perrie et al. (1973) that the RLC of myosin can be phosphorylated, but it took several years before the process was linked with activity dependent potentiation. Once the association was made, skinned fiber experiments were done, confirming that RLC phosphorylation increased Ca^{2+} sensitivity (Persechini et al. 1985). This increase in Ca^{2+} sensitivity by RLC phosphorylation has been the assumed mechanism of activity dependent potentiation for the past 35 years. In spite of this, activity dependent potentiation without RLC phosphorylation has been demonstrated (Rassier et al. 1999) leaving the possibility that other mechanisms may contribute to this mechanical response of skeletal muscle. It has also been reported that twitch potentiation is not entirely eliminated in mice lacking MLCK, the enzyme that phosphorylates the RLC (Zhi et al. 2005).

5 Excitation–Contraction Coupling

Comprehension of the mechanisms and implications of activity dependent potentiation is facilitated by an understanding of excitation–contraction coupling. In voluntary recruitment of motor units or electrical activation of individual muscle fibers, each muscle fiber is activated by a sequence of events that result in cross-bridge formation and force generation. This sequence of events is referred to as excitation–contraction coupling.

Excitation–contraction coupling begins with depolarization of the surface membrane by either neuromuscular transmission, or by direct electrical stimulation. Once the membrane potential reaches threshold for opening of the fast Na^+ channels, an action potential is generated and propagated along the surface membrane and down the transverse tubules. In addition to fast Na^+ channels and a variety of K^+ channels, the membrane of the transverse tubules contains modified Ca^{2+} channels known as dihydropyridine receptors. These receptors, which are packed more densely in the transverse tubules, are voltage sensors that detect the depolarization of the membrane associated with the action potential. The dihydropyridine receptors are structurally linked to half of the ryanodine receptors, which are the Ca^{2+} channels of the sarcoplasmic reticulum. Detection of transverse tubule depolarization leads to opening of the ryanodine receptors and release of Ca^{2+} from the terminal cisternae of the sarcoplasmic reticulum.

The release of Ca^{2+} from the terminal cisternae results in a rapid increase in [Ca^{2+}] in the myoplasm. There are several ligands in muscle that bind Ca^{2+}, so the increase in free [Ca^{2+}] is much less than would otherwise be expected for the amount of Ca^{2+} released. Of note, there are two regulatory molecules that bind Ca^{2+} in the range of concentrations achieved with activation. Troponin C, which is bound to the thin filaments in conjunction with troponin I and troponin T is one of these ligands. The binding of Ca^{2+} to troponin C results in exposure of binding sites for myosin on the thin actin filaments. This exposure of the binding sites is apparently achieved by physical movement of the tropomyosin away from these binding sites (Gordon et al. 2001). As long as Ca^{2+} occupies a given troponin C, the corresponding tropomyosin, which spans seven actin monomers, will be held away from the myosin-binding sites, and these seven sites will be available for cross-bridge formation. Myosin heads in the vicinity of these sites can undergo formation and dissociation of cross-bridges.

The increase in [Ca^{2+}] resulting from activation, also results in binding of Ca^{2+} to calmodulin, a regulatory protein that has several homologies with troponin C (Hiraoki and Vogel 1987). The binding of Ca^{2+} to calmodulin is important in the activation of several enzymes, but the most relevant in this chapter is MLCK. This topic will be further discussed in Sect. 6 below.

Contractile interaction of myosin with actin is terminated by the decrease in cytoplasmic free [Ca^{2+}] that results from the uptake of Ca^{2+} into the sarcoplasmic reticulum. This Ca^{2+} uptake is accomplished by a membrane pump known as sarco-endoplasmic reticulum Ca^{2+} ATPase (SERCA). The isoforms of SERCA that are important in skeletal muscle are SERCA1 (in fast-twitch muscle) and SERCA2 in slow-twitch muscle. In amphibian muscle and in fast-twitch muscle of small mammals, parvalbumin facilitates the decrease in free [Ca^{2+}] (Celio and Heizmann 1982; Rall 1996). Parvalbumin is a cytoplasmic protein that acts as a Ca^{2+} buffer.

The contractile response that we see as a twitch or tetanic contraction results from cyclic interaction of myosin heads of the thick filament with binding sites on the actin thin filaments, while Ca^{2+} is bound to troponin C. With the decrease in myoplasmic free [Ca^{2+}] to the resting level, the cross-bridge cycle is interrupted, and the number of cross-bridges in the strong-binding state decreases exponentially.

6 Mechanisms of Activity Dependent Potentiation

As mentioned above, once the relationship between RLC phosphorylation and activity dependent potentiation was identified, this has been the focus of attention with respect to a mechanism for this potentiation. Although considerable progress has been made in understanding this probable mechanism, further work is necessary. The basics of current knowledge about the mechanism of RLC phosphorylation, and the impact of this on contraction are described below.

6.1 Phosphorylation of the Regulatory Light Chains of Myosin

Activity dependent potentiation appears to be primarily a function of RLC phosphorylation (Grange et al. 1993; Sweeney et al. 1993). The evidence for this is fairly convincing, but as indicated above, there are a couple of observations that would argue that perhaps other mechanisms are involved. Phosphorylation of a protein is a typical method of providing short-term memory to any molecular system, and the contractile proteins would be no exception. Phosphorylation of the RLC allows the myosin to "remember" that it has recently been activated.

There is a strong correlation between phosphorylation of the RLC of myosin and activity dependent potentiation (Klug et al. 1982; Manning and Stull 1982). This observation gave the first hint that phosphorylation of this RLC was the mechanism responsible for activity dependent potentiation. This strong correlation is maintained during increase and dissipation of potentiation. Consistent with the notion that activity dependent potentiation is caused by RLC phosphorylation, it has been reported that such potentiation is limited or absent in slow-twitch skeletal muscle which has a lower activity of MLCK, and a higher activity of phosphoprotein phosphatase C (Moore and Stull 1984) an enzyme that dephophorylates RLC. In contrast with the usual observation of little or no activity dependent potentiation in slow-twitch muscle, Burke et al. (1976) reported a similar magnitude of potentiation with a faster decline or dissipation. This faster dissipation is consistent with the observed greater activity of the phosphatase. There may be important species differences in the fiber-type properties of skeletal muscle.

6.2 Myosin Light Chain Kinase

The fact that Ca^{2+} and calmodulin combine to activate myosin light chain kinase (MLCK) and that this enzyme was responsible for phosphorylating the RLC of myosin was known before the consequences of this phosphorylation were understood. Walsh (1983) had shown that calmodulin was involved in the activation of

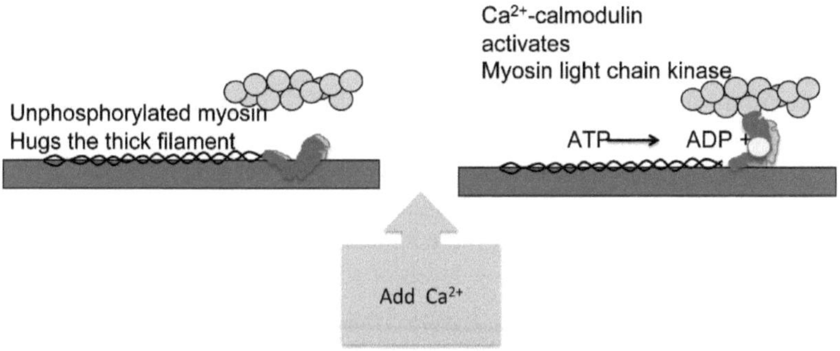

Fig. 4 Mechanism of regulatory light chain phosphorylation. On the left, the regulatory light chain is not phosphorylated, and Ca^{2+} concentration is at the resting level. With activation, $[Ca^{2+}]$ increases and Ca^{2+} binds to calmodulin (calmodulin binds $4Ca^{2+}$). The Ca^{2+}–calmodulin complex activates myosin light chain kinase, and this enzyme transfers an inorganic phosphate (*light dot*) from ATP to the regulatory light chain. The addition of the phosphate to the regulatory light chain causes the light chain to swing out away from the thick filament backbone

cardiac MLCK. Blumenthal and Stull (1980) showed that calmodulin needed four Ca^{2+} bound to activate MLCK. This scheme is illustrated in Fig. 4.

The $[Ca^{2+}]$ reached during a tetanic contraction is proportional to the frequency of activation (Chin et al. 1997). This fact is the likely reason that the force–frequency relationship resembles the force–pCa^{2+} relationship. Considering the relationship between frequency of activation and $[Ca^{2+}]$, and recognizing that Ca^{2+} binds to calmodulin in a manner similar to binding to troponin, it is expected that activation of MLCK is frequency dependent. The level of RLC phosphorylation achieved after a 2 s (200 Hz) tetanic contraction would be expected to be greater than that reached after 2 s of 10 Hz. In fact, the 2 s 200 Hz contraction elicits greater RLC phosphorylation than 10 s of 10 Hz stimulation.

6.3 Dephosphorylation of Regulatory Light Chains

When stimulation is discontinued, whether it was a single tetanic contraction or a series of submaximal contractions that caused potentiation, this potentiation will dissipate over the next few minutes of inactivity. This dissipation of potentiation can be monitored with infrequent twitch contractions, as shown in Fig. 3. Typically, after 4–8 min, the potentiation has disappeared, and the twitch contractions are not larger than they were prior to the conditioning contraction(s). The dephosphorylation of the RLCs is achieved by a phosphatase that seems to be unregulated (Sweeney et al. 1993). However, it has been observed by Decostre et al. (2000) that phosphatase may be inhibited by adrenaline. They observed that in the presence of adrenaline, posttetanic potentiation dissipated very slowly, and this corresponded with a very slow rate of dephosphorylation.

6.4 *Increased Ca^{2+} Sensitivity*

The specific mechanism by which phosphorylation of the RLCs could increase the active force of a submaximal contraction took a while longer to elucidate. However, Persechini et al. (1985) demonstrated that RLC phosphorylation resulted in increased Ca^{2+} sensitivity. The consequences of this increased sensitivity are illustrated in Fig. 5. It can be seen from this figure that the magnitude of relative enhancement of force would be expected to be greater at low levels of activation (for example, a twitch) than under conditions with higher [Ca^{2+}]. Increased Ca^{2+} sensitivity could explain the apparent intensification of the activation that is associated with activity dependent potentiation without affecting the time course of contraction (see Fig. 1).

It is one thing to realize that RLC phosphorylation increases Ca^{2+} sensitivity, but how does this occur? Increased Ca^{2+} sensitivity can occur due to either enhanced binding of Ca^{2+} to troponin or altered cross-bridge kinetics. It is hard to imagine that phosphorylation of the RLCs could result in enhanced binding of Ca^{2+} to troponin, which is located on a different filament, so the prime candidate for this mechanism is alteration of the cross-bridge kinetics.

6.4.1 Cross-Bridge Kinetics

Increased force generation by a muscle or muscle fiber can occur if there is greater force per cross-bridge, or there is a greater number of cross-bridges simultaneously

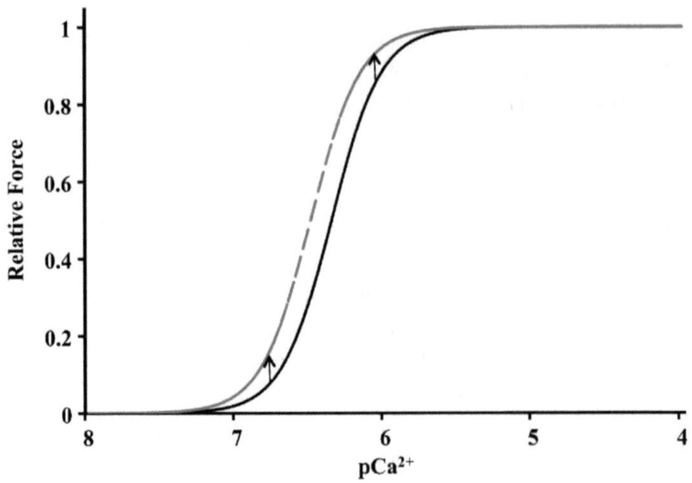

Fig. 5 Force–pCa^{2+} relationship. The relationship between Ca^{2+} concentration, expressed here as the negative log of [Ca^{2+}], is shown for a control condition (*solid line*) and after RLC phosphorylation. Phosphorylation of the RLCs results in an increase in Ca^{2+} sensitivity, indicated as a shift to the left. Sensitivity to Ca^{2+} can be quantified as the half-maximal [Ca^{2+}]. The vertical arrows indicate the increase in active force at two Ca^{2+} concentrations. Although the absolute increase is similar at the two spots on the graph, the relative increase is much greater at the low [Ca^{2+}]

engaged. More cross-bridges engaged at a given instant in time can be the consequence of either an increase in the rate of attachment or a decrease in the rate of detachment of cross-bridges.

Sweeney and Stull (1990) demonstrated in skinned rabbit psoas muscle that the increase in submaximal force that is related to increased RLC phosphorylation occurs with a proportional increase in stiffness and a proportional increase in myosin ATPase activity. They also demonstrated that the rate of force redevelopment after a quick release is affected by RLC phosphorylation; rate of force development is increased, giving an effect similar to that obtained with a higher cytoplasmic [Ca^{2+}].

These observations of Sweeney and Stull (1990) can be interpreted to indicate that the main effect of RLC phosphorylation is to increase the rate at which myosin heads attach to actin and transition to the strong-binding or force generating state. The observed proportional increase in stiffness indicates that the increase in active force associated with potentiation can be attributed to an increase in the number of cross-bridges formed, not an increase in the force per cross-bridge. The fact that ATPase activity or rate of splitting ATP also increased in proportion to the increase in force, provides evidence that the increased number of cross-bridges required increased energy. This is consistent with cross-bridge attachment time remaining constant. Attachment time of individual cross-bridges would be increased if the rate of detachment was slowed to accommodate the increased number of cross-bridges. A slow rate of detachment would result in more cross-bridges attached at a given instant without a proportional increase in use of ATP. An increase in the rate of attachment of myosin heads to actin is considered to be the primary mechanism by which RLC phosphorylation increases Ca^{2+} sensitivity. This mechanism is consistent with the increase in rate of force development after a quick release observed by Sweeney and Stull (1990).

6.4.2 Disordered Myosin Heads

The next step in understanding the mechanism by which RLC phosphorylation results in increased Ca^{2+} sensitivity, is to understand how this phosphorylation can affect the rate of cross-bridge formation. Levine et al. (1996) provided a hint for this mechanism. They reported direct effects of RLC phosphorylation on the structure of rabbit skeletal muscle myosin filaments. In the rested, nonphosphorylated state, myosin filaments appeared to have a uniform helical structure. This uniformity gave distinct layer lines in optical diffraction tests. This uniformity was lost when the filaments were exposed to MLCK, calmodulin and Ca^{2+}, and returned only when the filaments were treated with phosphatase, the enzyme that dephosphorylates the RLCs. This sequence of observations illustrates that the change in uniformity is most likely a consequence of RLC phosphorylation.

These observations suggest that in the relaxed non-phosphorylated state, the myosin heads are uniformly packed against the thick filament, but on phosphorylation of the RLCs, the myosin heads, still tethered by the S2 segment, have increased mobility; the S1 segment swings away from the thick filament backbone. This mobility is thought to bring the myosin heads in closer proximity to the thin filaments, increasing the probability of actin-myosin interaction. This proximity to the thin filaments is thought to enhance the rate of attachment of the myosin heads to actin when [Ca^{2+}] increases on activation. The increased rate of transition to the attached state allows more force at any submaximal [Ca^{2+}]. The increased Ca^{2+} sensitivity resulting from this mobility is thought to be the same as the increased Ca^{2+} sensitivity associated with longer sarcomere length (Konhilas et al. 2003). The increased Ca^{2+} sensitivity associated with longer sarcomere length is called length dependent activation (Lambert et al. 1979). This is discussed further in Sect. 8.1.

6.4.3 Ca^{2+} Also Causes Increased S1 Mobility

The increased mobility of the S1 head that results from RLC phosphorylation is similar to the increased mobility that results from increased [Ca^{2+}] (Podlubnaya et al. 1999). This means that effectively having a raised level of RL phosphorylation is like having a higher [Ca^{2+}]. The implication of these observations, and the interpretation presented above is that Ca^{2+} can have two mechanisms contributing to the force exerted by a muscle cell. At submaximal concentrations, Ca^{2+} binds to a proportion of the troponin C, exposing the corresponding actin molecules along that part of the thin filament. The S1 heads of myosin that are in close proximity to these activated actin molecules can interact. Ca^{2+} also binds to a proportion of myosin, increasing the mobility of the S1 heads. When [Ca^{2+}] is submaximal, this brings a higher proportion of the S1 heads to a close proximity of the actin. These similar mechanisms for Ca^{2+} and RLC phosphorylation can explain why RLC phosphorylation appears to result in an increased intensification of activation.

6.5 Alternative Mechanisms of Potentiation

Although RLC phosphorylation is generally considered to be the primary mechanism for activity dependent potentiation, other mechanisms have been proposed to be involved. These alternative mechanisms may work in conjunction with RLC phosphorylation, or they may be irrelevant. There are two viable supplementary mechanisms that have been proposed: (1) increasing peak free myoplasmic [Ca^{2+}] (Allen et al. 1989; Györke 1993; Abbate et al. 2002); and (2) sequential preponderance of oxidation then reduction potential in the cytoplasm (Posterino et al. 2003). These mechanisms have been proposed to contribute to staircase and/or posttetanic potentiation, but have not been rigorously studied. They will not be given further consideration in this chapter, but should be kept in mind as alternative or supplementary mechanisms of activity dependent potentiation. Any of these mechanisms could contribute to the activity dependent potentiation that has been observed without RLC phosphorylation (Rassier et al. 1999), and could otherwise contribute along with RLC phosphorylation to cause an enhanced contractile response following prior activity.

7 Mechanical Features of Activity Dependent Potentiation

7.1 Magnitude of Isometric Force Change

The magnitude of enhancement of the twitch response that can be obtained from prior activation is quite variable. Factors such as temperature, muscle length and initial twitch:tetanus ratio can influence the magnitude of staircase that can be expected. Stimulation of rat fast-twitch skeletal muscle at 37°C and close to optimal length can result in twitch enhancement by 20–90% during staircase at stimulation frequencies between 5 and 10 Hz. The magnitude of potentiation is decreased as temperature decreases and is increased at shorter lengths, decreased at longer lengths.

Incompletely fused tetanic contractions can also be enhanced by prior activation. When trains of five pulses are given, with two trains s^{-1}, for 7 s, a modified staircase response is seen. The magnitude of potentiation that results is inversely related to frequency of stimulation. This relationship is illustrated in Fig. 6. The dependence of potentiation of the frequency of activation is further explored below in Sect. 7.5.

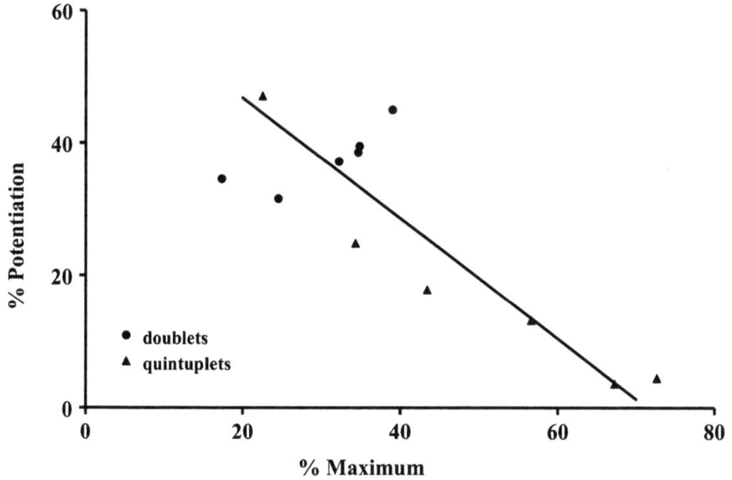

Fig. 6 Potentiation relative to active force expressed as % maximum. The potentiation observed after 7 s of repeated quintuplets or doublets at a variety of frequencies (see Fig. 7) is presented. % maximum refers to the initial active force (first contraction in the sequence) relative to maximal isometric force obtained with 200 Hz stimulation. Note that although doublets were at the same frequency of stimulation as the quintuplets, the higher relative active force obtained with longer stimulation is associated with a smaller magnitude of potentiation. From MacIntosh and Willis (2000), with permission

7.2 Twitch Shape

The twitch contraction time is not altered, or is shortened during staircase and posttetanic potentiation (Close and Hoh 1968; Desmedt and Hainaut 1968; MacIntosh and Gardiner 1987). The absence of an increase in twitch contraction time has been interpreted to indicate that potentiation is accomplished by an enhancement of the intensity of activation, not prolongation of the duration of activation (see Fig. 1). The rate of relaxation of the twitch is clearly accelerated during staircase (MacIntosh et al. 1994). It is not known if this acceleration is due to an increased rate of uptake of Ca^{2+}, or an increased rate of dissociation of cross-bridges, but the evidence for the mechanism of potentiation presented in Sect. 6.4.1 would suggest there is no change in the rate of dissociation, so a change in Ca^{2+} uptake seems more likely.

7.3 Time Course of Enhancement and Its Dissipation

Potentiation occurs very quickly Potentiation during staircase is evident within a tenth of a second and can continue to accrue for several seconds. Factors that influence this duration include temperature, frequency of contractions, muscle length and # of activation pulses per contraction. The magnitude of Ca^{2+} release can also influence the duration of positive staircase. Dantrolene is a drug, which attenuates Ca^{2+} release (Desmedt and Hainaut 1977) causing a decrease in the twitch: tetanus ratio. In the presence of dantrolene, staircase continues for a longer time and achieves a greater level of relative potentiation (MacIntosh 1991). A similar prolongation of the enhancement of contractile response associated with activity dependent potentiation is evident in fatigued muscle (MacIntosh and Kupsh 1987).

7.4 Dynamic Contractions Are Also Potentiated

Although the majority of work on activity dependent potentiation has been with isometric contractions, a few adventuresome investigators have considered the impact of prior activation on dynamic contractions (Grange et al. 1995, 1998; MacIntosh and Bryan 2002). It has been reported that shortening and velocity of shortening can be increased during sequential submaximal isotonic contractions. However, peak velocity is sometimes observed to increase (Grange et al. 1995). Maximal velocity for unloaded shortening does not increase, and may actually decrease.

7.4.1 Is Peak Velocity Slowed by Regulatory Light Chain Phosphorylation?

Maximal shortening velocity (V_{max}) is a function of the cross-bridge kinetics. In particular, the rate of dissociation of myosin from actin is thought to be a factor that influences maximal velocity (Huxley 1957; Metzger and Moss 1988). We have seen that during isometric contractions, the rate of cross-bridge dissociation is not apparently affected by RLC phosphorylation. If RLC phosphorylation increases the rate of attachment, and has no influence on the rate of detachment, then it would not be expected that maximal velocity would be affected by RLC phosphorylation.

In spite of this logic, there are several observations that indicate that the mechanism of activity dependent potentiation: RLC phosphorylation, is associated with a decrease in V_{max}.

One of the studies that has reported decreases in V_{max} is by Greenberg et al. (2009). These authors have used the in vitro motility assay to study the impact of RLC phosphorylation. In addition to slowing of V_{max}, they also reported an increased duty cycle of cross-bridge kinetics. However, they failed to see an increase in Ca^{2+} sensitivity, which they explain by the lack of a thick filament back-bone. They conclude that some effects of RLC phosphorylation must be independent of the thick filament back-bone. On the other hand, the constant relationship between ATPase activity and active force observed in intact fibers, would indicate that the reported increase in duty cycle in the in vitro motility assay is an artifact of the lack of a thick filament back-bone. Is the reported decrease in V_{max} also an artifact of this preparation? The fact that the duty cycle prolongation observed by Greenberg et al. (2009) was strain dependent also suggests this change in kinetics would not influence V_{max}. DeCostre et al. (2000) reported no change in V_{max} when RLC phosphorylation was increased. They were using the mouse extensor digitorum longus muscle in vitro at 20°C.

Crow and Kushmerick (1982) reported that RLC phosphorylation occurred as a consequence of tetanic contraction and that maximal velocity and the economy of force development also changed during a tetanic contraction. By association, it was concluded that RLC phosphorylation decreased the energy cost of the contraction and slowed V_{max}. It was later shown by Barsotti and Butler (1984) that these properties of sustained stimulation occurred independent of RLC phosphorylation. They concluded that RLC phosphorylation does not slow the turnover of cross-bridges. Sweeney and Kushmerick confirmed that V_{max} was not slowed by RLC phosphorylation in skinned rabbit psoas fibers (Sweeney and Kushmerick 1985).

The possibility that RLC phosphorylation may slow V_{max} has been recently revisited. Franks-Skiba et al. (2007) reported that RLC phosphorylation inhibits V_{max} in the presence of vanadate, a phosphate analogue. This observation lead to the suggestion that RLC phosphorylation may contribute to the fatigue-related depression of V_{max}, because [Pi] will be elevated in fatigue. Karatzaferi et al. (2007) extended this work to include acidosis. They assumed that their conditions: high [Pi], low pH and elevated RLC phosphorylation represented a fatigued muscle. The combination of acidosis, elevated [Pi] and RLC phosphorylation had a substantial effect, decreasing V_{max} to 60% of that observed under conditions mimicking a "rested" fiber. Of critical importance, these measurements were obtained at a temperature of 30°C, a condition approaching physiological temperature.

7.5 Frequency Dependence of Activity Dependent Potentiation

Although twitch contractions are typically used to evaluate activity dependent potentiation, it is clear from the information presented above that potentiation is evident for brief incompletely fused tetanic contractions. Figure 7 illustrates the staircase potentiation that occurs in response to sequential brief activations at a variety of frequencies. Clearly, potentiation is evident for intermittent stimulation up to 70 Hz (MacIntosh and Willis 2000) for in situ muscle (at 37°C). Figure 7 shows sequential brief contractions at 40, 50, 60 and 70 Hz, that have a clear dip in active force before showing obvious potentiation.

Theoretically, activity dependent potentiation should be associated with a leftward shift to the force–frequency relationship (see Fig. 8). The magnitude of contraction for all submaximal frequencies should be elevated but the maximal active force should not be affected. The fact that MacIntosh and Willis observed enhancement only up to 70 Hz indicates that other factors besides activity dependent potentiation are involved in these circumstances. The other factors would be related to fatigue, counteracting the tendency for potentiation, though acceleration of relaxation early in a period of repeated contractions probably also impacts staircase. This phenomenon can explain the initial decrease in active force in the first few contractions shown in Fig. 7. The possible contribution of fatigue in affecting this relationship is discussed further in Sect. 8.

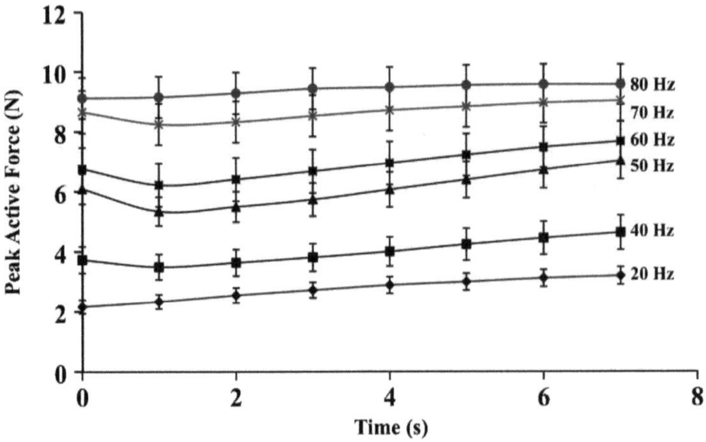

Fig. 7 Change in active force during repeated quintuplet contractions. Submaximal incompletely fused tetanic contractions were repeated twice per s for 7 s. Frequency of activation within each contraction is indicated on the right. Active force increased significantly by 7 s in all cases except 80 Hz. There was a clear initial decrease in active force before the staircase at the intermediate frequencies: 40–70 Hz. From MacIntosh and Willis (2000), with permission

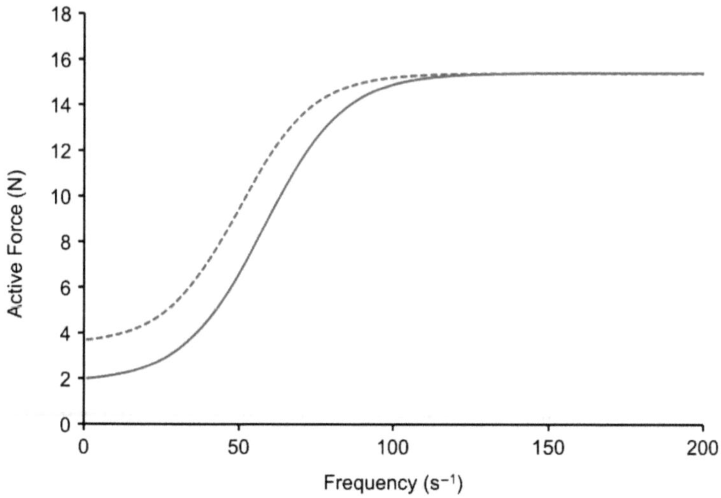

Fig. 8 Force–frequency relationship: activity dependent potentiation. The force–frequency relationship is shown for the control condition, no prior activation (*solid line*) and the potentiated condition, representing what would be expected if the entire force–frequency relationship could be evaluated after a single conditioning contraction (*dashed line*). Observed potentiation is typically less due to fatigue, but there is still a relatively greater effect of activity on enhancement of active force at lower frequencies

8 Interactions and Factors Affecting Potentiation

Assuming that the observation of disordering of the myosin heads on phosphorylation of the RLCs in isolated filaments has some corollary in intact fibers, then there are some predictions that could be made. If the effect of phosphorylation is simply to bring the S1 head to a position in close proximity to the actin, then a long length should have the same effect. Similarly, if there are other factors that affect the order/disorder of the myosin heads on the thick filaments, it would be expected that these other factors would have specific effects on the Ca^{2+} sensitivity and an impact on consequences of RLC phosphorylation.

It is known that Ca^{2+} sensitivity increases as sarcomere length is increased (Konhilas et al. 2002). This is consistent with the idea that proximity of the myosin head with the thin filament results in an increased probability of actin myosin engagement. This observation raises an interesting question. Does activity dependent potentiation still occur when the muscle is at a long sarcomere length? It should not, because Ca^{2+} sensitivity would already be high due to the proximity of the myosin heads to actin.

8.1 Length Dependence of Activity Dependent Potentiation

Several investigators have reported that activity dependent potentiation is more evident when tested at short muscle (sarcomere) lengths than when tested at long sarcomere lengths (Close and Luff 1974; Roszek and Huijing 1997; Rassier and MacIntosh 2002). Figure 9 illustrates this property of potentiation. Staircase progresses to a

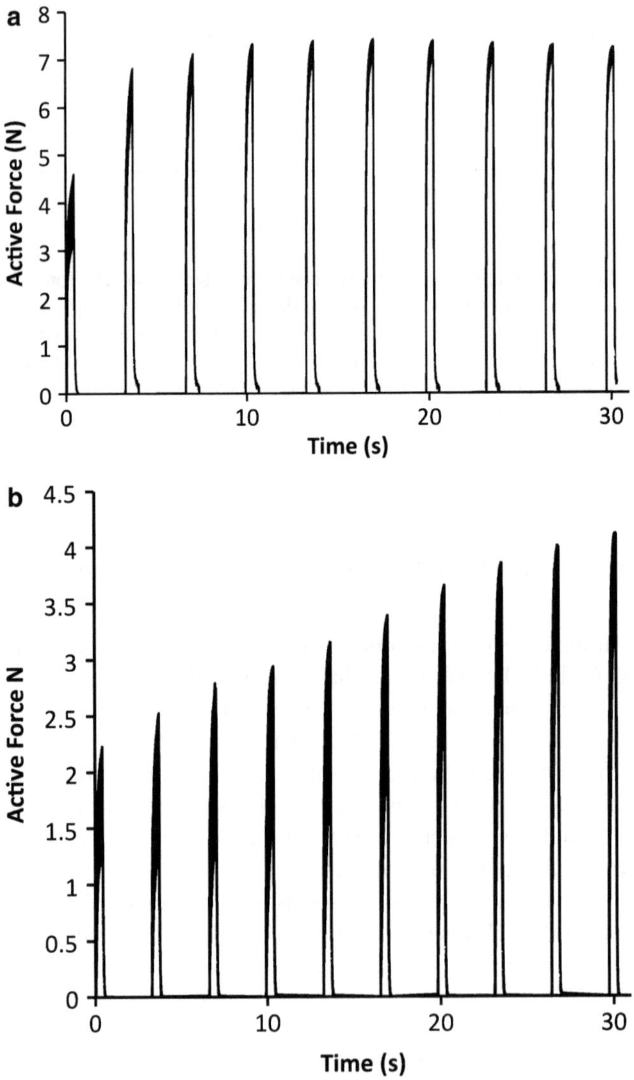

Fig. 9 Staircase at short and long lengths. Sequential incompletely fused tetanic contractions are shown at long (**a**) and short (**b**) lengths. Notice the greater final magnitude of potentiation with the contractions at the short length. Note also the greater fusion with the contractions at the long length, evident as dark shading within the upper parts of each contraction

greater magnitude and for a greater number of contractions in comparison with the same pattern of stimulation at a longer length.

Assuming the primary mechanism of potentiation is RLC phosphorylation, it is important to consider whether RLC phosphorylation can be affected by the sarcomere length of the muscle. Rassier et al. (1997) showed that the length dependence of potentiation was independent of RLC phosphorylation in rat gastrocnemius muscle in situ. They measured RLC phosphorylation in muscle frozen after 10 s of repetitive stimulation at 10 Hz at three different lengths. Potentiation during the 10 Hz stimulation was 118.5 ± 7.8, 63.1 ± 3.9 and $45.6 \pm 4.1\%$ at the short medium and long lengths respectively. RLC phosphorylation was not different between the lengths. It was not surprising that RLC phosphorylation was not different at the various lengths. Balnave and Allen (1996) have shown that the magnitude of Ca^{2+} transients is not affected by muscle length, so activation of MLCK should be similar with a given pattern of stimulation, regardless of length. Furthermore, when a conditioning contraction is elicited at one length and twitch contractions are evaluated at a variety of lengths afterwards, there is greater potentiation of the twitch measured at short lengths than at long lengths (Rassier and MacIntosh 2002). This test shows that the length at which the conditioning contraction is elicited is not a factor that affects the length dependence of potentiation.

The logical explanation for the length dependence of activity dependent potentiation, is that the impact of phosphorylation of RLC is diminished as the filaments are brought closer together; myofilaments are brought together by stretching the muscle. Since RLC phosphorylation increases the rate of transition to strong-binding by increasing the mobility of the myosin head, bringing the phosphorylated head closer to actin, it can be imagined that this mechanism would be less effective if the filaments were already close, as they would be at longer lengths. Yang et al. (1998) tested this hypothesis by osmotically shrinking the myofilament lattice at a short length. This shrinking of the lattice increased Ca^{2+} sensitivity and decreased the effectiveness of RLC phosphorylation at increasing the Ca^{2+} sensitivity. These observations indicate that the length dependence of activity dependent potentiation is related to proximity of the myofilaments. Figure 10 illustrates the relationship between sarcomere length and myofilament lattice spacing.

8.2 Temperature and Activity Dependent Potentiation

It has been reported that activity dependent potentiation is diminished at cold temperatures, and increases as muscle temperature is raised. This modulation of force response appears to be independent of myofilament lattice spacing. However, it has been reported that at low temperatures, the myosin filament is more disarrayed than at a warmer temperature (Xu et al. 2003). This means cooling the muscle has the same effect as RLC phosphorylation, increasing the mobility of the myosin S1 fragment. This may explain why RLC phosphorylation has less effect on Ca^{2+} sensitivity at room temperatures and cooler. It would be interesting to test Ca^{2+} sensitivity interactions between temperature, RLC phosphorylation, and myofilament lattice

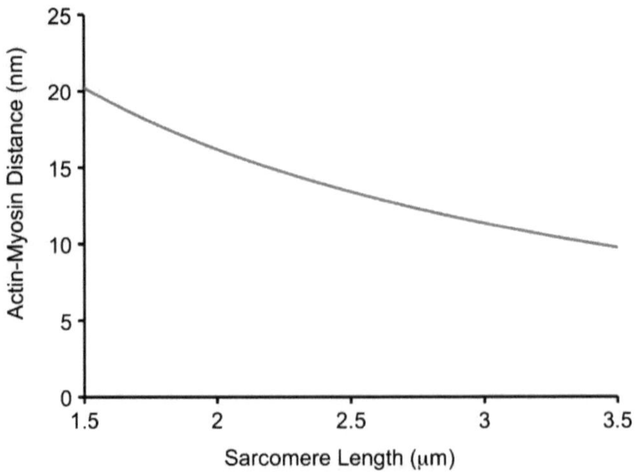

Fig. 10 Free distance between actin and myosin. The estimated distance between actin and myosin is shown across a range of sarcomere lengths. The effect of mobility of the myosin head is increased as you go from long to short lengths, because at long lengths, the myosin S2 segment is not far from the actin filament. Adapted from MacIntosh (2003)

spacing, similar to the observations of Yang et al. (1998) for length, RLC phosphorylation and myofilament lattice spacing.

An interesting observation may give us a clue of what would be expected with this test. If the relative enhancement of active force is compared for intact cells (posttetanic potentiation) and myosin RLC phosphorylation in skinned fibers, at a given (low) temperature, it is clear that the magnitude of potentiation is much greater in the skinned fibers. This greater force enhancement in the skinned fibers can be explained by the fact that skinning causes an expansion of the myofilament lattice (Irving et al. 2000) at a given sarcomere length. This would indicate that with an expanded myofilament lattice, the increased effectiveness of myosin head mobility has an added effect, even at room temperature. It is not clear if the two effects are purely additive, or if the evidence is incomplete. If the increased mobility of the myosin head that occurs on decrease of temperature is the same as the increased mobility associated with RLC phosphorylation, then maximal response from one of these mechanisms, should prevent the effectiveness of the other.

8.3 *pH and Mobility of the Myosin Heads*

Alkalosis has also been reported to increase the mobility of the myosin heads (Podlubnaya et al. 1999), similar to that obtained with RLC phosphorylation or reduced temperature. Acidosis decreases mobility of the myosin heads, similar to increasing temperature. This decreased mobility would decrease Ca^{2+} sensitivity as has been observed at room temperature and cooler (Stephenson and Williams

1985). Considering the interactions described above, this observation of a pH dependence of myosin head mobility gives reason to expect that acidosis would be less effective at decreasing Ca^{2+} sensitivity at a warm temperature, or a short length. It has been confirmed that acidosis is less effective at decreasing Ca^{2+} sensitivity when the muscle temperature is raised to 35°C. It might also be predicted that acidosis would be more effective at decreasing Ca^{2+} sensitivity in the presence of myosin RLC phosphorylation.

9 Interactions of Fatigue and Activity Dependent Potentiation

It is clear that the repetitive contractions that effect activity dependent potentiation also cause skeletal muscle fatigue. However, there is a stronger consensus for the cellular mechanisms of potentiation than for the cellular mechanism of fatigue. These two opposing processes can be simultaneously present, making it difficult or impossible to quantify the impact of one or the other, without measuring light-chain phosphorylation and/or the force–pCa^{2+} relationship.

It has often been suggested that the enhanced Ca^{2+} sensitivity that results from RLC phosphorylation serves to counteract the impact of fatigue (Rankin et al. 1988; Green and Jones 1989; MacIntosh and Rassier 2002). Considering that fatigue in a single cell can result from either less Ca^{2+} release, or decreased Ca^{2+} sensitivity, it seems logical to make this suggestion. In mammalian muscle at body temperature, it is thought that most mechanisms that can contribute to decreased Ca^{2+} sensitivity have minimal impact (Pate et al. 1995; MacIntosh 2003). Therefore, the primary mechanism of fatigue in mammalian muscle appears to be decreased Ca^{2+} release, resulting in lower peak cytoplasmic [Ca^{2+}]. Increasing Ca^{2+} sensitivity seems like a logical way to counter this decreased availability of Ca^{2+}.

It is of interest to consider if there is any advantage to the cell in having a system that requires large amounts of Ca^{2+} initially to activate the muscle, then has a system in place that decreases the requirement for Ca^{2+}. Why not start with contractile machinery that is more sensitive to Ca^{2+}? One possibility is that Ca^{2+} is needed at higher concentrations initially for other cell signaling mechanisms. Ca^{2+} has an important role in cell signaling, controlling a variety of enzymes and gene transcription processes. Further work is required to understand the implication of this possibility.

Considering the interacting effects of the factors described above: pH, temperature, Ca^{2+} and RLC phosphorylation, it is tempting to propose that the impairment of V_{max} attributed to RLC phosphorylation by Franks-Skiba et al. (2007) may actually be a consequence of the balance of factors affecting myosin S1 head mobility. At low temperature (increased mobility), acidosis decreases V_{max} (Edman and Mattiazzi 1981), but at high temperature (decreased S1 mobility) acidosis does not impair V_{max}. However, this impairment may become evident if you phosphorylate the RLC (increased mobility). This would be consistent with the observations of Franks-Skiba et al. (2007).

10 Less Potentiation with Inactivity

It has been demonstrated that in various types of muscle atrophy, activity dependent potentiation is diminished, or abolished (St-Pierre and Gardiner 1985; MacIntosh et al. 1988). The absence of potentiation under these circumstances could be a result of any of several mechanisms, including impaired RLC phosphorylation, counteracting effects of fatigue, or simply because atrophy is associated with a high twitch:tetanus ratio, removing the potential for enhancement. The absence of activity dependent potentiation corresponds with an absence of RLC phosphorylation (Tubman et al. 1996), confirming that the first of these possibilities is the most likely mechanism. Furthermore, it has been shown that denervation, an ultimate form of inactivity, results in a decrease in MLCK activity (Bozzo et al. 2005). This altered enzymatic activity provides an explanation for the absence of RLC phosphorylation.

In spite of this logical explanation for the lack of potentiation, Tubman et al. (1996) further investigated whether a high twitch:tetanus ratio might also be a factor that influences the loss of potentiation in atrophied muscles. Dantrolene sodium is a drug that is known to impair Ca^{2+} release via the ryanodine receptors of the sarcoplasmic reticulum. Administration of dantrolene to animals with atrophied gastrocnemius muscles caused a reduction in the twitch:tetanus ratio and restored staircase potentiation. Stimulation at 10 Hz for 10 s resulted in potentiation by 75%. There was not a significant increase in RLC phosphorylation during this staircase. This is an example of the substantial activity dependent potentiation without RLC phosphorylation. The circumstances of this study must favor an alternative mechanism (see Sect. 6.5).

10.1 Is Aging Just a Case of Less Activity?

It has also been noted that in aged rats and human subjects, activity dependent potentiation is diminished (Carlsen and Walsh 1987; Petrella et al. 1989; Hicks et al. 1991; Baudry et al. 2005). This attenuation of potentiation could be a consequence of changes similar to those associated with atrophy, just due to inactivity or it could be due to other factors, associated with aging. Although it might be considered that early fatigue might counteract an otherwise normal activity dependent potentiation, this seems unlikely. Muscles of old rats (and humans) tend to fatigue less than muscles of the young. Further research is needed to further investigate this property in muscle of old mammals.

11 Summary

One of the consequences of prior activation is an enhanced contractile response when activation is submaximal. This enhanced response occurs primarily due to phosphorylation of the myosin RLCs. The mechanism of this enhancement is thought to be due to an increased mobility of the myosin head, which brings the myosin head closer to the actin filament. This mechanism is less effective when the filaments are already close, as in long sarcomere lengths. Other factors that affect the mobility of the myosin head include: temperature and pH. These various mechanisms interact such that an increase in Ca^{2+} sensitivity due to increased myosin head mobility is more likely to occur if the Ca^{2+} sensitivity begins at a low level (immobilized myosin heads) or when the distance between filaments is relatively long.

References

Abbate F, Bruton JD, de Haan A, Westerblad H (2002) Prolonged force increase following a high-frequency burst is not due to a sustained elevation of $[Ca^{2+}]i$. Am J Physiol Cell Physiol 283:C42–C47

Allen DG, Lee JA, Westerblad H (1989) Intracellular calcium and tension during fatigue in isolated single muscle fibres from Xenopus laevis. J Physiol 415:433–458

Balnave CD, Allen DG (1996) The effect of muscle length on intracellular calcium and force in single fibres from mouse skeletal muscle. J Physiol 492:705–713

Barsotti RJ, Butler TM (1984) Chemical energy usage and myosin light chain phosphorylation in mammalian skeletal muscle. J Muscle Res Cell Motil 5:45–64

Baudry S, Klass M, Duchateau J (2005) Postactivation potentiation influences differently the non-linear summation of contractions in young and elderly adults. J Appl Physiol 98:1243–1250

Billeter R, Heizmann CW, Howald H, Jenny E (1981) Analysis of myosin light and heavy chain types in single human skeletal muscle fibers. Eur J Biochem 116:389–395

Blumenthal DK, Stull JT (1980) Activation of skeletal muscle myosin light chain kinase by Ca^{2+} and calmodulin. Biochemistry 19:5608–5614

Bozzo C, Spolaore B, Toniolo L, Stevens L, Bastide B, Cieniewski-Bernard C, Fontana A, Mounier Y, Reggiani C (2005) Nerve influence on myosin light chain phosphorylation in slow and fast skeletal muscles. FEBS J 272:5771–5785

Burke RE, Rudomin P, Zajac FE 3rd (1976) The effect of activation history on tension production by individual muscle units. Brain Res 109:515–529

Carlsen RC, Walsh DA (1987) Decrease in force potentiation and appearance of alpha-adrenergic mediated contracture in aging rat skeletal muscle. Pflügers Archiv 408:224–230

Celio MR, Heizmann CW (1982) Calcium-binding protein parvalbumin is associated with fast contracting muscle fibres. Nature 297:504–506

Chin ER, Balnave CD, Allen DG (1997) Role of intracellular calcium and metabolites in low-frequency fatigue of mouse skeletal muscle. Am J Physiol Cell Physiol 272:C550–C559

Close R, Hoh JFY (1968) The after-effects of repetitive stimulation on the isometric twitch contraction of rat fast skeletal muscle. J Physiol 197:461–477

Close RI, Luff AR (1974) Dynamic properties of inferior rectus muscle of the rat. J Physiol 236:259–270

Crow MT, Kushmerick MJ (1982) Myosin light chain phosphorylation is associated with a decrease in the energy cost for concentration in fast twitch mouse muscle. J Biol Chem 257:2121–2124

Decostre V, Gillis JM, Gailly P (2000) Effect of adrenaline on the post-tetanic potentiation in mouse skeletal muscle. J Muscle Res Cell Motil 21:247–254

Desmedt JE, Hainaut K (1968) Kinetics of myofilament activation in potentiated contraction: staircase phenomenon in human skeletal muscle. Nature 217:529–532

Desmedt JE, Hainaut K (1977). Inhibition of the intracellular release of calcium by dantrolene in barnacle giant muscle fibers. J Physiol 265:565–585

Edman KAP, Mattiazzi AR (1981) Effects of fatigue and altered pH on isometric force and velocity of shortening at zero load in frog muscle fibres. J Muscle Res Cell Motil 2:321–334

Franks-Skiba K, Lardelli R, Goh G, Cooke R (2007) Myosin light chain phosphorylation inhibits muscle fiber shortening velocity in the presence of vanadate. Am J Physiol Reg Int Comp Physiol 292:R1603–R1612

Gordon AM, Regnier M, Homsher E (2001) Skeletal and cardiac muscle contractile activation: tropomyosin "rocks and rolls". News Physiol Sci 16:49–55

Grange RW, Vandenboom R, Houston ME (1993) Physiological significance of myosin phosphorylation in skeletal muscle. Can J Appl Physiol 18:229–242

Grange RW, Cory CR, Vandenboom R, Houston ME (1995) Myosin phosphorylation augments force–displacement and force–velocity relationships of mouse fast muscle. Am J Physiol Cell Physiol 269:C713–C724

Grange RW, Vandenboom R, Xeni J, Houston ME (1998) Potentiation of in vitro concentric work in mouse fast muscle. J Appl Physiol 84:236–243

Green HJ, Jones SR (1989) Does post-tetanic potentiation compensate for low frequency fatigue? Clin Physiol 9:499–514

Greenberg MJ, Mealy TR, Watt JD, Jones M, Szczesna-Cordary D, Moore JR (2009) The molecular effects of skeletal muscle myosin regulatory light chain phosphorylation. Am J Physiol Regul Integr Comp Physiol 297:R265–R274

Guttman SA, Horton RG, Wilber DT (1936) Enhancement of muscle contraction after tetanus. In: Proceedings of the Society for Experimental Biology and Medicine 34:219–221

Györke S (1993) Effects of repeated tetanic stimulation on excitation–contraction coupling in cut muscle fibres of the frog. J Physiol 464:699–710

Hicks AL, Cupido CM, Martin J, Dent J (1991) Twitch potentiation during fatiguing exercise in the elderly: the effects of training. Eur J Appl Physiol 63:278–281

Hiraoki T, Vogel HJ (1987) Structure and function of calcium-binding proteins. J Cardiol Pharmacol 10:S14–S31

Huxley AF (1957) Muscle structure and theories of contraction. Prog Biophys Biophys Chem 7:255–318

Irving TC, Konhilas J, Perry D, Fischetti R, De Tombe PP (2000) Myofilament lattice spacing as a function of sarcomere length in isolated rat myocardium. Am J Physiol Heart Circ Physiol 279:H2568–H2573

Karatzaferi C, Franks-Skiba K, Cooke R (2007) Inhibition of shortening velocity of ksinned skeletal muscle fibers in conditions that mimic fatigue. Am J Physiol Reg Int Comp Physiol 294:R948–R955

Klug GA, Botterman BR, Stull JT (1982) The effect of low frequency stimulation on myosin light chain phosphorylation in skeletal muscle. J Biol Chem 257:4688–4690

Konhilas JP, Irving TC, De Tombe PP (2002) Length-dependent activation in three striated muscle types of the rat. J Physiol 544:225–236

Konhilas JP, Irving TC, Wolska BM, Jweied EE, Martin AF, Solaro RJ, De Tombe PP (2003) Troponin I in the murine myocardium: influence on length-dependent activation and interfilament spacing. J Physiol 547:951–961

Krarup C (1981a) Enhancement and diminution of mechanical tension evoked by staircase and by tetanus in rat muscle. J Physiol 311:355–372

Krarup C (1981b) Temperature dependence of enhancement and diminution of tension evoked by staircase and by tetanus in rat muscle. J Physiol 331:373–387

Lambert CR, Gladden LB, Stainsby WN (1979) Length-dependent activation of in situ canine skeletal muscle. Am J Physiol Cell Physiol 237:C38–C42

Levine RJC, Kensler RW, Yang ZH, Stull JT, Sweeney HL (1996) Myosin light chain phosphorylation affects the structure of rabbit skeletal muscle thick filaments. Biophys J 71:898–907

MacIntosh BR (1991) Skeletal muscle staircase response with fatigue or dantrolene sodium. Med Sci Sports Exer 23:56–63

MacIntosh BR (2003) Role of calcium sensitivity modulation in skeletal muscle performance. News Physiol Sci 18:222–225

MacIntosh BR, Bryan SN (2002) Potentiation of shortening and velocity of shortening during repeated isotonic tetanic contractions. Pflügers Archiv 443:804–812

MacIntosh BR, Gardiner PF (1987) Posttetanic potentiation and skeletal muscle fatigue: interactions with caffeine. Can J Physiol Pharmacol 65:260–268

MacIntosh BR, Kupsh CC (1987) Staircase, fatigue and caffeine in skeletal muscle in situ. Muscle Nerve 10:717–722

MacIntosh BR, Rassier DE (2002) What is fatigue? Can J Appl Physiol 27:42–55

MacIntosh BR, Willis JC (2000) Force–frequency relationship and potentiation in mammalian skeletal muscle. J Appl Physiol 88:2088–2096

MacIntosh BR, Roberge MC, Gardiner PF (1988) Absence of staircase following disuse in rat gastrocnemius muscle. Can J Physiol Pharmacol 66:707–713

MacIntosh BR, Grange RW, Cory CR, Houston ME (1994) Contractile properties of rat gastrocnemius muscle during staircase, fatigue and recovery. Exp Physiol 79:59–70

MacIntosh BR, Taub EC, Dormer GN, Tomaras EK (2008) Potentiation of isometric and isotonic contractions during high-frequency stimulation. Pflugers Arch 456:449–458

Manning DR, Stull JT (1982) Myosin light chain phosphorylation–dephosphorylation in mammalian skeletal muscle. Am J Physiol Cell Physiol 242:C234–C241

Metzger JM, Moss RL (1988) Thin filament regulation of shortening velocity in rat skinned skeletal muscle: effects of osmotic compression. J Physiol 398:165–175

Moore RL, Stull JT (1984) Myosin light chain phosphorylation in fast and slow skeletal muscles in situ. Am J Physiol Cell Physiol 247:C462–C471

Pate E, Bhimani M, Franks-Skiba K, Cooke R (1995) Reduced effect of pH on skinned rabbit psoas muscle mechanics at high temperatures: implications for fatigue. J Physiol 486:689–694

Perrie WT, Smillie LB, Perry SV (1973) A phosphorylated light-chain component of myosin from skeletal muscle. Biochem J 135:151–164

Persechini A, Stull JT, Cooke R (1985) The effect of myosin phosphorylation on the contractile properties of skinned rabbit skeletal muscle fibers. J Biol Chem 260:7951–7954

Petrella RJ, Cunningham DA, Vandervoort AA, Paterson DH (1989) Comparison of twitch potentiation in the gastrocnemius of young and elderly men. Eur J Appl Physiol 58:395–399

Podlubnaya Z, Kakol I, Moczarska A, Stepkowski D, Udaltsov S (1999) Calcium-induced structural changes in synthetic myosin filaments of vertebrate striated muscles. J Str Biol 127:1–15

Posterino GS, Cellini MA, Lamb GD (2003) Effects of oxidation and cytosolic redox conditions on excitation–contraction coupling in rat skeletal muscle. J Physiol 547:807–823

Rall JA (1996) Role of parvalbumin in skeletal muscle relaxation. News Physiol Sci 11:249–255

Rankin LL, Enoka RM, Volz KA, Stuart DG (1988) Coexistence of twitch potentiation and tetanic force decline in rat hindlimb muscle. J App Physiol 65:2687–2695

Rassier DE, MacIntosh BR (2002) Sarcomere length-dependence of activity-dependent twitch potentiation in mouse skeletal muscle. BMC Physiol 2:1–8

Rassier DE, Tubman LA, MacIntosh BR (1997) Length-dependent potentiation and myosin light chain phosphorylation in rat gastrocnemius muscle. Am J Physiol Cell Physiol 273:C198–C204

Rassier DE, Tubman LA, MacIntosh BR (1999) Staircase in mammalian muscle without light chain phosphorylation. Braz J Med Biol Res 32:121–129

Rosenblueth A, Morison RS (1937) Curarization, fatigue and Wedensky inhibition. Am J Physiol 119:236–256

Roszek B, Huijing PA (1997) Stimulation frequency history alters length-force characteristics of fully recruited rat muscle. J Electromyogr Kinesiol 7:161–177

Staron RS, Pette D (1993) The continuum of pure and hybrid myosin heavy chain-based fibre types in rat skeletal muscle. Histochemistry 100:149–153

Stephenson DG, Williams DA (1985) Temperature-dependent calcium sensitivity changes in skinned muscle fibres of rat and toad. J Physiol 360:1–12

St-Pierre DMM, Gardiner PF (1985) Effect of "disuse" on mammalian fast-twitch muscle: joint fixation compared with neurally applied tetrodotoxin. Exp Neurol 90:635–651

Sweeney HL, Kushmerick MJ (1985) Myosin phosphorylation in permeabilized rabbit psoas fibers. Am J Physiol Cell Physiol 249:C362–C365

Sweeney HL, Stull JT (1990) Alteration of cross-bridge kinetics by myosin light chain phosphorylation in rabbit skeletal muscle: implications for regulation of actin-myosin interaction. Proc Natl Acad Sci U S A 87:414–418

Sweeney HL, Bowman BF, Stull JT (1993) Myosin light chain phosphorylation in vertebrate striated muscle: regulation and function. Am J Physiol Cell Physiol 264:C1085–C1095

Tubman LA, MacIntosh BR, Rassier DE (1996) Absence of myosin light chain phosphorylation and twitch potentiation in atrophied skeletal muscle. Can J Physiol Pharmacol 74:723–728

Walsh MP (1983) Calmodulin and its roles in skeletal muscle function. Can Anaesth Soc J. 30:390–398

Xu S, Offer G, Gu J, White HD, Yu LC (2003) Temperature and ligand dependence of conformation and helical order in myosin filaments. Biochemistry 42:390–401

Yang Z, Stull JT, Levine RJ, Sweeney HL (1998) Changes in interfilament spacing mimic the effects of myosin regulatory light chain phosphorylation in rabbit psoas fibers. J Struct Biol 122:139–148

Zhi G, Ryder JW, Huang J, Ding P, Chen Y, Zhao Y, Kamm KE, Stull JT (2005) Myosin light chain kinase and myosin phosphorylation effect frequency-dependent potentiation of skeletal muscle contraction. Proc Natl Acad Sci U S A 102:17519–17524

Index

A

Acidosis, 333, 338–339
Acta Physiologica Scandinavica, 8
Actin
  γ-actin, 300, 304
  filaments, 1, 3, 5, 144
  light bands, I-bands, 179, 180
  molecules, 330
  myosin
    attachment, 64, 65
    interaction, 61–63, 263, 325
    interface, 69
    links, 232
  site, 43, 44
  striated muscles, 8
  titin
    attachment, 159, 160
    dynamic interaction, 111
Activity dependent potentiation
  aging, 340
  dantrolene sodium, 340
  definition, 321–322
    postactivation potentiation, 323–324
    posttetanic potentiation, 323
    staircase, 322–323
  dynamic contraction, 332–333
  enhancement and dissipation, 332
  fatigue interaction, 339
  frequency dependence, 334–335
  increased Ca^{2+} sensitivity, 328–330
  isometric force change magnitude, 331
  length dependence, 336–337
  MLCK, 326–327
  myosin RLC phosphorylation, 326
  pH and mobility, myosin heads, 338–339
  RLC dephosphorylation, 327
  sarcomere length, 335
  staircase/posttetanic potentiation, 330
  temperature, 337–338
  twitch shape, 332
  twitch–tetanus ratio, 340
Actomyosin
  ATPase, 216, 264
  bonds, 197, 204
  dissociation, 8
  interaction, energy economy, 48
Actomyosin interaction, energy economy
  actin site, 44–46
  actomyosin, 48
  ADP, 48
  AMDP, 48, 49
  ATP, 42
    hydrolysis, 46–48
    turnover, 50, 51
  ATPase, 42, 43
  cross-bridge model, 42, 43
  detachment rate constant, 44
  didactic two-state strain-dependent model, 43–45
  duty ratio concept, 43
  empirical relationships, 43
  energy cost, 41
  force production, 49
  force–velocity relationship, 42
  late *vs.* early events, 43
  mechanical properties, 53
  'molecule to man' integration, 53
  myofilament compliances, 44
  nice fit, force–velocity data, 46, 47
  non-force-producing attached state, 48–49
  parameters variation, 46, 47, 49
  physiological predictions, 52–53
  Pi binding and release, 51
  probability of attachment, 46, 50, 51
  skeletal muscle contraction, 42
  strain-independent model, 49, 50
  tuning, 52
  working stroke, 52

Adachi, K., 78
Adenosine triphosphate (ATP), 7, 42
  hydrolysis, 46–48
  turnover, 50, 51
Akimoto, Tsuyoshi, Dr., 78
Allen, D.G., 301, 324, 337
Ankyrin repeat proteins, 112
Anterior byssus retractor muscle (ABRM), 85
Asmussen, G., 217
ATP. *See* Adenosine triphosphate
ATP-induced cross-bridge movement
  amplitude
    distribution, 94, 95
    experimental condition, 95
    vs. two IP records, 94
  filament bare region, 95, 96
  myosin–paramyosin hybrid filaments
    ADP, 88
    amplitude distribution, 88
    ATP electrode, 87
    center of mass position, particle, 87, 88
    filament drift, 88
    laser-flash photolysis technique, 87
    synthetic filaments, 89
  reversibility
    M•ADP•Pi, 96, 97
    sequential position change, nine different pixels, 97, 98
Axelson, H.W., 227
Aydin, J., 285, 288, 290, 291

B
Bagni, A.M., 213
Bagni, M.A., 220, 237
Balnave, C.D., 337
Bansal, D., 309
Bárány, E.H., 8
Barsotti, R.J., 333
Bershitsky, S.Y., 263
Bianco, P., 238
Blix, M., 9
Blumenthal, D.K., 327
B-process
  biochemical-mechanical cycle, 58
  duty ratio, 58
  force production and time durations, 58, 59
  muscle force, 58
  ODE solution, isometric force and C-process
    apparent on-and off-rate, 64
    frictional force, 66
    Huxley's model, x-dimension, 64
    length displacement, 66
    myosin attachment and detachment, 64, 65
    one-dimensional equations, 65
    stiffness coefficient, 67
    total force, **N** myosin molecules, 64
  oscillatory muscles, 63
  power stroke, 58
  small sinusoidal length perturbation analysis
    A-, B-and C-processes, 60, 61
    elastic and viscous moduli, 61, 62
    force response, 60
    frictional force, 61, 62
    mathematical expression, complex modulus, 61
    myosin-actin interaction, 61
    negative viscous modulus, 63
    resultant Nyquist diagram, 60–61
    visco-elastic properties, 59–60
  strain dependency, myosin off-rate
    frequency dependency, 69–71
    lengthening and shortening phase, 68
    steady-state response, 69
Brandt, P.W., 63
Bressler, B.H., 214
Brown, L.M., 24
Bruton, J.D., 293
Buchthal, F., 231
Burke, R.E., 326
2,3-Butanedione monoxime (BDM), 194, 197–198
Butler, T.M., 333

C
Calcium
  binding, 160, 166
  concentration, 31, 32
  cross-bridges deactivation and force depression, 189
  dependence, 227–228
  intracellular calcium ([Ca^{2+}]i), 310
  maximum activated tension, 171, 173
  myofibrils
    activation and history dependence, 182, 183
    experimental protocol, 181
    force–SL relation and SL dispersion, 182, 184, 187
    solutions, 179
  release, 16, 31
  sensitivity, 328–330
Calpain 3, 112, 310
Calpain protease P94, 112

Campbell, K.S., 231, 237, 244
Caputo, C., 25, 33
Cardiac ankyrin repeat protein (CARP), 112, 113
Cardiomyopathy, 72
Cellular and whole muscle
  activity dependent potentiation (*see* Activity dependent potentiation)
  definitions
    postactivation potentiation, 323–324
    posttetanic potentiation, 323
    staircase, 322–323
    tetanic contraction, 320–321
    twitch, 320
  electromyography, 324
  excitation–contraction coupling, 324–325
  in situ method, 319
  intact myofilament lattice, 318
  in vitro method, 319–320
  in vivo method, 318–319
Chacko, V.P., 277
Chang, A.N., 174
Chronic obstructive pulmonary disease (COPD), 114–116
Colomo, F., 220
Connectin. *See* Titin
Contractile performance. *See* Striated muscles
Cooke, R., 48
Corsi, A., 173
Coupland, M.E., 251, 261
Crow, M.T., 333
Czapó, A., 8

D
Dantzig, J.A., 51
Davis, J.S., 263, 264
Day, S.M., 277
Decostre, V., 324, 327, 333
Delayed-onset muscle soreness (DOMS), 300
Denny-Brown, D., 230
Desmedt, J.E., 324
Diabetes ankyrin repeat protein (DARP), 112
Dihydropyridine receptor, 325
DMD. *See* Duchenne muscular dystrophy
Dobson, G.P., 277
Duchenne muscular dystrophy (DMD)
  characteristic, 311
  DAPC, 300, 301
  dystrophin gene, 300, 302
  extracellular calcium hypothesis, 301
  plasma creatine kinase, 302
Dystrophin-associated protein complex (DAPC), 300, 301

E
Edman, K.A.P., 9–11, 13–30, 32, 33, 35, 36, 220
Ekelund, M.C., 32
Endothermic force generation
  intact muscle fibers, 254
  mechanism, 263–264
  tension responses-temperature-jump, 255, 257
Engelhardt, 7
Environmental chamber (EC) system
  ATP-induced cross-bridge movement, 87–88
  carbon sealing film, 79, 80
  cylindrical compartment, 79, 80
  electron beam damage, 101
  glass capillary microelectrode, ATP solution, 81
Epstein, N.D., 264
Erdös, T., 8
Etoh, T., 90, 91
Excitation–contraction coupling, 324–325

F
Fenn-effect, 250, 265
Fenn, W.O., 12
Ferenczi, M.A., 263, 264
Film-sealed EC, 78
Finer, J.T., 5
Flitney, F.W., 194
Force redevelopment (kTR), 167, 170–172
Ford, L.E., 214, 263
Franks-Skiba, K., 333, 339
Frank-Starling law, 126
Fridén, J., 300
Fujita, H., 237
Fukami, A., 78, 79, 90, 91

G
Geeves, M.A., 70
Getz, E.B., 202, 232–235
Gilbert, S.H., 263
Gold particles
  colloidal, 83, 91
  cross-bridge attachment, 86
  data analysis, 84
Gordon, A.M., 9, 10, 133, 134, 143–147, 151, 155, 156, 158, 161
Granzier, H., 106
Greenberg, M.J., 333

Gunst, S.J., 226
Guttman, S.A., 324
Györke, S., 324

H

Hagbarth, K.E., 227
Hainaut, K., 324
Harada, Yoshiyasu, Dr., 90
Harrington, W.F., 89, 100, 263
Haugen, P., 236, 239
Herbst, M., 231
Herzog, W., 25
Hikikoshi, I.A., 5
Hill, A.V., 12, 13, 34, 42, 43, 46, 47, 145, 213, 244
Hill, D.K., 226, 235, 238, 239, 241
Hill, L., 24
Hill, T.L., 64
Himmelreich, U., 277
Hirst, D.G., 194
Holmes, K.C., 70
Howard, J., 43
Human cardiac muscles, 60, 71
Hutter, O.F., 303, 307, 311
Huxley, A.F., 42–44, 47, 64, 89, 100, 144, 262
Huxley contraction model, 100
Huxley, H.E., 11
Hydrated myofibrils, 81–82

I

Imaging plate (IP) system, 82–83
Incident electron dose, 82
Inorganic phosphate (Pi), 250, 254, 256
Insect flight muscle, 62
Intact muscle fibers, 2–3
  endothermic force generation, 254
  isometric tetanic contraction, 252, 253
  normalised tetanic tension data, 252, 253
  T-jump tension response, 252–254
  two state system, 254
Intra-molecular strain, 69–70
Isolated myofibrils, 126–127. *See also* Tropomyosin–troponin complex
Ito, H., 90, 91
Ito, Kazuo, 78

J

Japan Electron Optics Laboratory (JEOL) Ltd, 78, 90
Javadpour, M.M., 277
Jockusch, H., 306
Joyce, G.C., 217
Julian, F.J., 25

K

kACT. *See* Rate of force activation
Kad, N.M., 72
Kaiser, E., 231
Karatzaferi, C., 52, 333
Kawai, M., 63, 254, 263
Khakee, R., 300, 310
King, N.M.P., 245
Kinosita, K. Jr., 101
Kominz, D.R., 173
kTR. *See* Force redevelopment
Kumar, A., 305
Kushmerick, M.J., 333

L

Labeit, S., 106
Lakie, M., 227, 244
Lange, S., 113
Laser trap
  essay, 4, 5
  negative relationship, myosin off-rate and strain, 72
  probing, 58
  t_{on} measurement, 61
Leake, M.C., 110
Lee, E.J., 25
Levine, R.J.C., 329
Linari, M., 221
Linke, W.A., 106
Ljubimova, 7
Lombardi, V., 232, 233
Loram, I.D., 227
Lou, F., 25, 33
Lowey, S., 90
Lymn, R.W., 48, 210, 258

M

MacIntosh, B.R., 331, 334, 338
Maréchal, G., 159, 161, 188, 217
Margossian, S.S., 90
Marsh, B.S., 12
Maxwell model, 70
McNeil, P.L., 300, 310
Meiss, R.A., 226
Menke, A., 306
Microgrids, 79, 80
Minoda, H. Dr., 90
Mitov, M.I., 233

Monte Carlo simulations, 109
Morgan, D.L., 25
Morison, R.S., 324
Moss, R.L., 231, 237
Mouse cardiac muscle
ATP hydrolysis, 274
data normalization and statistical analysis, 272
enthalpy change (ΔH), 270
force transducer, 271
free energy conversion, 270
mechanical efficiency, 272
mitochondrial oxidative phosphorylation efficiency
ATP hydrolysis free energy, 276–278
glucose oxidation, 276
thermodynamic efficiency, 276
myothermic/heat measurement technique, 270
oxidative phosphorylation, 270
papillary muscle preparation, 271
recovery metabolism, 272
servo-controlled motor, 271
shortening velocity effects, 273
shortening velocity efficiency, 273, 274
thermodynamic efficiency (ΔG), 270
work production efficiency, cross-bridge
contraction protocol, 275
cross-bridge force-extension relationship, 276
energy use, 275
enthalpy, 274–275
skeletal *vs.* cardiac muscle, 275–276
Muscle ankyrin repeat proteins (MARPs), 112
Muscle contractile dysfunction, 282
Muscle LIM protein (MLP), 112
Muscle shortening and lengthening
crossbridge modelling
force response, lengthening, 215, 216
lengthening velocity, 216–217
Lymn-Taylor scheme, 215
ramp shortening, tension response, 215, 216
strain-dependent mechano-kinetic model, 215
stroke heads, 217
force–velocity (F–V) relation, 210
intact rat muscle fibers, 210
materials and methods, 210–211
mechano-kinetic model, 210
non-crossbridge contribution
elements stiffening, 220–221
experimental findings, 217–219
implications, 221
sarcomere instability, 219
thin filament deactivation, 220
ramp length change, tension response
P_1 transition, 214
P_2 transition and F–V relation, 212–213
ramp lengthening, 211
ramp shortening, 212
Muscle specific ring finger protein-1 (MURF-1), 113, 116
Muscle temperature perturbation
active isometric muscle force, 264
acto-myosin crossbridge, 263
Arrhenius plots, 264
ATP hydrolysis, 265
crossbridge cycle, 250
extrapolation, 263
Fenn-effect, 250, 265
force generation mechanism, 250
force transient, phases, 252
intact fiber experiments, 251
isometric force, temperature dependence
intact muscle fibers, 252–254
skinned fibers, 254–255
length step *vs.* tension recovery rate, 263
linear acto-myosin ATPase pathway, 261
mammalian muscle fiber, 250
mechanical recording system, 250
mechano-kinetic model, 262
muscle contraction, 249, 260
myosin-ATPase, 264
quick tension recovery, 262
skinned fiber experiments, 251
steady state force-shortening velocity curve, 264
temperature-jump technique, 250–251
temperature sensitivity, 263
tension responses and T-jump, 255–260
MyBP-C isoform, 113
Myofibrils, 1, 3–4
extraction, 148–149
force depression
average sarcomere length, 153
cross-bridge attachment, 160, 161
half-sarcomeres, 154, 155
isometric steady-state forces, 152–154
shortening magnitude, 152, 154, 159
stiffness, 150, 155
force enhancement, 150–152
micrograph, 147, 148
protocol, 149–150
stretch and shortening, 150

Myosin, 179–180, 189
  actin
    attachment, 64, 65
    filament interactions, 4, 5
    interaction, 61–63, 263, 325
    interface, 69
    links, 232
  filament, 4, 5
  head, 70
  RLC phosphorylation, 324, 326
Myosin light chain kinase (MLCK), 326–327
Myosin–paramyosin hybrid filaments
  ABRM myosin, 85
  advantages, 85
  ATP-induced cross-bridge movement, 87–88
  force and motion production, 89
  rabbit skeletal muscle, 84
  stability, cross-bridge position
    conventional electron micrograph, 85, 86
    distribution of distance, 86, 87
    gold particles, 85, 86
    IP record, 85
    mass position, particle, 86–87

N

Nagy, A., 107, 110
Nichol, J.A., 303, 307
Nyitrai, M., 48
Nyquist diagram, 60–61, 71

O

Ordinary differential equations (ODEs), 63–67

P

Page, S.G., 11
Pasternak, C., 307
Pate, E., 48
Permeabilized fibers, 2, 3
Perrie, W.T., 324
Perry, S.V., 90, 173
Persechini, A., 328
PEVK
  domain, 106, 238
  segment
    elasticity, molecular basis, 109–110
    E-rich motifs, 107
Pi. *See* Inorganic phosphate
Piazzesi, G., 52, 232
Pinniger, G.J., 216
Plaghki, L., 159, 161, 188
Portzehl, H., 8
Potma, E.J., 49, 50
Prado, L.G., 108
Puchner, E.M., 113

R

Rabbit skeletal muscle myosin
  ATP-induced cross-bridge movement
    amplitude, 93–95, 100
    filament bare region, 95, 96
    reversibility, 96–98
  bipolar thick filaments
    advantages, synthetic filament, 91
    gold particles image, 91, 92
    length and diameter, 91
    myosin–myosin rod mixture, 90
    myosin preparation, 90
    spindle-shaped thick filaments, 91, 92
  cross-bridge position, stability, 92–93
  cross-bridge recovery stroke
    attachment–detachment cycle, 98–99
    lever arm hypothesis, 100
    vs. power stroke, 99
  resumption of experiments, 90, 91
Rack, P.M.H., 217, 226
Rana esculenta, 195
Rassier, D.E., 337
Rate of force activation (kACT), 167, 170–172
Reggiani, C., 9
Regulatory light chain (RLC) phosphorylation
  dephosphorylation, 327
  dynamic contraction, 332–333
  fatigue interaction, 339
  increased Ca^{2+} sensitivity
    cross-bridge kinetics, 328–329
    disordered myosin heads, 329–330
    force *vs.* pCa^{2+}, 328
    S1 mobility cause, 330
  length dependence, 337
  MLCK, 326–327
  myosin light chain kinase, 326–327
  myosin RLC, 324, 326
  pH and mobility, 338–339
  pH and mobility, myosin heads, 338–339
  temperature, 337–338
RLC phosphorylation. *See* Regulatory light chain phosphorylation
Roots, H.R., 216
Rosenblueth, A., 324
Ryanodine receptor, 325

S

Sandow, A., 239
Sarco endoplasmic reticulum Ca^{2+} ATPase (SERCA), 325
Sarcomeres
  A-band movements and half-sarcomere dynamics
    displacement measurement, 136–137
    distance measurement, 134
  activation and force measurements, 129
  behaviour, 25
  elongation, 194, 196, 204
  force and sarcomere length (SL)
    activation period, 130, 132
    contractions, 132–134
    degree of SL dispersion, 131, 133
    relation, 133–135
    SL *vs*. time, 129, 130
    unstable behavior, 129, 131
  force–length relationship
    actin-myosin overlap, 160
    2,3-butanedione monoxime, 160
    cell-and whole muscle-level, 143
    contractile proteins, 156–157
    force creep, 145, 147
    force depression, 152–155
    force enhancement, 150–152
    force–time curve, 146
    frog skeletal muscle, 144, 145
    half-sarcomere length non-uniformities, 157–158
    history dependence, 158
    mid-segment, 147
    passive force, 158
    preliminary hypothesis, titin, 159, 160
    protocol, 149–150
    rate functions, 144
    single fiber forces, 146
    single myofibril extraction, 148–149
    sliding filament and cross-bridge theory, 144, 155
    troponin and tropomyosin, 159, 161
    "x"-distance, 144, 145
  half-sarcomere dynamics, 139
  instability, 219
  isolation, 127–128
  length dependent activation, 330
  M-line, 113
  myofibrils
    experiments, 137–138
    isolation, 126–127
    visualization, 127–129
  PEVK segment, 109, 110
  relation, force enhancement during stretch, 20–23
  relaxed muscle, 239
  residual force enhancement, 23–25
  stretch-induced muscle damage, 300
  striated muscle
    active shortening, 27
    length–tension relationship, 9–11
    organization, 1, 3, 4
  titin, 105, 106
Sarcoplasmic reticulum (SR)
  Ca^{2+}
    cold-acclimated UCP1 deficient mice, 288–291
    reduced myofibrillar Ca^{2+} sensitivity, 289–293
    Tfam deficient mice, 284–288
  protein, 111, 112
Sarkar, A., 110
Saupe, K.W., 277
Short range elastic component (SREC), 226, 236
Siemankowski, R.F., 48
Simmons, R.M., 262
Single intact frog fibers
  actomyosin interaction, 194
  critical force, tension and length
    normal Ringer and BDM treated fiber, 197–198
    relationships, 194
  cross-bridge means extension, 205
  degree of proportionality, 201
  fast stretch technique, 194
  filament compliance, 203
  force and sarcomere length response, 196
  force measurement, 201
  force potentiation, 204
  isometric plateau, 205
  isotonic Ringer, 203
  mean critical tension, 196
  methods, 195
  myofilament compliance, 202
  non-covalent chemical bond, 204
  ramp stretch technique, 194
  rupture force, 204
  sarcomere compliance, 202
  temperature effects, 200–201
  temperature potentiation and force generation, 205
  tonicity effects
    normal, hypo-and hypertonic ringer, 199
    tetanic tension, 198
    tetanus rise, 199–200

Single molecule preparation, 4
Siththanandan, V.B., 168
Skeletal and cardiac muscles
  actin-myosin bond, 241
  BDM, 237
  Ca^{2+} concentration, 226, 237
  canine sample, 242
  cross-bridge component, 236
  force response signals, 240
  inter-half-sarcomere interaction, 238
  mechanical history
    force response, 230
    muscle stiffness, 231
    myosin heads, 231–232
    recovery time-course, 230
    tension response, 229–230
    thixotropy, 231
  molecular mechanism
    calcium dependence, 227–228
    cross-bridge mechanism, 227
    myosin ATPase inhibitors, 228–229
  myocardium types, 240, 241
  myosin heavy chain effects, 242, 243
  N2B titin isoform, 241
  passive linear spring, 236
  physiological significance
    intact rat trabecula, 245
    myocardium stiffness, 244
    neuromuscular control system, 244
    sarcomere length inhomogeneity, 243
    short-range mechanical properties, 242–243
  rabbit psoas and rat soleus, 237
  ramp stretch force response, 241
  relaxed muscle
    actin-titin interaction, 238–239
    Feng effect, 239
    filament compliance, 240
    interfilamentary movement effects, 238
  shortening/re-stretch perturbation, 240
  short-range mechanical properties, 226–227
  short-range response
    advantage, 233
    biphasic force response, 232
    critical strain (Lc), 232
    elastic limit, 233
    'fitting by eye,' 234
    parameter estimation, 232
    piece-wise linear regression, 234
    stiffness estimation, 233
  short range stiffness, 226
  SREC, 226, 236
  static stiffness component, 237
  velocity-dependence, 234–236
Skeletal and cardiac myofibrils. *See* Tropomyosin–troponin complex
Skeletal muscles
  ageing process, 281
  cold-acclimated UCP1 deficient mice, 288–291
  force depression
    cross-bridge deactivation, 188–189
    force enhancement, 177
  history-dependent properties, 178, 185–186, 188
  intracellular activation-contraction pathway, 283–284
  muscle atrophy, 282
  myofibrils
    activation and history dependence, force production, 182–184
    dark-light intensity pattern, 179–180
    data analysis, 181
    homogenized muscle sample, 179
    isometric force, 182, 185
    MgADP activation, 183
    micro-needle displacement, 180
    multi-channel perfusion system, 180–181
    myosin, 179–180
    preparation, 178–179
    protocol, 181
    SL_{dis} *vs.* step change, 185, 187
    solutions, 179
    standard deviation (SD), 184, 186
  primary muscle disease, 281
  reduced myofibrillar Ca^{2+} sensitivity
    contractile function recovery, 289
    force production reduction, 291
    human motor unit, 290
    low-frequency fatigue, 290
    NADPH oxidase, 292
    prolonged low-frequency force depression (PLFFD), 290–291, 293
    rat FDB fiber, 292, 293
    superoxide dismutase (SOD), 292
  RyR1 channel complex, 294
  sarcomere length (SL), 178
  single muscle fibers, 282–283
  SL_{dis} and non-uniformity, 188
  sliding filament theory, 178
  Tfam deficient mice
    CASQ1 expression, 286
    cyclosporin A (CSA), 286–287

exercise intolerance, 284
fatiguing stimulation, 287, 288
genetic mutation, 284
inorganic phosphate ion (Pi), 285
mitochondrial membrane, 287
mitochondrial respiration, 288
mitochondrial transcription factor A, 284
reactive oxygen species (ROS), 287–288
tetanic contraction, 284–285
Skinned muscle fibers, 283
Sleep, J., 262
Smith, G.A., 262
Smith, N.C., 44
Solaro, J.R., 52
Specimen image
ATP and ADP application, 80, 83
data analysis, 84, 85
muscle contractile proteins, 81–82
position marking, individual cross-bridges, 83
recording, 82–83
Spindler, M., 277
SREC. *See* Short range elastic component
Stedman, H.H., 305
Sten-Knudsen, O., 236, 239
Stienen, G.J.M., 50, 233
Straub, F.B., 8
Street, S.F., 305, 307
Stretch-induced muscle membrane damage
calpain activation, 310
creatine kinase, 300
cytochalasin-D, 307
depolarization, 309
DMD
DAPC, 300, 301
dystrophin gene, 300, 302
extracellular calcium hypothesis, 301
plasma creatine kinase, 302
DOMS, 300
eccentric contraction, 299, 310
electrochemical gradient, 309
hypo-osmotic shock, 306
intracellular Na^+ and Ca^{2+}, 309
membrane permeability, 311–312
membrane strength measurement, 306–307
phospholipase A2 inhibitor, 310
popping sarcomere, 300
protein array, 308
ROS scavenger, 311–312
surface membrane, mechanical properties
calpain and proteolytic damage, 305
costamere, 304, 305
β-dystroglycan, 304
lateral connection roles, 305–306, 308
multicellular muscle preparation, 304, 307
muscle stretching and shortening, 302
myofibril Z discs, 305
skeletal muscle, 303
'spare' membrane, 304
tensile strength, 303
T-tubule network, 302
TRIM72, 311
Trypan blue/pyruvate kinase, 306
utrophin transfection, 308
Striated muscles
actin, 8
ATP, 7, 8
cardiac glycoside effects, 8
contraction mechanisms, 2
deactivation, active shortening
caffeine, 31
calcium induced contraction, skinned muscle fiber, 31, 32
depressant effect, 27, 29–31
free calcium concentration, 31, 32
movement effect, contractile system activation, 31
myosin bridges, 33
twitch period, 27, 28
electronic technique, 8
Engelhardt–Ljubimova discovery, 7
force enhancement, stretch
amplitude, 20
definition, 19
isometric tetanic force, sarcomere length, 21–22
myofilament, 23
oscilloscope records, intact muscle fiber, 21
residual force enhancement after stretch, 20, 23–25
sarcomere length *vs*. fiber width, 22
tetanus, single muscle fiber, 19–20
force reduction, loaded shortening
disappearance, period of relaxation, 26, 28
force deficit *vs*. coefficient of variation, 26, 27
length-controlled segment, 26, 29
non-uniform sarcomere behaviour, 25
single muscle fiber, tetanus, 25, 26
force–velocity relationship
active force, 12
biphasic shape, 12
cross-bridge models, 14

Striated muscles (*contd*).
degree of activation, contractile system, 16
empirical equation, 12–14
Gaussian position-dependence, 14
Hill's hyperbolic equation, 12
maximum speed of shortening, 15, 16
rectangular hyperbola, 12
single muscle fiber, 12, 13
unloaded shortening velocity *vs.* myofibrillar ATPase activity, 15
kinetic properties, individual muscle fibers
genetic information, 37
isomyosins identification, 36
segmental differences, 34, 35
skeletal muscle fiber, 36–37
speed of shortening, 34–36
length–tension relationship
bell-shaped relationship, 9
experimental approach, 11
isometric force, 9
length–tension curve, 9, 10
sarcomere length, 9
skeletal muscle, force production, 9
sliding-filament hypothesis, 9, 10
substantial variability, 11
tetanic force *vs.* sarcomere length, 11–12
myosin, 7, 8
relaxation process, 8
single filaments and molecules, 4–6
single muscle fibers, 2–3
single myofibrils and sarcomeres, 3, 4
slack test method
amplitude *vs.* time, 17, 18
braking force, 18
extended fiber length, 18
superimposed quick release records, 18, 19
superimposed records, plateau of tetanus, 17
titin molecules, 1
Stull, J.T., 327, 329
Sugi, H., 77–101
Sun, Y.-B., 33, 50
Sutoh, K., 83
Sweeney, H.L., 329
Szent-Györgyi, Albert, 7–8

T
Takagi, Y., 51
Tanokura, S., 90
Taylor, E.W., 48, 210, 258
Telethonin, 111. *See also* Titin
Tension responses-temperature-jump
crossbridge/AM-ATPase cycle, 258
endothermic force generation, 255, 257
force transient, 255, 256
forward rate constant, 258
isometric model simulation, 260, 261
Pi and MgADP effect, 255, 256
shortening simulation, 260, 262
strain effect, 257–259
Thermopile, 271
Tibial muscular dystrophy (TMD), 114
Titin
elasticity modulation, isoform diversity
Ca^{2+} responsiveness, 107
E-rich motifs, PEVK segment, 107
exon 49, 106
half-thick filament binding, 108
isoform size, 107
microarray, 107
N2BA titin isoform, 106–107
N2B titin, 106, 108
neonatal skeletal muscle, 107
titin gene, 106
elasticity, molecular basis
in vivo behavior, 110
persistence and contour length, 108
PEVK segment, 109–110
tandem Ig segments, 108–109
WLC model, 108
half-sarcomere dynamics, 139
layout, 106
protein complex, stress sensors
biomechanical sensor, 111
I-band, 112–113
M-line region, 113
MURF-1, 113
Z-disk, 111–112
sarcomere, 105
skeletal muscle diseases
COPD, 114–116
disuse atrophy, 116
hereditary myopathies, 114
spasticity, 114
stretch-release force hysteresis, 111
Tm–Tn complex. *See* Tropomyosin–troponin complex
Trentham, D.R., 48
Tropomyosin–troponin complex
actomyosin, 173, 174
Ca^{2+}-activated tension, 174
force activation and relaxation, 173
gelsolin treated/reconstituted systems, 173

irreversible Ca^{2+}-independent contraction, 168
isometric force, 169
kACT and kTR, 167, 170–172
maximal calcium activated tension, 170, 171
myofibrillar maximal isometric force measurements, 167
protein exchange assessment, 171
regulatory protein replacement, 166
release-restretch protocols, 170
removal and reconstitution protocol, 168
striated muscle contraction and relaxation, 165
Tm and Tn definitions, 165–166
Tm-reintroduced myofibrils, 167
Tm replacement, 168, 170, 173
Tm–Tn extraction-reconstitution
biochemical activity, 174
sarcomeres homogeneous, 166
SDS–PAGE gel electrophoresis, 172
tension generation, 170, 171
Tris–HCl SDS–PAGE gel analysis, 168, 169
Tsaturyan, A.K., 263
Tsuchiya, T., 20, 220
T-tubule system, 111, 112
Tubman, L.A., 340

U
Udaka, J., 116

W
Walsh, M.P., 326
Weber, H.H., 8
Weiss, R.G., 277
Westbury, D.R., 226
White, H.D., 48
Wild-type and mdx mice. *See* Stretch-induced muscle membrane damage
Willis, J.C., 331, 334
Woledge, R.C., 46, 52
Wormlike-chain (WLC) entropic spring, 108

Y
Yang, Z., 337, 338
Yoshioka, K., 173

Z
Zhao, Y., 263

MIX
Papier aus verantwortungsvollen Quellen
Paper from responsible sources
FSC® C105338

If you have any concerns about our products,
you can contact us on
ProductSafety@springernature.com

In case Publisher is established outside the EU,
the EU authorized representative is:
Springer Nature Customer Service Center GmbH
Europaplatz 3, 69115 Heidelberg, Germany

Printed by Libri Plureos GmbH
in Hamburg, Germany